PAINTING OF STEEL BRIDGES AND OTHER STRUCTURES

PAINTING OF STEEL BRIDGES AND OTHER STRUCTURES

CLIVE H. HARE

BRIDGESERIES

VNR Van Nostrand Reinhold
New York

Library of Congress Catalog Card Number 89-5676
ISBN 0-442-23499-6

Printed in the United States of America

Van Nostrand Reinhold
115 Fifth Avenue
New York, New York 10003

Van Nostrand Reinhold International Company Limited
11 New Fetter Lane
London EC4P 4EE, England

Van Nostrand Reinhold
480 La Trobe Street
Melbourne, Victoria 3000, Australia

Nelson Canada
1120 Birchmount Road
Scarborough, Ontario
Canada M1K 5G4

16 15 14 13 12 11 10 9 8 7 6 5 4 3 2 1

Library of Congress Cataloging-in-Publication Data

Hare, Clive H., 1941–
Painting of steel bridges and other structures / by Clive H. Hare.
p. cm.
Bibliography: p.
Includes index.
ISBN 0-442-23499-6
1. Painting, Structural. 2. Protective coatings. 3. Bridges, Iron and steel—Maintenance and repair. I. Title.
TT320.H29 1990
624′.28—dc19 89-5676
CIP

Contents

Figures

Tables

Preface

This volume is primarily concerned with a review of the various technologies involved in the design and use of protective coating systems for structural steel on highway bridges. Structural steel in such applications as storage tanks, ships, buildings, offshore drilling platforms, gantries, and many other structures, is, however, also coated with the same basic protective coating systems, using the same fundamental techniques.

It is hoped, therefore, that this volume will be of substantial value to individuals involved in any aspect of the painting process, no matter which industry he or she serves. Such paint-using industries include building construction, transportation, power generation and transmission, petroleum and mining, water treatment and storage, ship building and marine industries, paper production, chemical and other process industries, locks, dams and waterway construction and maintenance, and so on. In addition, this volume contains technical detail that should be of considerable assistance to personnel in the protective-coatings manufacturing industries and those chemical industries that supply and support the manufacturing of paints and protective coatings. This volume's value to rail and highway personnel involved in the maintenance of bridge structures and associated steel maintenance goes without saying. Although the majority of data is derived from domestic sources, foreign experience, particularly that of Western Europe, is well represented.

Basic data on practical corrosion science, paint technology, and paint-system design is provided along with valuable information on practical specification design, surface preparation, the painting process, application equipment, system service, longevity, and costs. Much of this data is not available elsewhere and is certainly not available in any other single source. For the reader interested in further review of specific topics, the comprehensive reference list and bibliography should provide a valuable key.

The work behind this volume was completed as part of a three-year study of the subject commissioned by the National Academy of Sciences and featured as the National Cooperative Highway Research Program Synthesis #136.

This volume, while it attempts to remain practical and understandable to both novice and expert alike, is as diverse as it is comprehensive. It addresses subjects as practical as the politics and psychologies of bidding practice and as technologically advanced as the latest research into the mechanisms by which coatings halt the corrosion process.

While every effort has been made to minimize subjectivity, interpretation and emphasis may reflect the author's own experience to some extent. Where this occurs, it is hoped that such emphasis reflects perception rather than prejudice.

This volume is the only handbook and compendium of all published data on the painting of bridges. It is hoped that it will also be a fundamentally important reference on painting in general.

ACKNOWLEDGMENT

This volume evolved from an American Association of State Highway and Transportation Officials (AASHTO) sponsored National Cooperative Highway Research Program (NCHRP) synthesis study on protective coatings for bridge steel. Because of the size of the present manuscript in relationship to the normal format of the NCHRP synthesis studies, the Transportation Research Board (TRB) elected to employ a condensed version of the study only. This appears as NCHRP Synthesis 136, published by the Transportation Research Board, Washington, D.C.

In view of the amount of valuable data contained in the full version of the dissertation, however, the TRB elected to allow the separate publication of the original text without abridgment.

Because this study was commissioned by the TRB and funded by the National Academy of Sciences, the author takes this opportunity to thank the TRB for permission to publish the full text of the study.

Chapter 1

Introduction

In many respects, the technology involved in the preparation and use of anti-corrosive coating systems for bridge structures was born in the middle of the nineteenth century. The first iron structure, the Wynch Bridge, built over the River Tees in England in 1741, was not painted and, after about sixty years, collapsed from corroded chains. The disaster pointed to the need for bridge maintenance.[1] The world-famous cast iron bridge at Coalbrookdale (Iron Bridge) in Shropshire, England, was not painted when it was built in 1779, but was coated with a bituminous varnish about nine years later. Painted many times since then, it still exists today.

As the century advanced and turned, demand for wider spans with greater tensile strength fostered the use first of wrought iron and later of steel-displacing cast iron in bridge construction. The Telford suspension bridge across the Menai Strait in Wales was an early (1817) wrought-iron bridge. Like many others at the time, after cleaning, this structure was protected by being dipped in boiled oil. Afterward, it was primed with a white lead/linseed oil paint.[2] Again, like other bridges of the time, the undersides were painted with a black bituminous coating. The practice of using heavy bituminous coatings on the underside of bridges appears also to have been employed in the United States. While evaluating the Longfellow Bridge in Boston, the author discovered two coats of bituminous material in some of the lowest layers of the paint-film strata removed from the bearings of this structure, which was built in 1907.

Although steel was in commercial production by 1866, the British Board of Trade banned its use on bridges until 1877. This ban appears to have resulted from British conservatism rather than from an appreciation of steel's greater vulnerability to corrosion as compared to wrought iron. Apparently, the latter fact was not established until the early 1900s. Because of the benefits inherent in the use of steel for larger spans and lighter support structures, and possibly because of the collapse of the Ashtabula Truss Bridge in Ohio and the bridge over the Firth of Tay in Scotland, steel came into its own in the 1880s when it was used in the construction of bridges in St. Louis, Brooklyn, and across the Firth of Forth in Scotland.

The paint system used on the 1,710 foot cantilever span of the Forth Bridge in Scotland is well documented, and in light of its similarity to systems used widely over the next seventy years, it is worth noting.

Before erection, the steel was scraped, wire-brushed, and given a hot or boiled linseed oil treatment. A red lead/linseed oil primer was then applied in the shop where possible. After erection, a second coat of the same paint was applied in the field. It was then recoated with a red oxide intermediate coat and one coat of a red oxide finish. In such areas as tube interiors, the red lead primer was recoated with two coats of white lead and oil.[3] In 1899, tests conducted by H. Smith indicated that one or two coats of red lead with at least two topcoats of iron oxide gave the best protection for iron structures.[4]

The Forth Bridge is also interesting because a program of continuous maintenance has been in effect since its erection. The effectiveness of this practice was highlighted in 1965 when the bridge was inspected and found to be sound and suitable for continued service.

With the exception of the questionable hot oil treatment, the paint system applied to the Forth Bridge is surprisingly similar to that used on steel bridges across the United States during the first half of the twentieth century. Red lead/linseed oil primers, red lead/iron oxide intermediate coats, and white lead and oil finish coats were often the standard paint systems used by authorities. Variations included the use of blue lead (lead sulphate) in finish coats and, later, the introduction of chromium green oxide as a pigment in finish coats and bodied linseed oil and alkyd modification of the finish-coat vehicles. The principal advantage of the red lead/linseed oil primer is that its good corrosion resistance and low surface energetics make it almost foolproof for use over surfaces with little or no surface preparation.

Other systems that were less readily accepted in the United States, but that were used with great success in Europe (e.g., the bridge at Muengsten, Germany, painted only six times since its erection in 1897), were based on the lamellar-barrier pigment micaceous iron oxide (lamella hematite).[5,6] An early example of the more modern, mixed-system concept (inhibitive primer/barrier finish) is the bridge over the Victoria Falls in Rhodesia. This span was originally painted with a red lead/linseed oil inhibitive primer, a micaceous-iron-oxide–based intermediate barrier coat, and an aluminum pigmented barrier topcoat.[7]

In the 1930s, the use of aluminum pigmentation in finish coats increased with the introduction of the phenolic varnishes. Also, zinc yellow (zinc potassium chromate) was introduced as a corrosion-inhibiting pigment and was used with synthetic varnishes and alkyds to provide an alternative inhibitive primer for steel. While this type of primer appears to have been used extensively on steel in other industries (e.g., shipbuilding),[8] its use on steel bridges has not been so widespread.

The last revolution of what we might call "the old era," was the replacement of red lead/white lead systems with the basic lead silico chromate (BLSC), "defense-in-depth" system.[9] Besides the clever marketing that fostered this change, the major force behind it appears to have been the somewhat improved stability of color and reduced chalking of the finish coats. The reduced weight of the paint also helped to sell this system. Although the orange color of the BLSC had to be overtinted, other colors were possible. In many cases, bridges in colors other than green (by now standard) were tried (e.g., blues, beiges, etc.). These colors were achieved using titanium dioxide and such tinting pigments as phthalocyanine blue, chrome yellow, and hydrated yellow oxide. Because of the reduced reactivity of BLSC and its effect on drying, alkyds were used in all three coats to modify the linseed oil. Specifications originally prepared by the pigment manufacturer, National Lead (now NL Industries), appear to have been adopted readily by many state highway departments, without extensive independent testing. These primers and midcoats have now evolved into AASHTO M-229 (TT-P-615). The extensive technical service and formulating assistance provided by National Lead probably contributed largely to the successful initiation of BLSC because then, as now, few state laboratories had formulatory capabilities.

In the 1950s and 1960s, the developing highway network was laid down, and numerous bridges still bear these BLSC systems (e.g., most small overpasses and such large structures as the Verrazano Narrows Bridge and several other of New York's giant spans). These systems appear to have provided good service in rural areas of the snowbelt, lasting as long as twelve years between repaintings. In heavy industrial areas, shorter service-life ranges (five to ten years) were more common. Comparisons with the red lead systems are difficult because of the gradually changing patterns of application and surface preparation that spanned this period of transition. It is probable that less BLSC-based primer was applied over the types of marginal substrates upon which red lead/linseed oil primers score. In many cases, BLSC paint systems were applied in the shop, while subsurfaces were either intact mill scale or sandblasted. On several large structures in the Northeast, hybridized systems employing red lead primers and intermediate coats and BLSC-based finish coats have been employed with success.

The defense-in-depth system was not as well accepted in Great Britain and Europe as it was in the United States. At least one critic thinks that fact is related to the system's unconvincing performance profiles compared to red lead and micaceous-iron-oxide–based systems (A. E. Claxton, Griffiths Bros, Wednesfield, England, personal communication, 1970).

In the 1950s, the remarkable advances that had begun in polymer chemistry during the previous two decades began to spawn a plethora of new coating vehicles. These were adopted by paint companies, which began to produce a bewildering variety of individual coatings, many with claims and counterclaims with regard to properties and performances. While many of these coatings fell by the wayside for one reason or another, many have remained on the market and have provided a level of performance heretofore unknown. Even in harsh environments, instead of thinking in terms of five to ten years between maintenance paintings, service is now discussed in terms of fifteen to thirty years. In some cases, even these estimates may turn out to be conservative.

Unfortunately, with so many new coating types now available for possible use in the protection of steel bridge structures, that number itself is geometrically increased when these individual coating types are utilized as part of full coating systems. Moreover, unlike the old lead-based systems in which oil and oil-modified vehicles were of basically similar types, the new coating systems are not only very different from the oil-based systems, but, as frequently as not, are also very different from each other. Curing mechanisms are no longer simply oxidizing in nature. Lacquers, two-pack chemically curing systems, and moisture-curing systems are now as often the rule in paint systems as they are the exception. An even more important factor, and one still not completely understood, is that one epoxy may be polymerically quite different from another. In other words, individual representations of a specific vehicle class (e.g., epoxies, urethanes, vinyls, etc.) can perform as differently from one another as can those from entirely different vehicle families. Even a relatively small family of vehicles, such as the vinyl chloride/vinyl acetate lacquers, will exhibit major property changes, as may be exemplified in a study of the changes in the adhesion characteristics of one resin as compared to another.[10] Among larger families of polymers, such as epoxies and urethanes, the property differences wrought by molecular engineering may be even greater.

In addition, paint formulating effects (e.g., changes in the pigment volume concentration, the critical pigment volume concentration and the ratio of one to the other, and the selection of pigments, solvents, and additives) can create radical differences in coatings—even those based on the same vehicle makeup. The problem becomes more complex when the factors that follow are taken into account.

1. The effects of high surface energetics
2. The demand for substrate preparation
3. The particular sensitivities of individual vehicles and vehicle classes to such phenomena as dehydrochlorination, ultraviolet (UV) deterioration, alkali-induced saponification, and so on
4. The sensitivities of certain coating types to untoward phenomena that develop from formulation and/or application effects (e.g., lateral adhesion failure in zinc rich films, etc.)

Further complicating the problem is the fact that in many cases those who formulate and advocate these systems do not completely understand the pitfalls of the various high-performance systems in field usage today.

Suddenly, paint technology has become less of an art and more of a science. Therefore, if they are to do their job properly, the individuals responsible for compiling modern bridge paint systems must be not only corrosion engineers, but also paint and polymer chemists. Small wonder then that there have been numerous instances of catastrophic coating system failures on bridges and other structures since modern painting systems have come into frequent use.[11] Inevitably, as the number of available paint systems increases, so does the statistical probability of system design error.[12]

If a technological revolution in coating science has occurred, it has occurred none too soon. Paralleling the rapid rise of the new coating systems has been an alarming increase in the deterioration of bridge structures. There are probably several reasons for such deterioration. Many of the older structures were designed fifty or more years ago when bridges were not subjected to today's heavier and increased traffic loads.[13] Moreover, increased, often indiscriminate use of deicing salts and increased air pollution by oxides of sulphur have added significantly to the severity of the environments in which many of these structures must perform.

Neglect has probably been the major factor behind the deterioration of bridges during the past thirty years. After the surge of bridge building in the 1950s and 1960s, it was inevitable that more maintenance would be needed as these newer structures joined those already in maintenance cycles in the 1970s. The consequent squeeze for the available bridge-maintenance dollar could only be met by an increase in available maintenance funds in the 1970s. Unfortunately, not only did this increase fail to materialize, but funds were actually cut back. Public works spending declined from about 4% of the gross national product in 1965 to 2% in 1982. By 1982, total spending for public works had decreased about 19% since 1965 (adjusted for 1972 dollars).[14] While federal funds from the 1983 gas tax increase should augment sorely depleted state coffers for bridge maintenance, in many cases the funding increases have been "too little, too late." As a result, state engineers now are making difficult decisions concerning which structures should be saved and which should not.

Statistics from the New York/New Jersey metropolitan area that were revealed at the First World Bridge Congress in 1982 showed a dramatic increase in bridge postings during the last decade and alarming projections for future closings.[15] Today, almost one-half of the bridges in the United States have been described as being either obsolete or in some manner deficient, and almost one-quarter of U.S. bridges are either officially posted or closed.[14]

Of course, the increased cost of bridge maintenance only makes the problem worse, particularly since many structures have been allowed to deteriorate to the extent that spot repair is impractical and entire structures must be stripped of all old paint. The cost of such radical blast-cleaning operations is very high. In certain urban environments where the old paint systems include lead-based paints, renewal of the paint system involves the initial removal of these old paints from the bridge and the costly disposal[16] and containment[17] of blast debris. In these cases, the process may become incredibly expensive.

Consequently, the need to ensure that every maintenance dollar is spent efficiently is very real. In bridge maintenance, this translates into achieving adequate coating system service at the lowest possible cost per year. In effect, it means that the best coating system for any specific bridge is that system that will give the longest possible service before additional maintenance is required and that will cost the least in the first place. Because both low cost and long service life are rarely possible with a single system, great care must be taken to ensure that the initial cost of expensive, high-performance coating systems and costly surface preparation is returned by way of an extended service life.[11]

In view of the complexity of the design procedures involved and the need for care and intelligent decision making, all those responsible for the design of bridge paint specifications should be well trained in all facets of the technology and have an appreciation of the pitfalls that lie between the design and the final delivery of any bridge paint maintenance program.

It is hoped that this volume will provide a general understanding of the subject and identify sources in which more detailed treatments of particular topics may be found.

Chapter 2

Corrosion Control by Coatings

THEORY OF CORROSION

Unfortunately, the metallic state of engineering metals is an unstable condition. Once reduced from their ores, these metals continually attempt to rid themselves of the energy pumped into them during the smelting process and, in so doing, return to a more stable state by combining with their environment to form oxides or other compounds similar to the ores from which they were derived.[1] This oxidative degeneration is the corrosion process. The stability of any particular metal in the metallic state will depend on the energy originally absorbed to achieve this state: the more energy required, the more unstable the metal will be; the less energy required, the less readily the metal will corrode.[2]

Corrosion is an electrochemical process that can be readily demonstrated as follows: If zinc is electrically connected to iron (see Figure 2-1) by means of an ammeter and then both metals are immersed in an electrolyte such as salt water, a measurable electrical current will flow between the two metals. This demonstrates that the corrosion process is an electrochemical reaction.[3] In fact, a galvanic cell that is not very different from a battery has been set up. The recorded current flow indicates the passage of electrons from the zinc surface through the electrical connection to the iron surface. The electrons are released as the zinc dissolves into the electrolyte as positively charged zinc ions. At the iron's surface, the electrons react with water and dissolved oxygen to form negatively charged hydroxyl ions. These ions then combine with the zinc ions to form zinc hydroxide. It is important to note that while the two metals are connected, only the zinc (the anode of the cell) corrodes; the cathode (iron) does not.

If the ammeter is replaced with a voltmeter, a voltage or potential difference will be seen to exist between the two metals. If the zinc is replaced with magnesium, this potential difference will increase and the magnesium will corrode at a greater rate than did the zinc. This can be verified by again replacing the voltmeter with the ammeter and observing the increased corrosion current generated.

Alternatively, if the zinc is replaced with copper, the corrosion current is seen to flow in the opposite direction, and now the steel becomes the anode and corrodes. This is because copper is electrochemically more stable than iron; that is, it has a more electropositive potential and is, therefore, cathodic to iron. A gold/iron cell having a still greater potential difference between the two metals would have produced more intensive iron corrosion. Depending upon the exact nature of the electrolyte all metals may be classified in terms of their electrochemical potential with respect to each other.

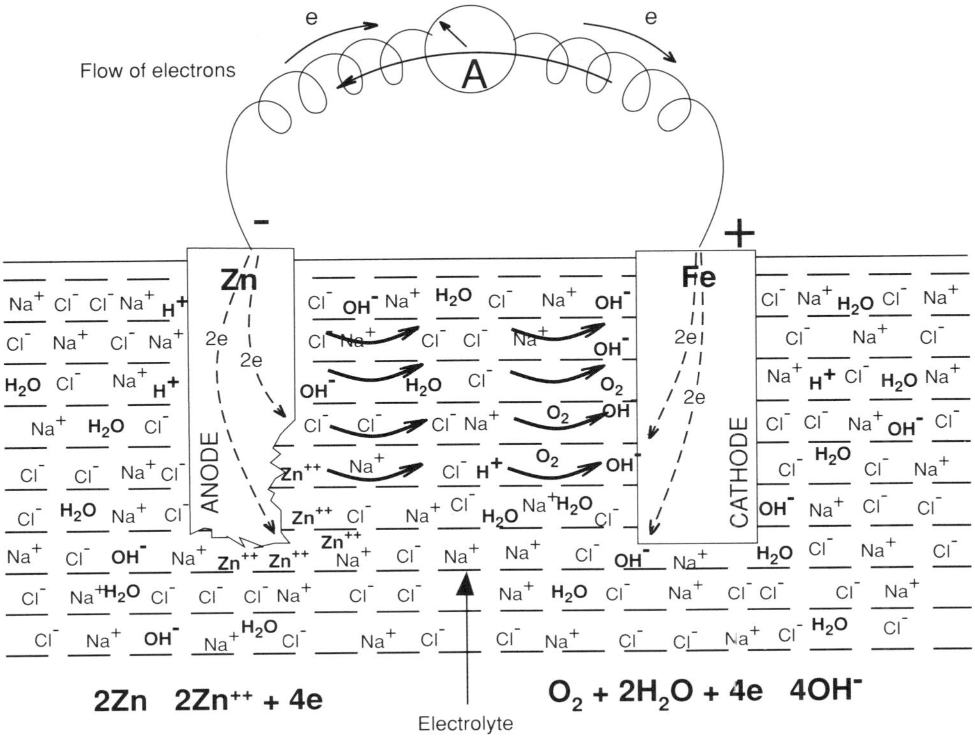

Fig. 2-1. Electrochemical Corrosion Cell

Such a listing is seen in Table 2-1. The relationship between the potential difference (E) and the current flow (I) can be predicted by Ohm's Law,

$$I=\frac{E}{R} \tag{2-1}$$

where R is the overall electrical resistance of the cell made up of the resistance of the two metals, their connector, and the electrolyte in which they are immersed. Expanding R, the equation becomes

$$I=\frac{E}{R_a+R_c+R_e} \tag{2-2}$$

where R_a is the resistance of the anode, R_c is the resistance of the cathode, and R_e is the resistance of the electrolyte.

From Equation 2-2, it may be realized that the current (I) may be reduced not only by decreasing the potential difference between the metals (E) but also by increasing (R), the electrical resistance of the cell.[1,4] Because the electrical conductivity of metals is so much higher than electrolytic conductivity, control of cell R_e is more readily accomplished than the control of R_a and R_c. This may be demonstrated in Figure 2-1 by substituting tap water, which has a lower ionic content, for the salt solution. In this

Table 2-1
Electromotive Series of Metals

The Galvanic Series of Metals in Sea Water	
Magnesium	Active
Zinc	
Aluminum	
Mild Steel	
Lead	
Tin	
Nickel (Active)	
Brass	
Copper	
Nickel (Passive)	
410 Stainless	
Titanium	
304 Stainless	
Silver	
Gold	
Platinum	Noble

case, the overall current flow is decreased markedly, and the anode corrodes much more slowly. If distilled water is used, corrosion may stop almost entirely.

The resistance of the cell in Figure 2-1 might also be raised by electrically separating the two metals. In this case, electrical contact would be broken, and the overall cell would cease to exist. The rate of zinc corrosion would be reduced as minor heterogeneities of the zinc surface, each with slightly differing electrochemical potentials, form myriad minor, local galvanic cells. The same thing would happen on iron, which would no longer be totally cathodic but instead would be made up of a patchwork of tiny galvanic cells, the anodes of which would corrode, while the cathodes did not.

Because bimetallic cells on properly designed bridges are the exception rather than the rule, this type of local cell corrosion is most commonly involved in the deterioration of steel bridges. Many factors, including heterogeneities in the steel, heterogeneities in electrolytes, variations in the access of electrolytes to the steel, and varying degrees of contamination all add to small, localized potential differences along the steel surface and lead to localized corrosion.[5] (Mill scale is particularly cathodic to steel and produces quite substantial potential differences.) All this notwithstanding, the basics of the corrosion process above apply and only the magnitude of the potential differences involved and the resultant corrosion current levels are different from bimetallic cells.

Fortunately, corrosion rarely proceeds at the same rate at which it began. Rather than being linear (Figure 2-2), the rate usually decreases with time and is modified by the formation of films on the anode and/or cathode that may increase the resistance of local cells or modify the electrochemical potentials of the particular surface on which they form. Equation 2-2 then becomes

$$I = \frac{E}{R_a + R_c + R_e + R_{af} + R_{cf}} \qquad (2\text{-}3)$$

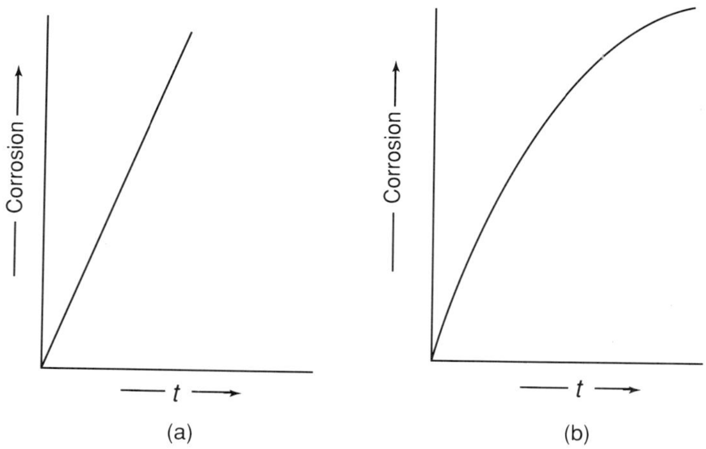

Fig. 2-2. Effects of Electrode Films (Polarization) on Corrosion Rates: *a,* linear rate—no polarization, or the protective films are removed or dissolved on formation; *b,* parabolic rate—the polarizing effects of anode and cathode films decrease the rate of corrosion with time.

where R_{af} is the resistance of the anode film and R_{cf} is the resistance of the cathode film. In local cells with little potential difference between anode and cathode areas, the potential of the anodic metal when coated with corrosion product may actually increase so as to become cathodic with respect to the original cathode. Consequently, the original cathode becomes the anode of the cell and itself corrodes. Such continual reversals of polarity lead to the general corrosion of the entire steel surface, a process that is common to unpainted steel in a corrosive environment.

Whether by the reduction of potential difference or the increase in electrical resistance, the effect of the growth of electrode films, known as polarization, inevitably results in a diminution of the magnitude of current flow and, therefore, corrosion.[6]

The permanence of such polarizing films is often of critical importance. As a metal corrodes (i.e., dissolves into the electrolyte) an equilibrium is eventually set up between its tendency to go into solution, thereby increasing the concentration of its ions in the electrolyte, and an opposing tendency for those ions to coat and polarize the metal as films of corrosion product, thereby reducing the concentration of its ions in solution. In an electrolyte with low metal ion concentration, corrosion of the metal will be greater than one with a higher metal ion concentration. Where metal ion concentrations are nonuniform, this electrolytic heterogeneity will itself produce a potential difference between sites on the same metal surface that are in contact with these different concentrations. Areas of metal in contact with an electrolyte having low metallic ion contact will become anodic and corrode, while areas in contact with solutions of higher concentration will become cathodic until equilibrium is restored and the metal ion concentration at the anode equals that at the cathode. Problems occur when the attainment of equilibrium is prevented by continuous depletion of metal ion concentration at the anode. For example, when any area is continuously stripped of a polarizing film, corrosion will inevitably be sustained at a more or less linear rate and metal consumption will become excessive because electrolytic homogeneity and cell equilibrium are never achieved. In most cases, removal of a polarizing film is related either to abrasion within the corrosive environment or relative movement between the electrolyte and the metal. On bridge structures, the phenomena might be seen as a result of faulty drainage. When water continually spills over a steel section, it prevents the accumulation of rust and continuously exposes new steel for renewed attack (see Figure 2-3).

Similarly, a potential difference can be set up if a metal is in contact with an electrolyte that has local differences in levels of oxygen (oxygen concentration cell). In this case, equilibrium can only be achieved by increased oxygen consumption by way of the cathode reaction. Areas of metal in contact with electrolyte having high oxygen content will become cathodic to areas in contact with less oxygenated electrolyte. These will corrode until homogeneity of oxygenation of the electrolyte is achieved. Again, serious problems will occur if oxygen is continuously replenished at the cathodic areas and not at the anodic sites. In such cases, severe continuous corrosion at the anode can consume the metal.

Polarizing oxide films of corrosion product on some metals, such as aluminum, are so adherent that they effectively stifle all corrosion.[7] On steel, the corrosion product is usually porous and loosely adherent and provides little permanent protection (see Figure 2-2). Artificial polarizing films may be induced to form on steel surfaces, however, by the use of certain corrosion inhibiting moieties that may be introduced into the electrolyte.[6] Chromates, nitrates, molybdates, borates, phosphates, and certain amines and organic lead salts are typical of such species and are thought to function either by fostering the adsorption of passivating films of oxygen or oxides on the metal surface[8] or by thickening and reinforcing the naturally occurring oxide layer.[9]

Certain high-strength, low-alloy steels containing nickel, chromium, and copper (e.g., A588 Weathering Steel) as they corrode themselves form more adherent oxide films, which are thought to prevent further corrosion in much the same way as the oxide coating on aluminum prevents attack.[10,11] Additional evidence would suggest that this can only be accomplished in certain environments, and that environments with high chloride ion concentration are not conducive to protection by such alloying techniques.

Any discussion of the corrosion process would be incomplete without consider-

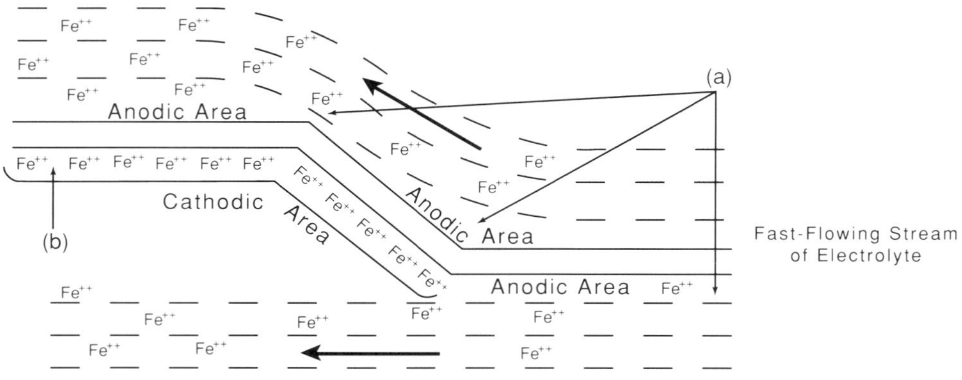

Fig. 2-3. Localized Removal of Anode Film Maintains Linear Rate of Corrosion—Metal Ion Concentration Cell: *a,* areas unsheltered from the flow of water are stripped of all ionic accumulation, preventing polarization. Areas are, therefore, permanent anodes and continue to corrode at a more or less rapid rate; *b,* surfaces sheltered from the flow of water accumulate metal ions and are polarized, thus acting as effective cathodes.

ation of the effects of the relative areas of anodes and cathodes on the intensity of the attack. In Figure 2-1, both electrodes are represented with identical surface areas. If, however, the zinc anode area is diminished to one-tenth the size of that depicted, then the corrosion current generated by the same large cathode and falling on the smaller anode will be ten times as intense as before. Alternatively, if the anode is made ten times larger in surface area than the cathode, then the same corrosion current will be dispersed over a larger area and the intensity of the attack will be correspondingly diminished. The effect is often referred to as the catchment area principle[12] and can be put to practical use in the design of systems that will reduce corrosion intensity (see Figure 2-4).

CORROSION CONTROL BY COATINGS

In the control of corrosion by protective coatings, one of three mechanisms may be used: the inhibitive primer, the sacrificial primer, or the barrier coating.[13] In practical composite paint systems, combinations of such techniques may be employed. Care and intelligence is required in the design of such systems to ensure that combinations produce a compatible whole without the various components working against each other.

The Inhibitive Primer

The inhibitive primer functions by modifying the anode and/or cathode reaction by means of artificial polarization of the electrode. In effect, it decreases the potential difference between the anode and the cathode and/or increases the electrical resistance across the cell.[14] As a result, it can only function when applied directly to the substrate.

The primer derives its inhibitive properties from the pigment. Historically, these have been either a partially soluble hexavalent chromium salt, such as zinc potassium chromate, strontium chromate, or zinc tetroxy chromate, or basic lead compounds, such as red lead.

The chromates are known as direct inhibitors and will function irrespective of the vehicle type with which they are combined as long as the volumetric ratio of the pigment to nonvolatile vehicle is such that small quantities of moisture may slowly penetrate the coating. Essentially, such pigments act as a reservoir of inhibitive ions. These ions dissolve slowly in minute amounts of water that penetrate the coating film. They are then carried to the metal interface where they may effectively modify the electrical resistances by fostering the formation of passive films (see Figure 2-5). Much study has been done on metallic inhibition by chromates. For a comprehensive review see Leidheiser.[15]

Solubility of the pigment is, of course, most important and dictates both the pigment loading levels and the type of vehicles in which the pigment is employed. Too soluble a pigment can produce not only severe blistering problems but can also rapidly exhaust itself as a reservoir of inhibitive ions.[16] On the other hand, a pigment that is too insoluble will not produce enough inhibitive ions quickly enough to maintain the passive film on the metal substrate. This may result in the establishment of galvanic cells between large passivated areas of metal and small spots where passivation has broken down. This could lead to aggressive corrosion of the pitting type.[12]

Lead pigments, sometimes referred to as indirect inhibitors, are thought to function by way of a reaction with acidic derivatives of specific vehicle types, such as certain

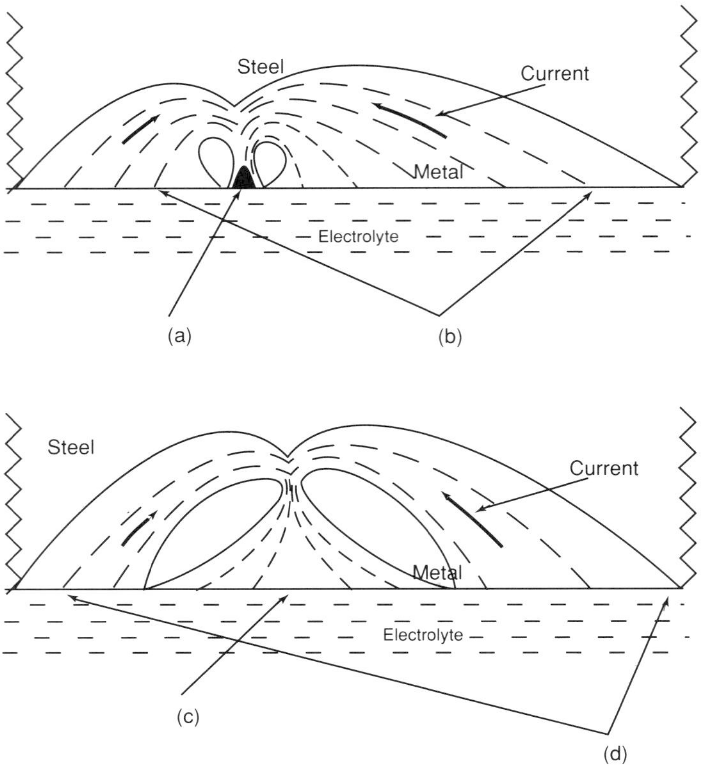

Fig. 2-4. Catchment Area Principle: *a,* large corrosion current falls on small anode area resulting in a high current density at anode and intensive corrosion; *b,* large cathode area produces a large corrosion current; *c,* smaller corrosion current falls on large anode resulting in a lower current density at the anode and much less intensive corrosion; *d,* smaller cathode area produces a smaller corrosion current.

vegetable oils and vegetable oil–modified vehicle species (e.g., alkyds, epoxy esters, uralkyds, etc.).[17] The resultant lead salts of long chain mono- and di-carboxylic acids dissolve in water in much the same manner as the chromates and again function by modifying the nature of the metallic substrate. Mayne and others have postulated that such inhibition involves the thickening of the naturally occurring air-formed oxide layer on the steel surface.[18] While it is true that red lead is inhibitive in water,[19] possibly by virtue of its basicity, this is probably no more than a temporary condition because it converts to an insoluble carbonate in air. It also has been postulated that red lead may also function by scavenging corrosive, soluble sulphate ions by reacting with ferrous sulphate contamination and producing insoluble lead sulphate.[20] Although such reactions probably do occur, the evidence supporting Mayne's position remains very persuasive. As in many vehicle systems, red lead is far less effective than it is in oxidizing vehicles, such as linseed oil.[21]

BLSC,[22] while ostensibly both a lead pigment and a chromate pigment, owes its success more to its basic lead characteristic than to the hexavalent chromium anion.[23] It is believed that lead chromate is too insoluble to provide the levels of performance derived from the pigment,[24] although both lead chromate and lead molybdate have been found to provide some measure of improved corrosion inhibition when evaluated as prime pigments against other conventional prime pigments, such as red oxide, chromium green oxide, and titanium dioxide.[25]

Since the early 1980s, both lead and hexavalent-chromium compounds have become a source of some toxicological concern, which has severely curtailed their use.[26] These concerns have spurred the development of a wide range of alternative pigments with reduced toxicological profiles. Many of the inhibitive pigments that are currently available are as follows:

Lead pigments
- red lead
- lead suboxide
- basic lead silico chromate
- dibasic lead phosphite
- tribasic lead phosphosilicate

Chromate pigments
- zinc potassium chromate (zinc yellow)
- basic zinc chromate (zinc tetroxychromate)
- strontium chromate
- (calcium, barium, and lead chromate are also possible inhibitors)

Borate pigments
- calcium borosilicate
- barium metaborate
- zinc borate
- (sodium, aluminum, and magnesium metaborate are also possible inhibitors)

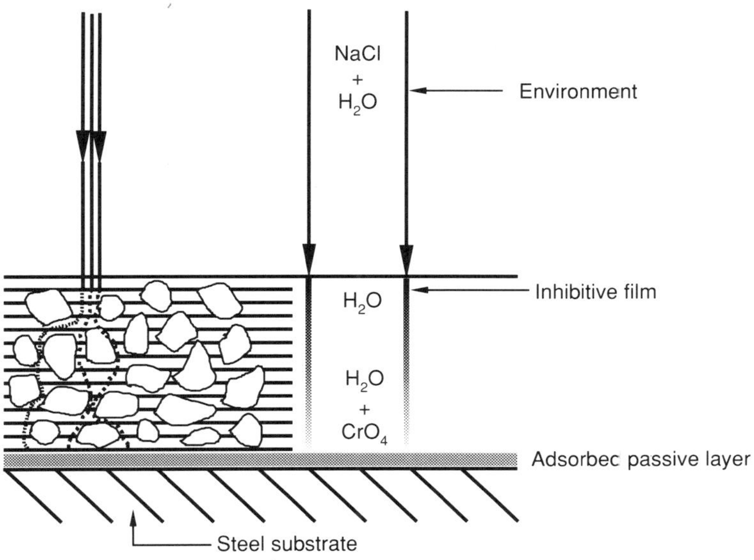

Fig. 2-5. Steel Protection by Inhibitive Primers. Film is slightly permeable to water but not to ionic solutions. Inhibitive ions dissolved from pigment by pure water are adsorbed onto the steel to form a surface less prone to corrosion.

Phosphates
- zinc phosphate (dihydrate)
- zinc phosphate (tetrahydrate)
- basic zinc phosphate
- basic aluminum zinc phosphate
- basic zinc molybdate phosphate
- zinc hydroxy phosphite
- zinc phosphosilicate
- calcium barium phosphosilicate
- calcium phosphosilicate
- calcium strontium phosphosilicate
- zinc barium phosphate

Molybdates
- basic zinc molybdate
- basic zinc calcium molybdate

Others
- zinc ferrite
- calcium ferrite
- zinc nitrophthalate
- lead nitrophthalate
- tungstates

These pigments may be loosely grouped together under their various chemical identities (e.g., borates[27] and borosilicates,[28] phosphates,[29] phosphites,[30] and phosphosilicates,[31] molybdates,[32] and ferrites[33]). Most are salts and compounds of calcium, zinc, strontium, aluminum, and barium.

Some controversy exists concerning the acceptability of barium compounds as nontoxic materials.[34] None of the new inhibitors is found to be as efficient as either the basic lead pigments or the chromates.[35] Basic pigments, such as zinc oxide,[36] magnesium oxide,[37] and calcium silicate,[38] also have been shown to be valuable auxiliary pigments in inhibitive primers. Early independent test data of many of the nontoxic inhibitors were quite disappointing, but this could be related to the primitive nature of early formulations.[39] As experience with such pigments has grown, better performance profiles have been achieved, although they still are not up to par with the conventional pigments. Several states and authorities currently are evaluating coatings based on such materials on existing bridges.

The Sacrificial Primer

Sacrificial primers are based on the same general principle as are sacrificial anodes used in the cathodic protection of steel hulls, buried pipe, and so on. Their function is similar to galvanized films and metallized zinc films. They set up generalized cells, in the presence of which all local cell activity on the steel surface ceases; the steel becomes totally cathodic and is thereby protected from corrosion.[40,41] The anode, by definition, must be electrochemically negative compared to steel and so, for all practical purposes, is always zinc metal (see Figure 2-6).

Alloys of zinc and aluminum, which provide cathodic protection to steel, have been used in metallizing (thermal sprayed coatings)[42] and in hot-dipped coatings.[43] In coatings, zinc is dispersed as a pigment throughout the paint film and is applied directly to the steel surface being protected. Pigment loadings must be sufficiently high so that the

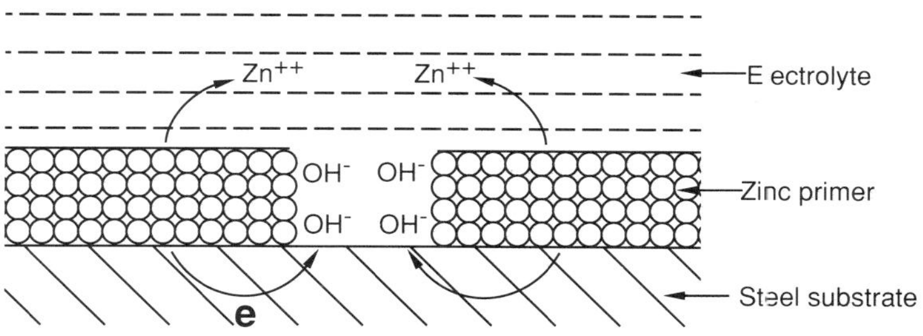

Fig. 2-6. Steel Protection by Zinc Rich Primers. The presence of strongly electronegative zinc-pigmented coating short-circuits all local cell activity on steel. The steel becomes totally cathodic to the zinc anode. When the zinc corrodes, the steel will not corrode even at bare spots. It is mandatory that the zinc coating is in electrical contact with the steel surface. Therefore, the steel must be stripped of all contamination.

effects of such highly dielectric vehicles as organic binders, which separate particles of zinc from each other and from the steel substrate, are minimized, but not so high that cohesive integrity of the primer film is compromised.[44]

Although experimentation has been done with several different particle types, sizes, and shapes,[45,46] the zinc form most widely used in such primers is the spherical zinc dust with mean diameters ranging from four to approximately eight microns.[47] As such, conductivity occurs by the tangential contact of individual spheres of zinc with one another and with the steel substrate. This tangential contact, along with the paucity of vehicle, the electrically insulative effects of substrate contamination, and the high surface energy and reactivity of certain inorganic vehicle species dictate an excellence of surface preparation and a mil profile not necessarily required by the inhibitive primers.[48] As with inhibitive primers, however, sacrificial primers must be employed as the first coat of any coating system.

Zinc rich primers may be either organic or inorganic. Radically different formulating parameters, curing mechanisms, and performance profiles give authority for such characterization.[13]

Organic Zinc Rich Primers

The organic zinc rich primer is quite properly considered an organic coating. It is pigmented at a pigment volume concentration (PVC) that equals or exceeds the critical pigment volume concentration (CPVC) (the level at which the pigment packing is tightest and there is just enough vehicle to surround each pigment particle with a monomolecular layer of vehicle and to fill in the interstices between the pigment). If the primers are formulated much beyond the CPVC compositions, they lose interstitial vehicle and become porous and weakly cohesive; if they are formulated much below the CPVC level, the particles of zinc are separated so that the electrical resistance of their films becomes too high to maintain cathodic protection. Consequently, there is a relatively limited formulating window within which such compositions must be controlled.[49] Various attempts have been made to identify this window by specifying weight loadings of zinc between various minima and maxima, generally between extremes of 80% and 95% by weight of the dry film.[50–53] Little success has been realized from such effort, however, and such specifications tend to be unreliable in practical use with organic zinc rich primers. In addition to including many excellent compositions, organic zinc rich primers may also include coatings that perform poorly and exclude still other materials with excellent performance characteristics. Specification difficulties of this type are hardly surprising in view of the complex paint chemistry involved in the formulation of organic zinc rich primers. These physiochemical factors include the many and various effects of the specific gravity and dielectric constant of the vehicle, the presence of wetting agents, auxiliary pigments, thixotropes, anti-gassing agents, and the effects of CPVC, zinc particle size, and distribution. The subject has been discussed in several papers.[54–57]

Moreover, although formulations may be satisfactorily adjusted for zinc content, variations resulting from untoward application procedures may produce widespread differences in zinc content of an organic zinc rich film on any given job.[58] Zinc is a heavy pigment and tends to settle rapidly in paints despite the use of thixotropes. Improperly homogenized material may result in vehicle-rich material from the top of the can producing films of overly bound and nonconductive product, while overly pigmented material from the bottom of the same can produces weak underbound films (see Figure 2-7). From somewhere in the middle of the same can might come material fortuitously having the correct balance of pigment and vehicle to produce the kind of film that should have been produced from the fully homogenized product. It is for this reason that manufacturers recommend continuous agitation (usually by in-pot mixers) of zinc rich materials both prior to and throughout the application process.[59]

By virtue of the type of organic binder[47] employed individual organic zinc rich primers may themselves be categorized as follows (all are SSPC Paint 20 Type II):

- One- or two-pack
  - high molecular weight linear epoxy (Phenoxy™*)
  - chlorinated rubber
  - silicone
  - epoxy ester
  - moisture-curing polyurethane
  - uralkyd
  - polystyrene
- Two- or three-pack
  - epoxy/polyamide
  - vinyl

As with other organic coatings, the type of binder used will have a marked effect on the coating's performance and will, to some considerable extent, dictate the properties of the individual product. Epoxy/polyamide, phenoxy, chlorinated rubber, and epoxy esters are widely used as binders for organic zinc rich coatings. Each has its own advantages and disadvantages. Medium molecular weight epoxy/polyamides have probably achieved the widest use in organic zinc rich primers. They exhibit a property profile typical of a chemically curing epoxy thermoset and provide strong adhesive films.[60,61,62] They are two-component materials and require the complete mixing of two or sometimes three components (if the zinc is packaged separately) before application. Because mixing ratios sometimes depart from the desired one-to-one by volume ratio typical of most epoxy/polyamide materials, in inexperienced hands use of these materials can be problematic. Phenoxy (high molecular weight linear epoxy) systems, usually single-pack ready-mixed zinc primers, perform well but, unless heavily modified with other polymers (e.g., phenolics), may be too solvent resistant for good intercoat adhe-

*Phenoxy is a trademark of Union Carbide.

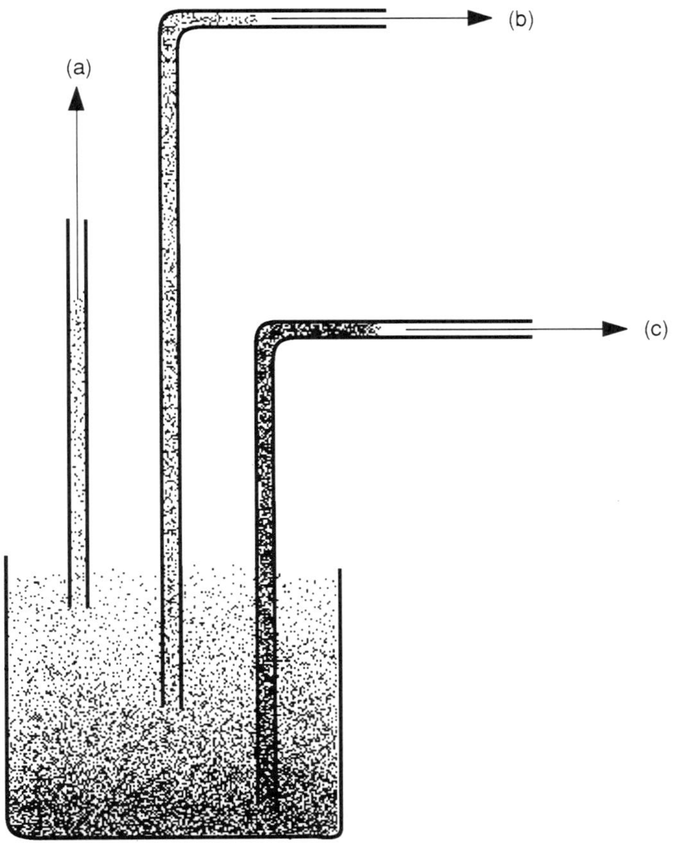

Fig. 2-7. Effects of Heterogeneity (Insufficient Mixing) on the Applied Zinc Rich Film: *a,* Material from upper layers of a nonhomogenized can has too little zinc to maintain galvanic current; no cathodic protection; *b,* material taken from middle layers of a nonhomogenized can has the correct amount of zinc—enough to provide cathodic protection, but not so much that physical properties are compromised; *c,* material from lower layers of a nonhomogenized can carries too much zinc; film is very porous and physical properties are wanting.

sion with certain topcoats, such as vinyls.[63,64] Pure phenoxy zinc/vinyl systems are possible, but require a WP-1 (DOD-P-15328D) tiecoat.[65] Chlorinated rubber zinc rich primers are much more readily recoated and are also currently in use.[66,67] The propensity of this resin to undergo dehydrochlorination, which is catalyzed by the presence of zinc, makes for a dangerous single package composition without a high degree of stabilization. Epoxy esters, although employed in Japan,[68] have not often been used on steel bridge structures in the United States. Extensive use has been made, however, of this type of system[69] in the automobile industry. Epoxy-ester–based zinc rich paints have better adhesion to poorly prepared substrates than do other zinc rich systems. Although little negative field data has been reported, however, the alkali sensitivity of this type of vehicle (varying radically with individual vehicle compositions) has apparently dissuaded the widespread investigation of this type of system. In a properly functioning zinc rich system, the evolution of alkali at the steel cathode is inevitable and should attack the ester groups of the epoxy ester vehicle.

Inorganic Zinc Rich Primers

Inorganic zinc rich primers have been produced with phosphates, titanates, and silicates, although the last are by far the most important commercially. Inorganic zinc silicate primers may be categorized by virtue of their vehicle type, although it seems likely that once fully cured, all zinc silicate films are more or less similar in composition, irrespective of their composition at the time of application.[70]

The two basic types, the alkaline silicate–based zincs and the alkyl silicate–based zincs, cure by somewhat different mechanisms. The alkaline silicates, typified by the silicates of sodium,[71] potassium,[72] lithium,[73] and quaternary ammonium silicate,[74] cure in the presence of zinc. The curing mechanism involves the initial evaporation of water and the hydration of the silicate to silicic acid by way of carbonic acid from the atmosphere or the application of a postcuring solution (such as a weak acid). The silicic acid then reacts with ionized zinc on the zinc particle to form a three-dimensional zinc–oxygen–silicon matrix. This matrix is further crosslinked over several months by the slow neutralization of residual vehicle alkalinity by additional acidity from the air.[63,70,75]

In the alkyl silicate zinc primers (usually prepared from partially hydrolized ethyl silicate or silicates of higher alkyl homologs or cellosolve silicates),[76] solvent evaporation leads to the hydrolysis of the alkyl silicate to silicic acid and the pertinent alcohol (which also evaporates). A reaction between the silicic acid and zinc ion, essentially similar to that described above with the alkaline silicates, follows and the formation of a zinc–oxygen–silicon matrix results.[70]

Historically, inorganic zinc rich primers have been two-component materials, wherein the zinc is added to the silicate mix just before application. In the case of the alkaline silicate, reaction between the solvent carrier (water) and the zinc itself mandates the two-pack nature. In the alkyl silicates, however, various techniques have been employed over the last decade to produce single-component zinc rich products.[76,77] In general, the industry consensus remains that the two-pack inorganic zincs are more reliable for high performance than are the single-pack systems, and cure rates are inevitably faster.

On bridge structures, ethyl silicate–based zincs are by far the most common inorganic primers. None of the alkaline silicates has achieved the same usage, although inorganic zincs based on a colloidal silicate have been used in California.[78] If anything, the alkaline silicates outperform the alkyl silicates, although surface preparation and recoating requirements are perhaps slightly more demanding with these systems than with the ethyl silicates.

An attempt has been made to classify zinc rich coating systems under SSPC-Paint 20.[79] A breakdown of the various coating types outlined above, under this specification, is as follows:

- Alkaline silicates
  - post-cure (SSPC Paint 20 Type IA)
    - sodium silicate
  - self-cure (SSPC Paint 20 Type IB)
    - potassium silicate
    - lithium silicate
    - sodium/lithium silicate

quaternary ammonium silicate
Alkyl silicates (self-cure)
two-pack (SSPC Paint 20 Type IC)
ethyl silicate
cellosolve silicate
ethyl silicate (butyral modified)
one-pack (SSPC Paint 20 Type IC)
polyol silicate
amine-modified colloidal silicate
silane-modified alkyd silicate
sodium-methoxide-modified alkyl silicate

The newly formed inorganic zinc rich film, irrespective of its derivation, is not only cohesively strong but is extremely adhesively secure on well-prepared surfaces. While some controversy exists concerning the mechanism of such strength and adhesion, and while it remains a possibility that the tremendous adhesion typical of these systems is solely a result of secondary valency attractions, the most reasonable explanation seems to be that the primary valency-bonded matrix involves iron ions from the steel substrate as well as zinc ions from the pigmentary surface.[75] Consequently, zinc, iron, and silicate would all be closely bound together in a single large matrix.

Unlike organic systems, the zinc silicate matrix is electrically conductive[70] and the possibilities of highly dielectric encapsulation of some zinc metal in this case are absent. Lower zinc loadings without the switch-off phenomena (brought about by too low a PVC in organic systems) are possible, although performance and longevity of protection (anode life) are directly related to the amount of zinc employed.[80]

As the cured inorganic zinc rich primer ages, the zinc continues to react with the environment, producing a variety of zinc corrosion products (e.g., hydroxides, carbonates, oxychlorides, etc.) depending on the environment.[81,82] As this happens, the resultant corrosion current continues to render the steel totally cathodic and protected. Eventually, the zinc corrosion product expanding from the surface of the zinc particles in the film occupies the interstitial areas between the zinc particles, thereby sealing the primer film completely and producing a dense, extremely hard, abrasion-resistant film. Thus, polarization occurs, and the mechanism of the protection of the steel is gradually converted from one that is primarily sacrificial to one with considerable barrier properties. If the coating in this state is scratched and the steel exposed, sacrificial protection will locally recur until the cut is sealed. In this case, however, only the zinc at newly exposed sloping sides of the cut (rather than the entire surface) will afford a newly bared anode area to protect the exposed steel at the base of the cut. Because anode/cathode area ratios are smaller than would be the case were a scribe to be made shortly after application, the intensity of protection is reduced. Where large abrasions occur and expose considerable steel, anode areas may be too small to prevent steel corrosion.

For cathodic protection to be at all viable, a continuous film of electrolyte is needed across both anodic and cathodic areas. Only a thin surface film is necessary; atmospheric service, however, particularly in dry environments, often precludes this. In this case, steel protection cannot be offered at any bared area.

In most cases on bridge structures, films are recoated fairly quickly after application and long before the primer films are sealed. The penetration of solvents from the topcoat into the primer creates a vapor pressure that produces bubbles or "blowholes" in the finish.[83] The problem is reduced as the inorganic system ages and becomes sealed, and it does not occur in postcured alkali silicates in which sealing is artificially induced during postcuring. (The surfaces, however, require careful cleaning of excess curing residues before coating.) Claims are also made that the vinyl butyral-modified alkyl silicate zincs with smaller particle-sized zinc dusts are more resistant to topcoat bubbling.[84] With organic systems, the problem occurs in the more porous films and may be related to PVC/CPVC ratios. Several techniques have been explored to reduce or even eliminate the problem.[81] The most effective technique seems to be the use of a tiecoat such as the vinyl butyral wash primer (DOD-P-15328D). Other less reliable techniques appear to be the use of thin mist coats to seal the zinc before the application of the full topcoats or the use of "zot coats" (radically diluted finish coats) that are applied before the fully concentrated finish. A more complete dissertation on the effects of topcoating of zinc rich films on performance may be found in the Steel Structures Painting Council (SSPC) report on the subject.[85]

Zinc rich primers are often deliberately extended with nonzinc pigments. These extenders have included either (1) inert materials such as asbestos, mica, or silica to control application, rheology, mudcracking tendencies, gassing or, in organics, PVC/CPVC ratio or (2) reactive pigments such as red lead and lead chromate, which may improve film formation and prolong pot life (inorganic systems) or improve durability.[41] The hexavalent chromates of lead and zinc have also been found to improve durability. It has been postulated that the effect of such inhibitors is to control the rate of anode consumption.[45] One of the most common reasons to modify zinc systems with other pigments is to tint the zinc in order to differentiate easily between the blasted steel substrate and newly primed areas during application. Because the color of the unmodified zinc primer and the white blasted steel are similar, pigments such as lead chromate, yellow iron oxide, lampblack, and red iron oxide have been used. Selection of the particular pigment requires care, especially in inorganic systems (particularly single-pack inorganics) because both stability and performance may be affected. Certain conductive pigments are said to be of value as zinc replacements in sacrificial primers. Careful performance testing of such "extended" systems is advocated, with particular attention being paid to the levels of zinc replacement, before such primers are employed on a commercial basis.

The Barrier Coating

Unlike the inhibitive primer and the sacrificial primer, the third type of coating used in corrosion control, the barrier coating, is used both as a primer film and as a finish coating over other systems.[13] It is typified by an inert pigmentation at lower PVC levels than either of the other two systems and produces a tighter, more cohesive film morphology with significantly lower permeability to water, oxygen, and, in particular, to ionic material.[16] The barrier system is exemplified by a wide range of coatings, such as coal tar enamels, low-build vinyl lacquers, epoxy and aliphatic urethanes, and coal tar epoxies.

Pure barrier systems are often simple composites of multiple coats of virtually the same product. Examples are the Bureau of Reclamation's VR-3 vinyl system[86] (while rarely used on bridge structures, this has been used extensively on locks, dams, and the interior linings of potable water storage tanks), and the two-coat 16 mil U.S. Corps of Engineers polyamide cured, low molecular weight epoxy/coal tar system, which is suitable for the most severe environments.[87] On bridge structures, the coal tar epoxy system, which has poor UV resistance, has been used with success on the steel of the expansion bays beneath the road deck. Corrosion control is often the most difficult there due to the continual access of chloride solutions from leaky joints.[88]

Despite the reduced transmission properties of the barrier system, it has been shown that normal film thicknesses of even the best barrier coatings allow more than sufficient oxygen and moisture vapor through their continuum to support the cathode reaction.[89] Undoubtedly, this coating concept is extremely efficient in maintaining corrosion control even under the most demanding conditions, including immersion in both fresh and salt water, burial in soils, and the like, and service in highly corrosive chemical environments.

The mechanism of protection appears to rely on the ionic impermeability of these films[88,90] (see Figure 2-8). This property ensures that any moisture that penetrates the film's continuum and reaches the steel substrate is of such high electrical resistance that the electrical conductivity of the electrolyte is low and the transfer of corrosion current between anode and cathode is minimized (i.e., I is minimized by maximizing R_e in Equation 2-2). In effect, the barrier coating acts like a continuous filter, allowing only pure water through the film and preventing, or at least severely hindering, the migration of all ionic species. It also seems possible that very thick barrier films may so significantly reduce all moisture and oxygen transmission as to render protection by the simple exclusion of moisture.

Much theoretical investigation of the mechanisms involved in moisture penetration through barrier films has been conducted by Mayne and his coworkers at Cambridge University.[91,92] Changes in the electrical conductivity of free unpigmented polymer films immersed in ionic solutions were found by Mayne to vary either directly (D-type films) or indirectly (I-type films) with the conductivity of the solutions in which they were immersed. D-type films were found to take up both water and ionic species from solution, whereas I-type films took up only water. It was found that polymer films were typically made up of areas of one type of film interspersed with areas of the other.

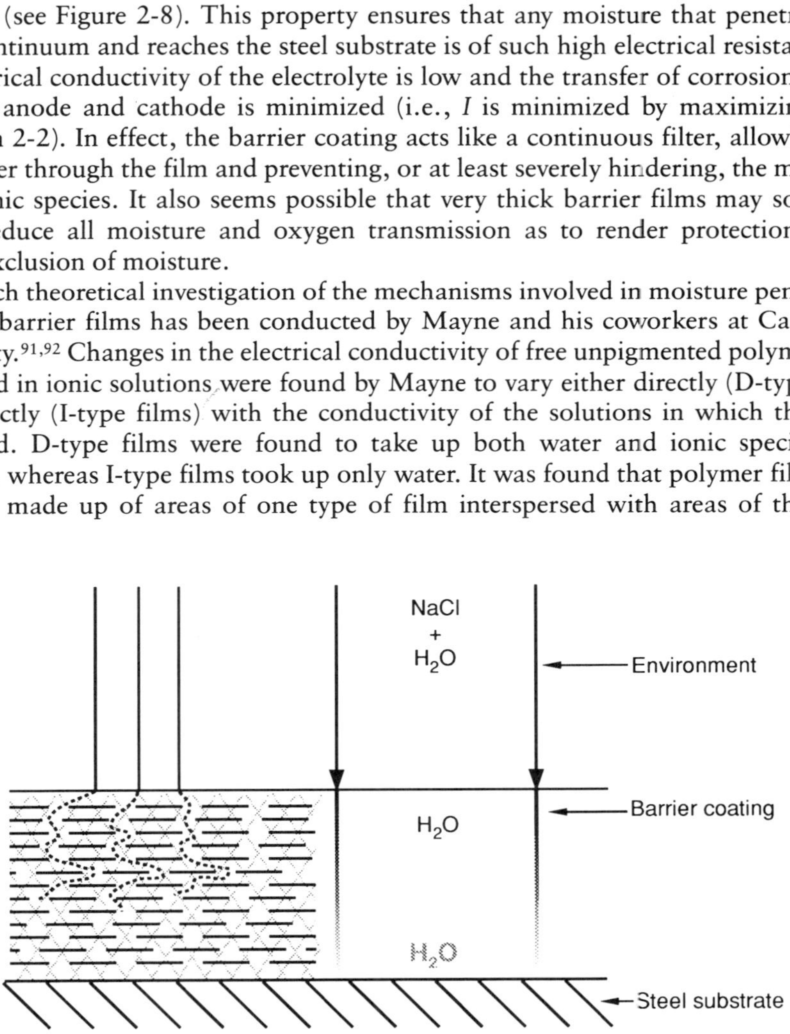

Fig. 2-8. Steel Protection by Barrier Coatings. Highly impermeable films allow no ionic penetration and only slight oxygen and water penetration. Any water reaching the steel contains no ions and has a very high electrical resistance. Hence, there is no external passage of current between local anodes and cathodes and no steel corrosion is present under the film.

Mayne concluded that polymeric films were heterogeneous in crosslink density and that the D-type areas were areas of lower crosslink density that allowed the transmission of ionic species. The I-type areas, on the other hand, allowed transmission of water only. Furthermore, the D-type areas were subsequently associated with corrosion of the substrate.[93] Thus, it may be concluded that uniformity and degree of crosslinking are important to satisfactory barrier systems. More recently, Callow and Scantlebury have demonstrated dyeing techniques to identify these D-type areas of film.[94] It may be that a similar homogeneity in thermoplastic systems is also related to diminished ionic uptake, although work with a chlorinated rubber system would not support this.[95] Other studies have related the effectiveness of barrier films to molecular and/or morphological uniformity and compactness.[96]

As a result, in the formulation of the barrier system, it is essential that all ion-producing pigments are replaced by inert materials. For example, the inclusion of inhibiting pigments in the barrier film and, in some cases, in films beneath a barrier system is counterproductive, especially where ionogenic (ion-producing) pigmentations of high solubility are employed in systems for use in high-humidity environments or in fresh-water immersion service.[16] Under these conditions, osmotic gradients are likely to be set up resulting in film breakdown by excessive blistering.[97] Similarly, the entrapment of ionogenic species beneath the film may short circuit the necessary high electrical resistance of moisture getting through the film, thereby nullifying the functioning mechanism of the system.[98] It is for this reason that surface preparation with true barrier systems must be totally exacting: residual chlorides and sulphates are devastating.

Flat, platey pigments are extremely valuable in the barrier film because they assist in preventing moisture and oxygen penetration by increasing the path of diffusion through the film.[99] Glass flake, micaceous iron oxide, graphite, and mica are all employed in such coatings. Where possible, wide use has been made of aluminum flake in barrier systems, and experimental evaluations of this type of pigment against other lamellar pigments have shown its value.[100] High-solids epoxies[101] and moisture-curing urethane barrier systems[102] employing aluminum have recently been achieving an increasing share of the bridge-coating market. Because of their high build and good wetting properties, several of these materials have been marketed as coating systems for poorly prepared surfaces.[101] Still all manufacturers encourage optimized surface preparation where maximized performance is required and do not recommend the systems' use over highly pitted steel surfaces, preferring a blasted surface. It might seem that the effects of aluminum pigments in films of relatively high build would reduce water penetration sufficiently to achieve some degree of protection when used over marginal substrates.[103] The technique, however, seems fraught with dangers and is not considered wise practice from a theoretical standpoint unless such factors as environmental restrictions or bridge design limit uniformity of excellence in surface preparation and lead pigmentation is prohibited.

Judicious pigmentation, whether in type or amount, must also include consideration of ways to maximize the electrical resistance of the paint film.[104]

Vehicle selection is also important and is generally best confined to those materials with inherently low moisture- and oxygen-transmission rates and high dielectric constants. Polymers bearing carboxylic groups, hydroxyl groups, and backbone ester groups do not generally produce as good a barrier system as polymers having carbon and carbon-oxygen backbones and aromatic structures.[98,105,106] Paradoxically, polar groups, such as hydroxyl and carboxyls, are more suited to the maintenance of good adhesion, and their absence in many barrier systems is a further mandate for excellence in surface preparation.

Vinyls, chlorinated rubbers, and coal tar systems make excellent barrier vehicles. Although epoxies and urethanes do not make quite as excellent barrier vehicles as do vinyls and chlorinated rubbers because of their somewhat higher moisture and oxygen transmission properties,[107] they are very soluble, particularly when pigmented with barrier pigments, because they are formulated at much higher solids and this allows for the deposition of much thicker films.[108,109,110] Oil and alkyd systems, with their overabundance of backbone ester groups, do not make good barrier vehicles because of their vulnerability to alkaline saponification. Latex systems, like alkyds, do not make good barrier coatings[111] and are best employed as inhibitive systems. Even then, multicoat systems are required and their use in snow-belt and in coastal locations will probably be less successful than in drier interior environments. In latex systems, the mechanisms of film formation[112] do not encourage the deposition of continuous films, and it has been shown experimentally that latex systems are more permeable to salt solutions than other coatings.[113] Careful formulation can maximize performance of latex systems, but while aluminum pigmentation and new resins such as vinylidene chloride[114] terpolymers and certain special proprietary acrylic latices[115] improve impermeability somewhat, this type of coating still falls short of the solvent cast vinyls, chlorinated rubbers, and other high-performance solvent-cast systems in impermeability.

The tendency of vinylidene chloride to undergo dehydrochlorination (producing chloride ions in situ) is a concern despite claims made by manufacturers[116] regarding the positive effects of polymer block spacing. The unique acid-buffered formulation technology of these new systems is also claimed to assist the stabilization of these paints.[117]

Film thickness is critical and, all other things being equal, the thicker the barrier coating system, the longer and more successful will be the protection afforded.[105,118] Unfortunately, while economic necessities foster the trend toward fewer thick coats, in general, barrier systems give better performance when the same film thickness is built up from three or four successive thin coats rather than two coats with twice the film thickness or a single coat with four times the film thickness.[119] Pressures on the industry from environmental groups are also spurring much development work in the field of high-barrier coatings.[120]

In general, high-build barrier systems are of higher PVC and, morphologically, are less compact than low-solids barrier systems. This inevitably demands some compromise in impermeability, although this may be made up with increased thickness. High-build systems used as intermediate coats between zinc rich primers and low-build finish coats are quite popular.

In addition to reduced impermeability, high-build barrier systems of higher than normal PVC may also suffer from their solvent system. These materials are often formulated with high levels of thixatrope and slow solvent systems. In cases in which such solvent systems involve the use of water miscible solvents (e.g., ethylene glycol mono ethyl ether acetate), problems have also been observed in premature immersion service. Insufficient drying before immersion or use in environments of high humidity may lead to water inbibition and consequent blister formation.[121]

Composite Coating Systems

Compromises in impermeability may be acceptable from high-build barrier systems when such systems are used over zinc rich primers, particularly those of the inorganic type. In this case, the finish is no more than the front line of defense and is supported by the galvanic protection of the zinc rich system. It may be argued that the only reason for a finish coat on a properly functioning zinc rich system is aesthetics; therefore, high-build barrier coats with greater permeability make perfectly sensible finish coats for such systems.

It may also be argued that in practical circumstances, the tight, relatively moisture impermeable barrier systems may actually contribute to problems when used over improperly cured zinc rich systems. Many zinc rich and other primer films require either water or oxygen to completely cure. If recoated too rapidly with highly impermeable finish coats, the curing process in the primer may be virtually halted by the presence of the topcoat, and the undercured state may be "locked in," precluding the attainment of optimum physical properties in the primer.

When using coating systems involving barrier topcoats, it is necessary to verify complete cure (and/or solvent release of the primer) before the primer is recoated. This is particularly important when the drying of the primer is slow or its cohesive strength is likely to be limited, as is very likely with zinc rich systems. Too rapid a recoat time with too impermeable a film has been cited as cause for intrafilm splitting (lateral cohesive failure) in several paint system failures on bridges bearing inorganic zinc/vinyl systems (W. M. Medford, personal communication, 1984). On the other hand, in a laboratory study by Gelfer and Vondorsten, exceptionally rapid recoating of zinc rich films has resulted in no obvious problems.[122] Lateral cohesive failure is a problem to which all zinc systems may be heir.[54] It may also occur when lacquer-type primers (phenoxies and chlorinated rubbers), particularly those containing slow solvents, are recoated with thermosetting coatings before solvent evaporation is complete and the primer's strength is consequently optimized. Similar failures have also been seen when epoxy-ester zinc rich coatings have been recoated with barrier systems before oxidation is sufficiently advanced. One major manufacturer requires at least seven days curing before recoating inorganic zincs with vinyls (D. Miller, Tnemec Company, personal communication, 1984). Other manufacturers, although less conservative, have tended in recent years to increase the recommended cure time of newly applied zincs before recoating. Inevitably, the time required for such zincs to acquire a full cure will vary from job to job, based not only on temperature and relative humidity but also on thickness of the applied primer.

Determining how long inorganic zincs must cure before recoating is the subject of much research within the industry. The coin hardness test has been used historically, although work at the Georgia Institute of Technology would indicate that a methyl ethyl ketone (MEK) rub test is more reliable.[123] Results in this case have been correlated to cure kinetics measured with infrared (IR) spectroscopy. The subject is also discussed elsewhere by Starr, Oechsle, Lanning, and Ginsberg.[124]

The design of functioning composite paint systems is often more exacting than the design of the individual coatings themselves. It is for this reason that individual elements of a complete coating system should be selected from the same manufacturer. Only he is totally aware of the particular properties of individual elements (paint films) and only he will have examined the interactions of those elements in a total composite system. Coating system failures in the field are not as likely to result from a defect in a single element of the coating system as they are from a reaction between one element and another. What is more, such effects are less likely to be identified in the laboratory as the system is being designed than they are in the field when the system is put into service.

Such multicoat design considerations are now well appreciated at the design stage. In fact, the concepts involved are so basic that it is difficult to determine whether the system design precautions respond to problems that were first predicted from theory, derived from laboratory practice, or identified as a result of field problems. Some problems (usually related to application) are so obvious that laboratory experimentation would probably reveal them long before field difficulties were encountered. Examples of such failures are (1) the attack of the aromatic and oxygenated solvent systems of finish coats on newly applied oxidizing systems, (2) the bubbling of vinyl topcoats over highly porous zinc films,[125] or (3) the alligatoring of highly inflexible finish coats over inappropriately flexible primers.[97] Other design considerations might be predicted from theory. These include (1) the need to use nonmetallic primers to insulate metallic finish coats bearing electropositive pigmentation (copper or stainless steel) from steel substrates to avoid possible galvanic substrate pitting; or (2) the complete degradation of an oleoresinous coating system by cathodic alkali in the areas of expansion bays subjected to continual drenching in salt water.

Probably the majority of other design considerations result from difficulties originally encountered in the field. These are most often longer-term problems that result in failure after the system has been in service for a number of months or even years. Design considerations prompted by these problems include (1) the inadvisability of applying alkyds and oil-based finish coats directly over zinc surfaces,[83] (2) the necessity of using a washprimer tiecoat in recoating unmodified phenoxy zinc rich primers,[63] and (3) intercoat adhesion difficulties with epoxy systems that result from amine carbonate contamination of slow-curing (or noninducted) primers when recoated without carbonate removal. Unfortunately, it seems that most problems that indicate a need for further design control considerations are encountered in the field. Perhaps this indicates a lack of foresight and understanding on the part of even the best system design engineers.

While there are certainly instances of failure caused by carelessness in coating system design and application, it seems that in many, if not a majority, of cases, such failures are the result of a deficiency in laboratory testing procedures. During the experimental evaluation of potential coating systems, more attention to all aspects of design, use, and application procedures might help to identify system design problems before they occur in the field.

Chapter 3

Surface Preparation

With the revolution in bridge-painting systems that began in the 1960s inevitably came a corresponding revolution in surface-preparation techniques. While oil and long-oil alkyd systems were rather tolerant of the type of surfaces left by simple hand- or power-tool cleaning techniques, the newer high-performance coating systems were not; to work satisfactorily, they demanded clean surfaces free of contamination. As the emphasis in coating systems gradually changed from straight inhibitive systems to zinc rich and barrier systems, so the importance of mill scale removal increased, and the elimination of residual chlorides and sulphates became more critical. As environmental regulations prohibiting the use of the more efficient inhibitive pigments based on lead and hexavalent chromium spurred on the experimentation with such nontoxic inhibitors as borates, phosphates, and molybdates, the greater sensitivity of these inhibitive systems to chloride-ion depassivation again encouraged the abandonment of the older preparatory techniques in favor of those producing a physically and chemically purer surface.

By the late 1960s, ten states had abandoned hand- and power-tool cleaning in favor of some form of abrasive blasting.[1] Generally, this was merely brush-off blasting (SSPC-SP-7), although one state, Louisiana, adopted a white-blast standard for a while and shortly thereafter changed to a commercial SSPC-SP-6 standard as a more cost-effective procedure. Even so, many states continue to employ oil and alkyd systems, applying these over an SSPC-SP-6 or -7 type surface. By the 1980s,[2] almost all states were employing abrasive blasting in some form in maintenance operations as well as in new construction. Many states were employing several alternative coating systems (usually two): a conventional lead-based oil/alkyd system, and a high-performance system, most often an alkyld silicate-based inorganic zinc rich system. This dichotomy produced two standards of abrasive blasting: brush-off or commercial for the alkyds, and commercial or near-white for the zinc systems.[3]

Those states that retained the oil-based systems have found advantage in improved performance with a higher level of surface preparation. This is certainly true in new construction where entire steel surfaces are blasted. As is discussed below, problems still can occur in spot maintenance, during which old films around the periphery of a spot-blasted area may be severely damaged without actual removal. Recoating can only further weaken the old paint integrity. Despite the wariness that exists in some circles over high-performance coating systems (probably originating in the many reports of catastrophic paint failures that have occurred with the use of these systems), the selection of low-performance oil paints for shop application over high-quality blasted surfaces is somewhat surprising. The better control that the shop application of these high-performance systems provide should produce a consequent improvement in performance and better reliability than is possible in the field.

The improved performance of the simple oil/alkyd systems over better prepared surfaces cited by many states is hardly surprising. The beneficial effects of improved surfaces on paint-system performance has long been recognized, although over profiled substrates, the application of films of sufficient thickness becomes more critical. Since the early work of Hudson,[4] numerous other studies[5–9] have demonstrated the importance of good surface preparation, so that now, equating surface preparation with performance is axiomatic. Data from a Federal Highway Administration (FHWA) study evaluating the effectiveness of a wide range of specialty coatings for use over rusted steel substrates indicates that none is capable of producing a performance profile better than that of most bridge-paint systems on properly prepared steel substrates.[10]

SCIENTIFIC JUSTIFICATION FOR SURFACE PREPARATION

In the day-to-day discipline of steel maintenance, the basic reasons for the benefits of good surface preparation are often lost. Before proceeding to a discussion of the various methodologies involved in surface preparation, then, it serves to review the scientific justification for the process.

The four principal reasons for surface preparation are as follows:

1. To ensure a uniform substrate that is as close as possible to the theoretical model conceived by the specifying paint-system design engineer
2. To ensure adhesion of the paint system to the steel by removing all loose and tightly adherent contamination and foreign matter (including rust scale and mill scale) from the steel surface, and so freeing up reactive sites on the metal surface that may be covered or chemically occupied by such contamination in order that subsequent reaction with the paint may take place
3. To improve adhesion through greater primer/substrate reaction density by increasing the number of reactive sites per given square centimeter of surface area (i.e., by increasing the real surface area per apparent surface area)
4. To ensure that the mechanism by which the paint system protects the steel is neither hindered nor prohibited, either chemically or electrically, by the presence of such soluble moieties as chloride and sulphate ions on the steel surface.

Normalization of Substrate

Although its basis is not strictly scientific, the normalization of substrates is important. In writing a bridge-paint specification or developing a paint formulation or paint system in the laboratory, the design engineer conceives and uses steel in a condition that may be quite different from conditions in the field. Two-hundred feet above the water, all types of conditions not present in the theoretical concept of the surface for the paint system may prevail. These conditions may include heavy rust scale, mill scale, salt contaminations, road dirt, pigeon droppings, old paint films and paint film residues, and many other conditions that may be isolated or unique to a particular structure. Adequate surface preparation can transform these many different surfaces into one that more or less closely approximates the surface on which the paint system was developed in the laboratory. Patterns of pitting, design geometrics, and exposure conditions may compromise the good theoretical paint specification, but a good surface preparation can give the system a fighting chance from the start.

Removal of Interference Material

Multicoat paint systems are an arrangement of individual entities that are linked together by a system of adhesive forces, which anchor successive coats to each other, and cohesive forces, which vertically and laterally hold the single paint films together. Because the primer film or first coat is adhesively anchored to the steel substrate, the substrate itself must necessarily be considered part of the paint-system composite. As a result, the characteristics and quality of this surface are as important to the whole as are the type and quality of each paint film that comprises the paint system.

Most paint films are quite different from a steel substrate. This creates a need for special consideration if we are to maintain adhesion of the first paint coat (or primer film) to the steel.

Paint films adhere to steel by one of two basic mechanisms. In the first, the binder or glue that holds the paint film together actually reacts chemically with the steel substrate. This is known as *primary valency bonding* (see Figure 3-1). It is the most satisfactory kind of bonding, but it is still rare with paint systems and is confined to such materials as conversion coatings,[11] wash primers,[12] and inorganic zinc systems.[12a] More commonly, paint films adhere to steel by way of physical attractions or *secondary valency bonds* (hydrogen bonds, Van der Waal's forces, etc.), which are the same kind of forces that prevent us from pulling two pieces of plate glass apart (see Figure 3-2). The strength of a primary valency bonded wash primer has been measured to be as much as six times that of a secondary valency bonded polyvinyl acetate lacquer.[13]

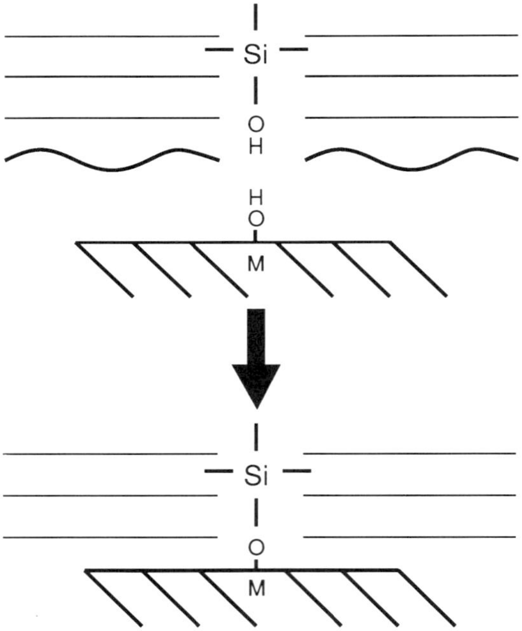

Fig. 3-1. Primary Valency Bonding. Paint vehicle forms chemical bonds with substrate.

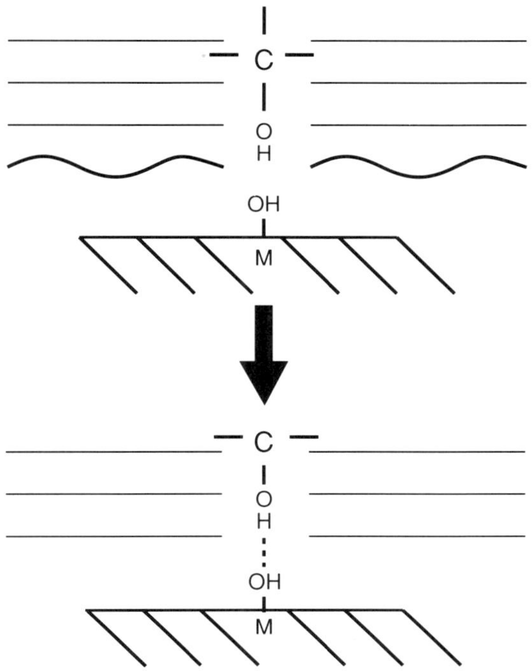

Fig. 3-2. Secondary Valency Bonding. Paint vehicle is held to the substrate only by a strong physical attraction.

Unfortunately, for either type of adhesive bonding to occur, the paint film must approach the steel substrate very closely. As distances between attracting sites on the steel and the coating exceed five angstroms (approximately three times the diameter of an oxygen atom), the forces that pull them together decrease very rapidly. On this scale, a particle of road dirt is like a boulder and will totally prevent the paint from ever reaching the steel. The paint may stick to a dirt layer, but unless the dirt is tightly adherent to the steel, the entire system remains loose and will peel away as the system is placed into service. In service, the system is continually stressed by a variety of stresses—first internal, and then external. External stresses (e.g., expansion and contraction of the paint system with temperature differentials, imbibition and desorption of water, traffic vibrations, impacts, etc.) can be very severe, but not as severe as internal stresses—those initial severe demands that are exerted on the system as newly applied films with active solvents first wet the existing film and then lose solvent and/or polymerize, shrinking both vertically and laterally.

To achieve such proximity to the substrate, the paint film must wet the substrate. Under ideal circumstances, this wetting should not be difficult. As a general rule, liquids of low surface tension readily wet solids of high surface energy, and paint films with surface tensions of 20 to 70 dynes per centimeter have little problem wetting surfaces of high energy, such as steel (about 400 dynes per centimeter) (see Figure 3-3). Unfortunately, a smear of oil one molecule thick on a steel surface can transform the surface-free energy of that steel from 400 dynes per centimeter to less than 25 dynes per centimeter so that a paint of, say, 40 dynes per centimeter will never wet the surface. Paint films of low surface-free energy, say, 20 dynes per centimeter, will still wet that

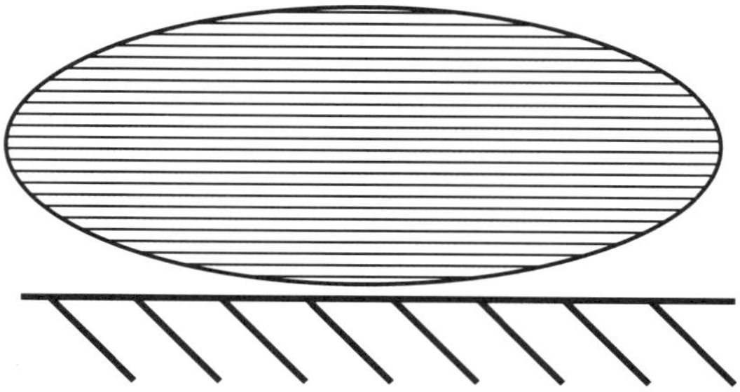

(a)

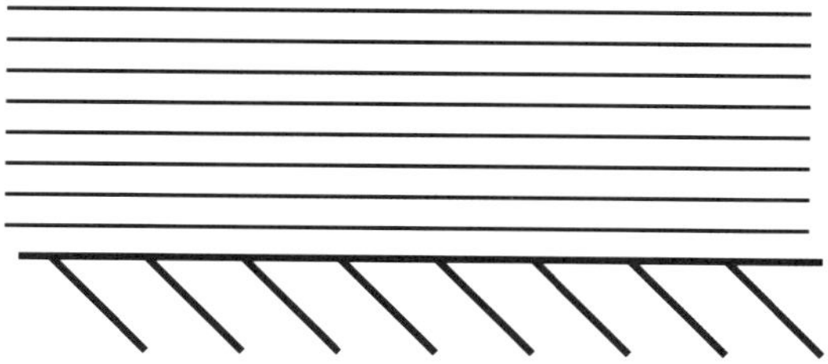

(b)

Fig. 3-3. Wetting and Surface Energetics: *a,* high-surface-tension liquid will not wet low-energy substrate, nor will it approach surface close enough for good adhesion; *b,* low-surface-tension liquid will wet high-energy substrate; the approach will be close enough to allow adhesion.

surface, however, and this is the basic reason that for years, the low-surface-energy oil paints have been successfully used over all types of ill-prepared steel surfaces. Unfortunately, most of the newer, higher performance systems (the inorganic zincs, the vinyls, etc.) have higher surface energy and thus cannot tolerate the same poor-quality surfaces as did oil paints. Therefore, as we demand higher performance and longer service lives from our systems, the need for better surface preparation also increases.

A steel surface may be imagined as a plane on which, with some regularity, there are sites that are more conducive to reaction with complementary sites on a paint vehicle. This holds true whether adhesion is by means of primary bonding reactions or secondary valency attractions (see Figure 3-4a). Unfortunately, on the in-service steel surface, these reactive sites are often occupied by oxide deposits and scales or screened from potential coatings by dirt or other contaminants that prevent or at least impede access of such coating vehicles (see Figure 3-4b).

Oxide deposits may be loosely adherent hydrated oxides (rust), tightly adherent oxide scales formed during the high-temperature rolling of steel (mill scales), or heavy, adherent scales of oxides formed beneath paint films and made up largely of magnetite (rust scales). While application of the paint films over loose rust may produce more immediate failure than application over mill scale, both mill scale and heavy rust scale will eventually delaminate from the steel, producing an equally untenable bond. By

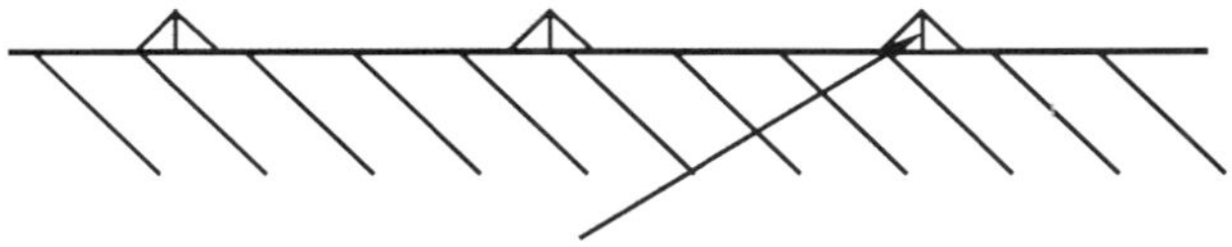

Fig. 3-4a. Stylized Representation of Steel Surface

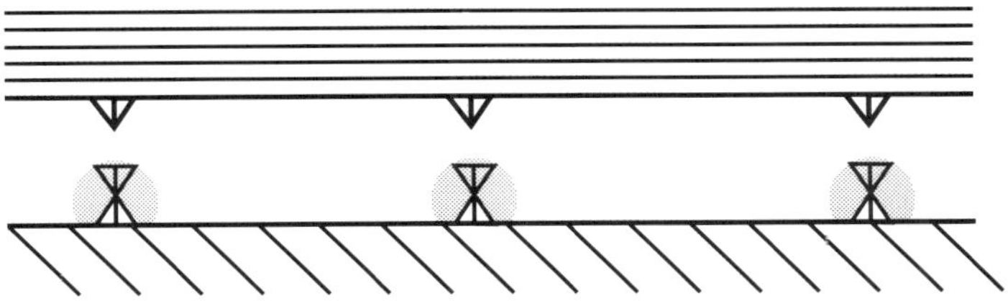

Fig. 3-4b. Rusty Steel Surface. Reactive sites are occupied by rust, and paint will not bond.

stripping these heavy oxide films from the steel surface, surface preparation presents highly active steel surfaces with newly bared reaction sites that may then be satisfied in reaction (either chemical or physical) with reactive groups on the vehicle of the paint system (see Figure 3-4c).

As is discussed in Chapter 2, corrosion of steel generally occurs because of local heterogeneities in the makeup of the steel surface. One common source of such heterogeneities is contamination, which must be removed. Mill scale, for example, is electrically less active than steel and will actually form the cathode of a galvanic cell with bare steel, intensifying the corrosion at areas where the mill scale has been broken. Similarly, rust may act as a contaminant and, by preventing or impeding access of a corrodent to the steel over which it lies, may artificially change the electrochemical potential of the steel beneath the rust, setting up concentration cells that again may induce corrosion. Other impurities, such as accumulations of road dirt, sand, pigeon droppings, refuse, and the like, will produce similar problems.

With some types of surface preparation, the removal of mill scale and rust is not accomplished easily or thoroughly. And while the resultant surfaces may seem improved, they may bear hidden problems. Whenever mill scale, rust, and other contaminants are beaten or forced into the steel by certain types of hand-tool cleaning or, in

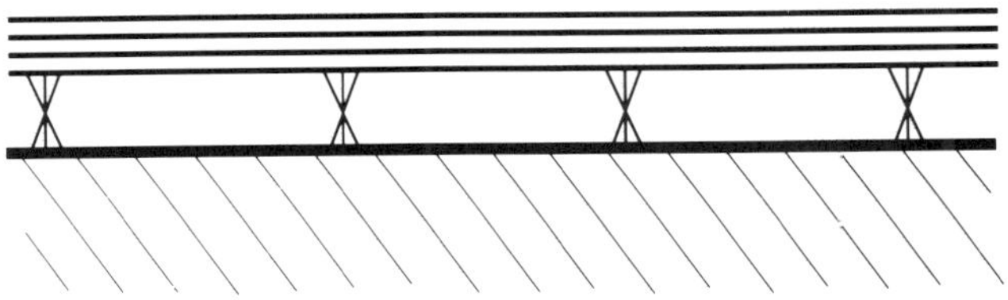

Fig. 3-4c. Well-Prepared Steel Surface. Rust stripped to expose reactive sites that can subsequently attract complementary sites on paint vehicle, and paint will bond.

some cases, even by shot-blasting the surfaces, the resultant steel may be left in a contaminated state with increased heterogeneities[14,15] (see Figure 3-5). Scanning electron microscopy has shown, in fact, that all abrasives, including grit, leave trace amounts of impurities in the steel. Alumina gives the cleanest surface and, generally, nonmetallic abrasive gives better paint performance than do metallics.[16] On the other hand, where the steel is blasted with angled grit so that the surface is cut instead of peened, these effects are significantly reduced. Also, where mill scale and rust scale are incompletely removed and beaten into the steel, the reaction sites discussed above are incompletely bared and adhesion of the subsequent coating is deficient.

Increasing Surface Area

Where steel surfaces are prepared by physical or chemical scarification, such as sandblasting or pickling, there is a further advantage. In such removal, the contour of the steel surface left after scarification is transformed from a flat plane into one that is broken up into numerous peaks and valleys (more so during grit-blasting than shot-blasting). This, in turn, increases significantly the real surface area of the steel over any given square centimeter. As a result, the density of those reactive sites is increased over any given square centimeter of surface.[17] Because there are more potential sites upon which to anchor a paint film vehicle (whether such anchorage is by way of primary valency bonding, physical attraction, or secondary valency bonding), the overall adhesion of the system is improved (see Figure 3-6).

Removal of Water-Soluble Salts

It has been shown[18] that the inhibitive primer functions by the formation of protective layers on the steel surface beneath the paint film. While chloride and sulphate ions in service environments are powerful depassivators, thereby nullifying the effectiveness of passive films by preferential adsorption onto the steel at the metal surface, the low ionic permeability of properly designed primers ensures a favorable ratio of inhibitor (derived from the paint film) to corrosive ionic species (derived from the environment) at the interface. Such passivating primers continue to function well. On the other hand, if the surface to which the paint is applied were contaminated with chlorides or sulphates

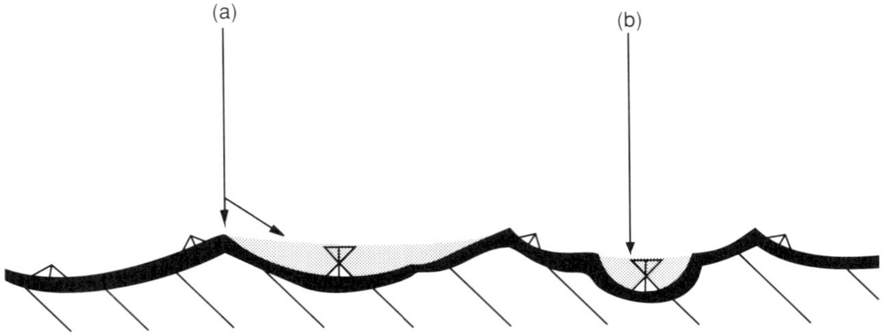

Fig. 3-5. Peened Surface with Rust-Scale Residues Forced into Steel: *a,* heterogeneities with severe differences in electrochemical potentials; *b,* reactive sites still occupied by rust scales.

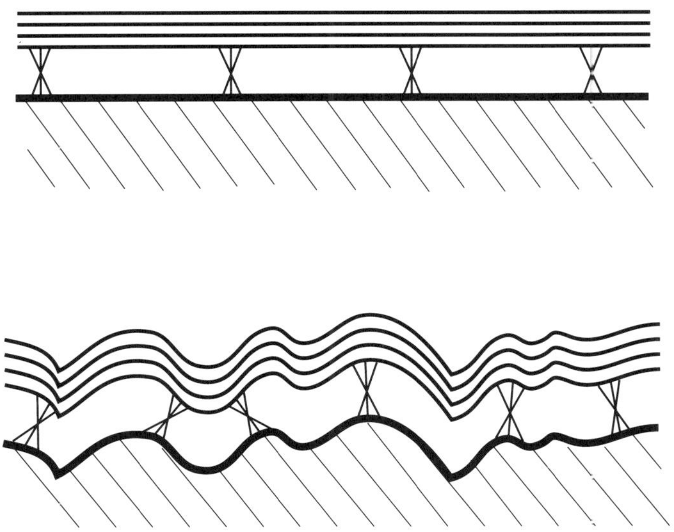

Fig. 3-6. **Effect of Surface Scarification on Adhesion.** Increased real surface area per apparent surface area increases number of reactive sites available for reaction with paint vehicle.

before the application of the primer, the low ionic permeability of such primers serves little purpose. In this event, water penetrating the coating finds a rich reservoir of soluble corrosive ionogenic (ion-producing) matter at the steel surface (beneath the paint). This effectively prevents passivation by the inhibitive ions derived from the paint and ensures rapid underfilm deterioration of the substrate, despite the effectiveness of the best inhibitive primer formulations (see Figure 3-7).

A similar scenario exists with barrier systems. Here, of course, there is less penetration of water, virtually no ionic penetration, and the film contains no soluble inhibitor. Protection is afforded solely by the high electrical resistance of any moisture penetrating the coating to the metal interface. The presence of corrosive ionogenic species at the interface, therefore, dramatically increases the conductivity of any moisture reaching the substrate, which readily corrodes in the presence of a favorably conductive electrolyte[19,20] (see Figure 3-8).

Chlorides are particularly common on bridges in the snow belt and those near the coast (see Figure 3-9). Equally damaging to paint systems, however, are the soluble sulphates that predominate in industrial areas (see Figure 3-10). These result from contamination of the steel with gaseous oxide of sulphur. Schikorr has shown a relationship between the corrosion of steel and the levels of sulphur dioxide in the air.[21] Similarly, Mayne[22] has shown that paint life may be related to levels of iron-sulphate contamination of rust on a corroding surface. A periodicity in contaminant levels by season that was shown in both studies is related to the burning of high-sulphur fuels in winter. Rust absorbs large quantities of sulphur dioxide, and sulphate "nests" grow beneath the rust pits on the steel. As iron sulphate is oxidized and hydrolized, the resultant rust occupies a much larger volume than did the parent metal, and this results in the upheaval and undercutting of the coating around the pitted area and the growth of the nest itself. Moreover, the resultant area of coating breakdown tends to absorb and retain water longer than areas of sound coating (see Figure 3-11), thus encouraging

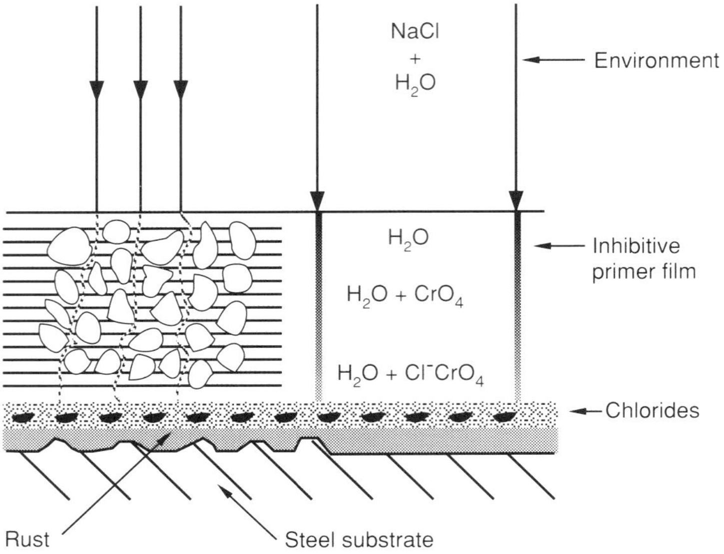

Fig. 3-7. Prohibitive Effects of Chlorides on Inhibitive Primer Protection. Unremoved salt contaminating steel beneath paint film dissolves in water that penetrates film and is preferentially adsorbed onto steel surface instead of inhibitive ions. No passive film is set up and steel corrodes beneath film.

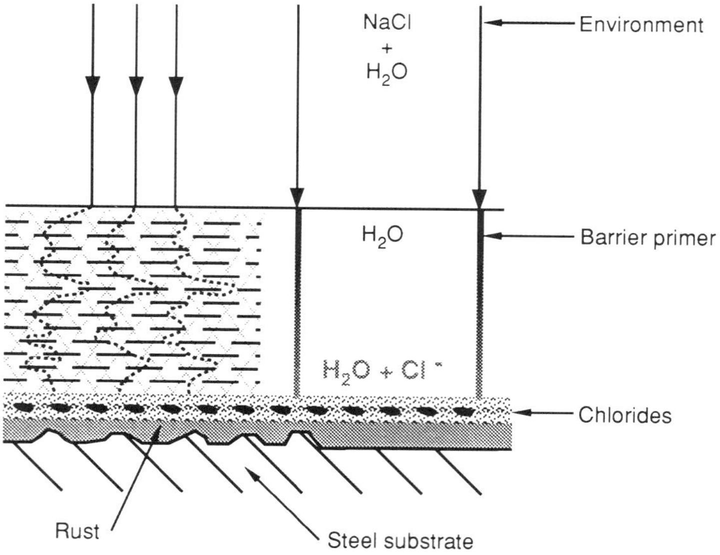

Fig. 3-8. Prohibitive Effects of Chlorides on Barrier Coating Protection. Coating still prevents the penetration of ionic solutions, but pure water, penetrating to substrate, dissolves chloride ions at the substrate and becomes electrically conductive and corrosive. Steel corrodes and the paint eventually blisters.

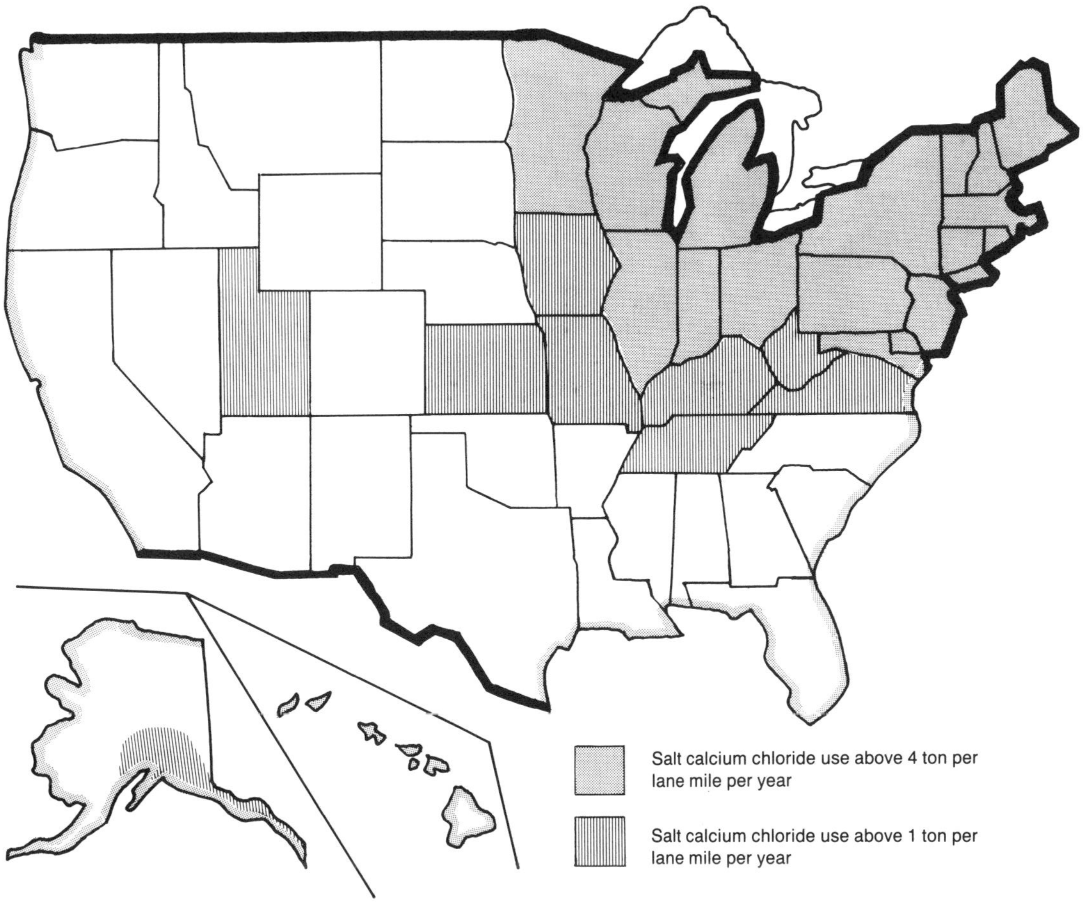

Fig. 3-9. High Chloride Environments Related to Use of Deicing Salts and Marine Proximity

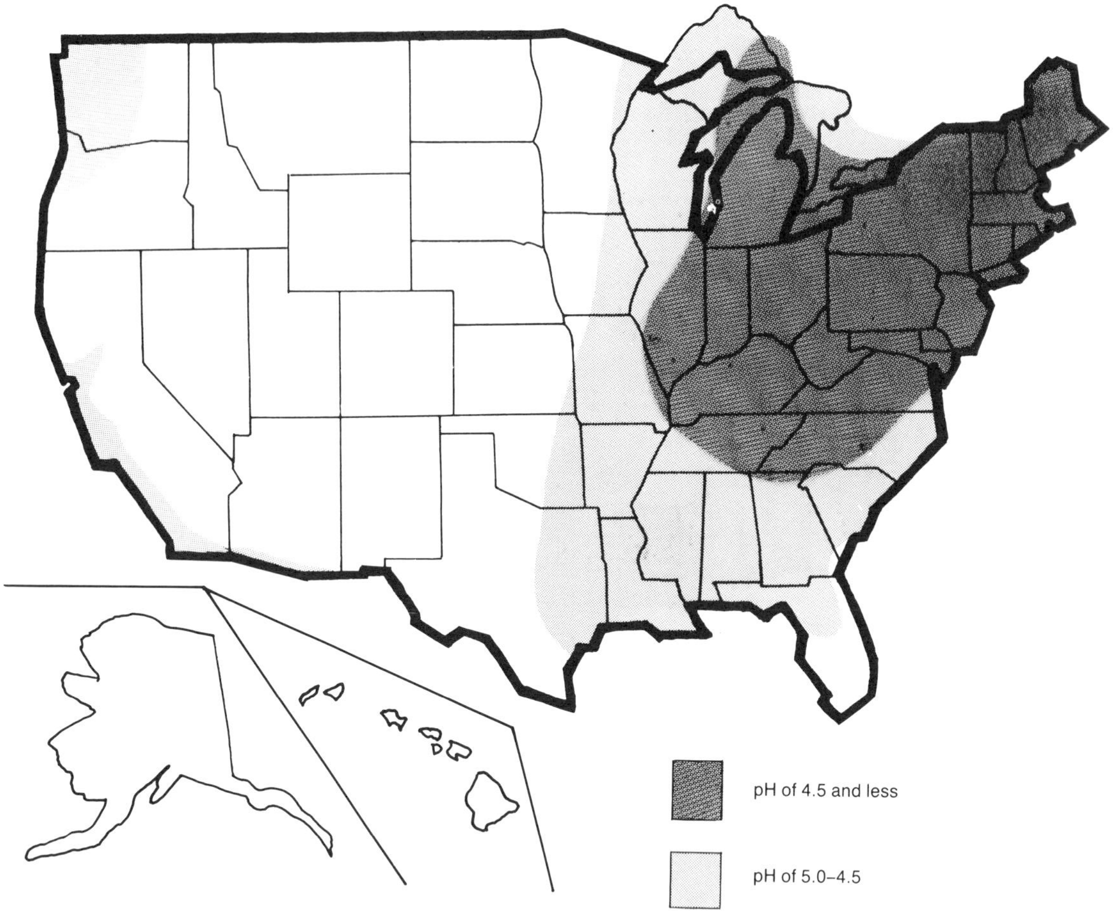

Fig. 3-10. Mean pH of Precipitation in the United States (late 1970s)

Fig. 3-11. Drying of Water Wet Surfaces Is Much Slower in the Corroding Areas Than It Is Where Paint Is Intact

further deterioration. Since the iron sulphate nest is beneath the rust, it is not readily removed with the rust; because of the salt's relatively low solubility in water, it is only solubilized with difficulty.

Igetoft[23] reports that there is a body of data that shows that in noncoastal areas sulphates may be more of a problem than chlorides. However, this is unlikely to be the case during the winter on bridge structures in the snow belt, where chloride contamination from the use of deicing salts is considerable. In industrial areas, or in the northeastern United States in the summer, acidic air pollution could well increase the levels of sulphate contamination above the levels of chlorides. There is some evidence that synergistic effects of chloride and sulphate contamination may produce significantly increased corrosion.[24]

SURFACE PREPARATION METHODOLOGIES

Several organizations across the world have established standards for surface preparation.[25-33] Some of these are confined to the specifications of blast-quality standards only,[26,28] while some cover other types of surface preparation.[25,30] A comparison of such standards is shown in Table 3-1. The list that follows outlines attempts made by SSPC to categorize the initial surface conditions of both new and painted steel surfaces.

UNPAINTED

A. Steel surface covered completely with adherent mill scale with little, if any, rust (ASTM D2200 or SSPC-Vis-1 Rustgrade A).
B. Steel surface that has begun to rust, and from which the mill scale has begun to flake (ASTM D2200 or SSPC-Vis-1 Rustgrade B).
C. Steel surface from which most of the mill scale has rusted away or from which the mill scale has begun to flake (ASTM D2200 or SSPC-Vis-1 Rustgrade C).
D. Steel surface from which the mill scale has rusted away and where pitting is visible ASTM D2200 or SSPC-Vis-1 Rustgrade D).

PAINTED

A. Paint almost intact; some primer may show; rust covers less than .1% of the surface (ASTM D 610 or SSPC-Vis-2 Rustgrades 8–10).
F. Finish coat somewhat weathered; primer may show; slight staining or blistering; after stains are wiped off, less than 1% of area shows rust, blistering, loose mill scale, or loose paint film (ASTM D610 or SSPC-Vis-2 Rustgrades 6–8).
G. Paint thoroughly weathered, blistered, or stained; up to 10% of surface is covered with rust, rust blisters, hard scale, or loose paint film; very little pitting visible (ASTM D 610 or SSPC-Vis-2 Rustgrades 4–6).
H. Large portion of surface is covered with rust, pits, rust nodules, and non-adherent paint. Pitting is visible (ASTM D610 or SSPC-Vis-2 Rustgrades 0–4).

None of the specifications details all methodologies, and most are more concerned with the visually apparent quality of the blast rather than either surface area or freedom from ionogenic material. In many specifications, surface area and mil profile are controlled separately from the control of blast quality. Few current U.S. specifications consider the chemical cleanliness of the substrate, and, in general, an awareness of the importance of the removal of chlorides and sulphates has not assumed the same importance here that it has in Europe at this time. These deficiencies are thoroughly considered by McKelvie in a paper reviewing present surface preparation standards.[34] The establishment of a reliable quantitative field test is difficult, however, and even in Europe viable standards remain elusive. Subcommittee TC35/SC12 of the International Standards Organization is presently at work designing a more comprehensive international standard. The scope of this work is described by McKelvie.[35]

The preferred surface-preparation technique, therefore, is that which satisfies all the four requirements of (1) normalization of the surface, (2) exposing the reactive steel, (3) increasing the real surface area of the steel, and (4) removing all ionogenic materials. Surprisingly few techniques currently in use are totally satisfactory (see Table 3-2).

Hand- and Power-Tool Cleaning

There are few techniques that fall under hand-tool (SSPC-SP-2) or power-tool (SSPC-SP-3) cleaning techniques that satisfy any one of the four basic requirements (see Figure 3-12). By definition, hand and power tools were never intended to remove more than loose

Table 3-1
Comparison of Surface Preparation Standards

Cleaning Method	SSPC Standard	Description[a]	NACE	Swedish Standard	British Standard	Japanese Standard
						(Primary)
Solvent cleaning	SSPC-SP-1	Complete removal of oil, grease, wax, dirt, and other contaminants by cleaning with solvents, vapors, emulsions, alkalis, or steam. Interior environments of low humidity only	NACE #5			
Hand-tool cleaning	SSPC-SP-2	Removal of loose rust, loose mill scale, and paint by manual labor with wire brushes, hand scrapers, sanders, etc.	NACE #6			
				St-2[b]		Pt-1[b]
Power-tool cleaning	SSPC-SP-3	Much the same as above but utilizing power tools such as chippers, descalers, grinders, etc.	NACE #7			Pt-2[b]
				St-3[b]		
						Pt-3[b]
Brush-off blast	SSPC-SP-7	Blast cleaning of all except tightly adhering residues of mill scale, rust, and old coatings, exposing numerous, evenly distributed areas of underlying metal	NACE #4			Sd 1 & Sd 2
				Sa 2	Third quality[b]	
Commercial blast	SSPC-SP-6	Sandblasting until at least two-thirds of each element of surface area is free of all visible residues	NACE #3			
					Second quality[b]	
Near-white blast	SSPC-SP-10	Blast cleaning until at least 95% of each element of surface area is free of all visible residues	NACE #2	Sa 2.5		Sd 2 & Sh 2
White-metal blast	SSPC-SP-5	Complete removal of all visible rust, paint, mill scale, and foreign material by wheel or pressure blasting using (wet or dry) sand, grit, or shot. Suitable for all the most severe environments, including immersion service	NACE #1	Sa 3	First quality	Sh 3 & Sd 3

[a]For complete specification description, see original SSPC document.
[b]Misalignment is intentional because some of the specifications fall between those for SSPC or NACE.

Table 3-2
Effectiveness of Surface Preparation Techniques in Attaining Satisfactory Substrates

Cleaning Methodology	Normalization of Substrate	Removal of Interference Material	Increasing Surface Area	Removal of Water-Soluble Salts
Hand-tool cleaning	Poor	Poor to fair	Poor to fair	Poor
Power-tool cleaning	Fair	Fair	Fair	Poor to fair
Brush-off blast cleaning	Fair	Fair	Good	Poor to fair
Commercial blast cleaning	Good	Good	Excellent	Good
Near-white blast cleaning	Very good	Very good	Excellent	Very good
White-metal blast cleaning	Excellent	Excellent	Excellent	Very good
Water blasting	Good	Good	Poor	Fair
High-pressure water blasting	Very good	Good to very good	Poor	Good
Wet abrasive blasting	Very good to excellent	Very good to excellent	Excellent	Excellent

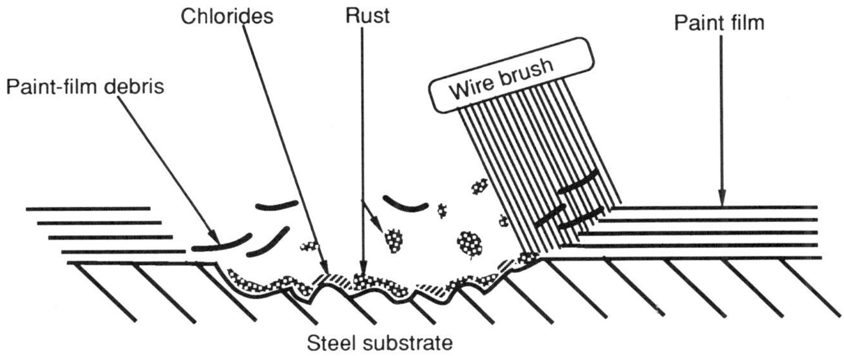

Fig. 3-12. Hand- and Power-Tool Cleaning. Wire bristles only partially scarify the steel surface and remove loose paint, rust, and dirt. They do not remove tight paint, rust and mill scale, nor embedded ionic contamination. This technique causes little air pollution but is ineffective and inefficient.

rust, paint, and the like. Many hammers, chisels, and other tools do more damage than good because they force as much corrodent into the steel as they remove. They are, however, valuable in removing heavy rust scale prior to blasting. Blasting through heavy scale to bare steel is extremely time consuming and impractical.

Hand-held brushes, even when used correctly, are hardly effective, and, with modern labor costs, they are certainly not cost effective. Power brushes are more effective but leave much to be desired in the quality of the surface they provide. They are also less cost effective than blasting (see Figure 3-13). Power grinders are similarly limited, but serve reasonably well, perhaps better than abrasive blasting, in the feathering of the peripheral edges of paint films surviving spot-cleaned areas. Neither hand- nor power-tool cleaning methods is suitable for the removal of tight mill scale and neither technique was ever really suitable for much more than spot cleaning because of their low production rates when used properly. One serious drawback of this type of cleaning (because of the slow productivity and weight of the equipment) is operator fatigue, which inevitably will further compromise the quality of the preparation.

Needle descaling (see Figure 3-14), possibly one of the most widely employed power-tooling techniques at this time, tends to force corrosion product into the steel. Despite these limitations, needle guns are very useful for such intricate areas as rivet heads, bolts, and seams, and for removing weld spatter.

New power tools, such as the 3M Roto-Peens and Scotch Brite "Clean & Strip" discs and wheels,[36] remove tightly adherent material and provide scarification of the same quality as abrasive blasting.[37] Roto-Peens tend to rip the steel very aggressively and often require a secondary modification of the profile with 3M pads in order to reduce the profile and irregularities. This author has found production speeds to be somewhat slower than with needle descaling because of this double process. The major drawback of these tools on bridge structures, however, appears to be their limited adaptability to complex shapes and angles, and, despite the better surfaces they impart, they are less universally adaptable than are needle descalers. The frequency with which the Roto-Peen flaps and 3M pads must be replaced also makes these newer power tools more expensive to operate from a material-consumption standpoint. Notwithstanding these drawbacks, where steel configurations allow the use of these techniques, the resultant life-cycle costs may be lower with the 3M equipment. This is because they provide

Fig. 3-13. Typical Surface Quality Left after a Thorough Power-Tool Cleaning Operation

better surface preparation as compared to other power tools. This, in turn, allows the application of higher performance systems and raises the service-life expectancy of conventional systems.

Fig. 3-14. Needle Descaling of Intricate Areas

Dry Abrasive Blasting

Dry abrasive blasting is in far more universal use on most areas of bridges. In the field, certain designs (lattice, trusses, etc.) are hardly conducive to maximum efficiency as far as the use of abrasive is concerned. Still, the resultant surfaces are nonetheless of the same quality as, for example, are the webs of continuous beams.

Although the quality of the surfaces produced will depend on both the type and size of the abrasive and the condition of the steel, for the most part, with the exception of the brush-off blast (SSPC-SP-7), blasting fulfills the first three requirements for surface preparation noted above. This fulfillment improves as the quality of blasting increases from commercial (SSPC-SP-6) through near-white (SSPC-SP-10) to white-metal (SSPC-SP-5) (see Figure 3-15). Igetoft[23] reports less fulfillment of the fourth requirement (the elimination of soluble salts) as the blast improves. But he also finds greater differences in the reduction of water-soluble salts with the condition of the original surface (surfaces bearing intact mill scale being much lower in soluble salt contamination than rusted surfaces, particularly those bearing deep pits). Supporting this, Calabrese and Allen found that industrially corroded steel recorroded more rapidly after blasting than did new steel and that the corrosion product contained both chlorine and sulphur.[38]

In dry abrasive blasting, one of two principal methodologies is employed: centrifugal blasting (see Figure 3-16) or air blasting (see Figure 3-17). Centrifugal blasting,[15,39,40,41] normally a shop procedure, involves the passage of steel members through a field of high-velocity abrasive particles that are produced by several electrically driven rotating bladed wheels that hurl the abrasive at the steel. Portable units[42,43] that move over the steel surface, have been employed in the ship-building industry, but they are rarely used on bridge structures in the field because of the complex geometries that are involved. The increased availability of centrifugal-blast equipment has played a signifi-

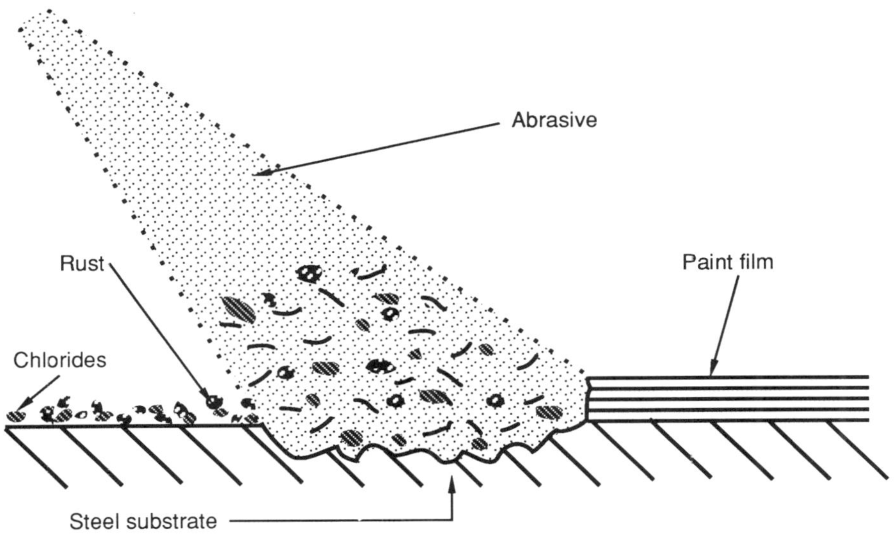

Fig. 3-15. Dry Sandblasting. Provides an excellent scarification of steel surface, removing paint, rust, steel, and most contamination. May not entirely remove chlorides, particularly from pits. May produce hazards from blast debris and is dirty, with much waste product.

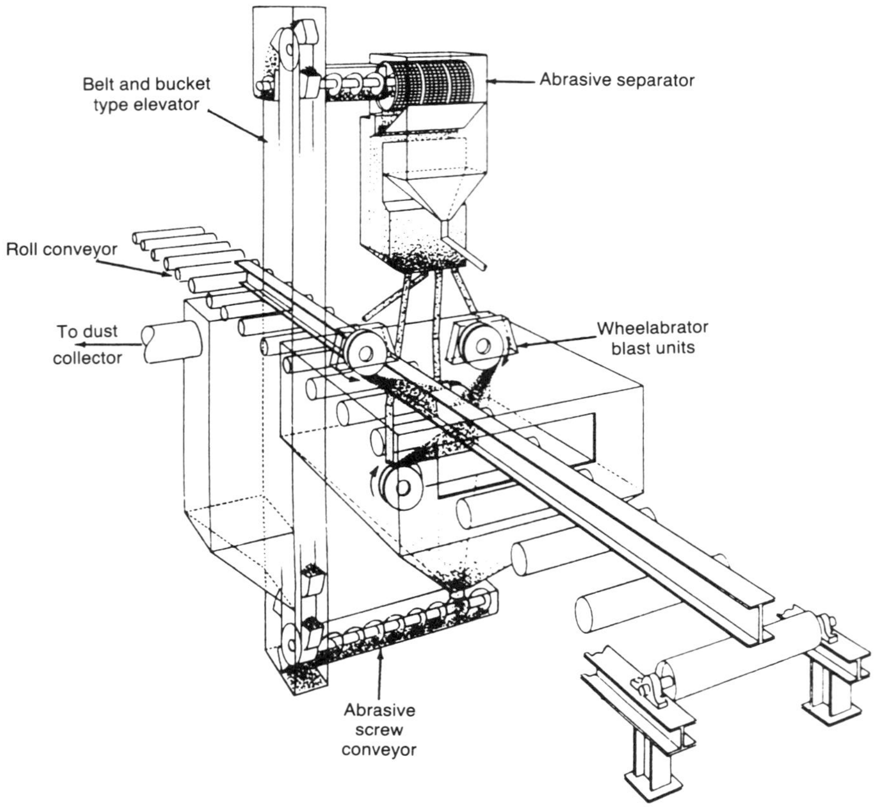

Fig. 3-16. Centrifugal Blasting Methodology. (Courtesy Wheelabrator-Frye, Inc.)

cant role in the increased use of abrasive-blasting preparatory techniques, which was a prerequisite to the adoption of the high-performance coatings. Because the abrasive is normally reusable, the more expensive blasting media[44] may be used in centrifugal blasters; and because steel in fabrication is usually clean, with few contaminants except rust and mill scale, metal shot or, preferably, shot and grit mixes can be employed. Care must be taken to ensure that recycled abrasive is continually stripped of contaminants (oils, etc.), as these can severely affect the quality of the blast and the production. In general, centrifugal blasting saves much time and money compared to air blasting.

In the field, air blasting is normally used where abrasive is propelled at high velocity from a venturi nozzle in a stream of compressed air.[45,46] In such open blasting, collection and recovery of abrasive is difficult; consequently, lower-priced abrasives are employed. Silica sand, coal slag, garnet, and crushed flint are all used. However, in some areas, silica sand is prohibited because of toxicological effects (silicosis). In areas where abrasive recovery is possible, steel grit or aluminum oxide may be used.

In field blasting, corrosion of steel during the interval between the blasting and priming operations is always possible and can be a costly complication. In humid environments or in areas where fog or mist can develop rapidly and unexpectedly, blasting/priming schedules can be disrupted by the newly blasted steel "turning." The problem is always more sensitive in areas where deeply pitted steel has been blasted because of salt contamination. This readiness to corrode attests to the presence of corrosive electrolytes (thin films of sulphate and chloride solutions) and exemplifies the difficulties in removing such contamination by blasting. Zinc shot[47,48] and other abrasives in which zinc is mixed[49] or, more effectively, where abrasives are coated with zinc,[50] have been found to provide much longer noncorroded intervals between blasting and priming. The zinc embedded in the steel surface gives a limited but satisfactory cathodic protection to the blasted steel for as long as several days. Adhesion of subsequent coatings is said to be good.[51] The major limitation of the zinc-modified blast appears to be high cost, which relegates the use of this technique to areas where its advantages are absolutely necessary. Clark[52] has shown that costs become increasingly competitive as surface preparation becomes more difficult and as time losses from climatic impact become more frequent. The technique has not as yet been used widely in the United States.

In a recent study by Peart and Flounders,[53] several types of common abrasives were evaluated for a variety of properties, including purity, pH, chlorides, and sulphate content, as well as cutting rates and breakdown characteristics. The chemical purity of slag, sand, and abrasives, in general, is of considerable importance because of the dangers of deposition of impurities on the steel surface. Copper abrasives are likewise of concern because of the effects of the deposition of electrochemically cathodic (with reference to steel) copper metal on the steel surface and the possible corrosion that this can cause.

Not only is the type of grit important in abrasive blasting, but both shape and size are directly related to the typical anchor pattern or mil profile that is provided to a given surface.[54,55] Round shot produces a craterlike surface topology with smooth rims; grit-

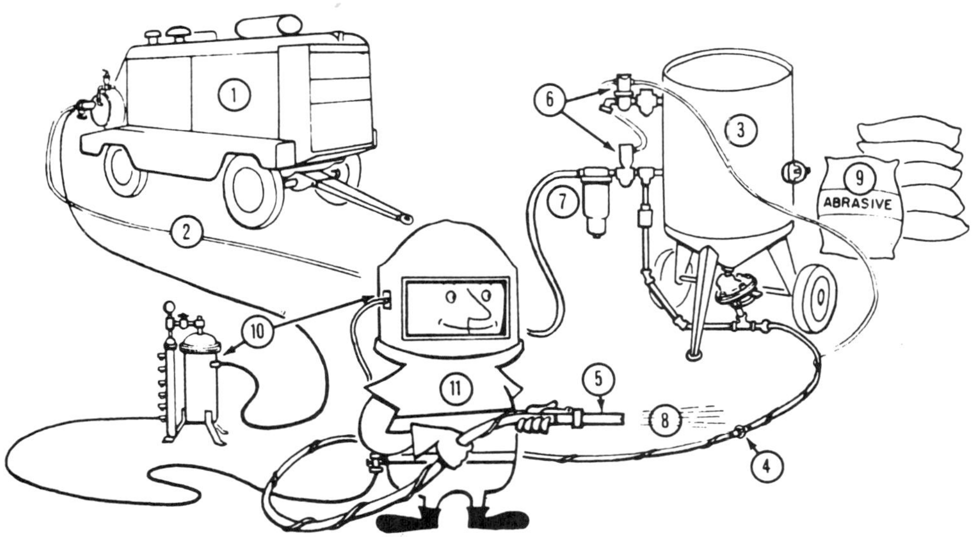

Fig. 3-17. Air Blasting Methodology. The components of a good air-pressure blast system are: (1) compressor giving an adequate and efficient supply of air; (2) air hose, couplings, and valves of ample size; (3) a portable, high-production sandblast machine; (4) the correct size antistatic sandblast hose (with externally fitted quick couplings); (5) high-production venturi nozzle; (6) pneumatic remote-control valves for safety and cost savings; (7) moisture separator; (8) high air pressure at nozzle; (9) correct type and size of abrasive; (10) air-fed helmet and air purifier (in good working order); (11) a well-trained operator. (Courtesy Clemco Industries Corp., San Francisco, California)

abraded surfaces are more irregular, particularly when nonmetallic abrasives are employed. Coal slag, for example, is one of the most widely employed abrasives in maintenance painting. It produces highly irregular surfaces compared to those produced by steel grit, which has a more orderly array of superimposed craters.[56] Some disagreement among authorities exists concerning the relationship of anchor-pattern depth to film thickness. Some suggest that coating-film thickness should be equal to the anchor pattern plus 3 or 4 mils;[46] others feel that the film thickness should be related to the anchor pattern by a factor of at least three times.[57] Others, including the SSPC, conclude that the relationship of film thickness and anchor pattern is less important.[56] Nevertheless, it is clear that there should always be a safety margin of film thickness above the peaks of the anchor pattern to avoid early corrosion. This is complicated by the fact that on surfaces blasted with steel shot or grit, strange needlelike hackles are often seen to protrude above the steel by as much as 5 mils.[56] Nonmetallic abrasives do not appear to produce hackles. It is important that film thickness be measured over the anchor pattern peaks and not the troughs. On the other hand, it is equally important that the film thicknesses are not so great compared to the anchor pattern that there is insufficient keying to maintain adhesion as the coating system works in service (i.e., undergoes internal stresses and service stresses that are developed in the curing and the response of coating systems to temperature, humidity, and vibration).

The inability of the dry-abrasive-blasting process to rid steel of ionic contamination is, of course, a more severe problem in maintenance applications than it is in new construction. On old corroded bridges ionic contamination is not only likely to be greater, but surfaces also are less conducive to easy contaminant removal. The presence of such impurities on steel may be verified by several indicators: potassium ferricyanide paper or 1 : 10 phenanthroline showing the presence of soluble iron; also, silver or mercuric nitrate solutions may be used in spot tests for chlorides, and so on. The limitations of the dry-blasting process and the susceptibility of corroded spots to ionic contamination are also clearly visible when newly blasted areas of spot-corroded steel are allowed to stand without coating (see Figure 3-18). In this case, it is always in the old pitted areas (now newly blasted), where the old coating had deteriorated, that corrosion of the new surface is first seen.[3]

Water Blasting

Water blasting (hydroblasting)[53,58,59] is very effective in removing chloride contamination and quite effective in removing sulphates (see Figure 3-19). At 10,000 to 20,000 pounds per square inch (psi), the technique will also remove old paint, rust, and loose scale. It will not, however, entirely remove all tightly adherent mill scale (except at very high pressures) or mechanically scarify the surface. At such high pressures, the technique also involves considerable operator fatigue and is dangerous. Therefore, pressures nearer 5,000 psi are more often the norm in bridge maintenance.

Wet Abrasive Blasting

A more satisfactory surface preparation is achieved if abrasives are fed into the water stream.[60,61] These wet abrasive techniques satisfy all of the criteria mentioned above for good surface preparation (see Figure 3-20). They suffer only from the messy sludges that result (increasing the hazard to operators, and, on high structures, to traffic below) and from a tendency of the steel so blasted to corrode before paint can be applied.

Fig. 3-18. Effect of Chloride and/or Sulfate Nests in Corroded Areas on the Rapidity at Which the Steel Recorrodes after Blasting

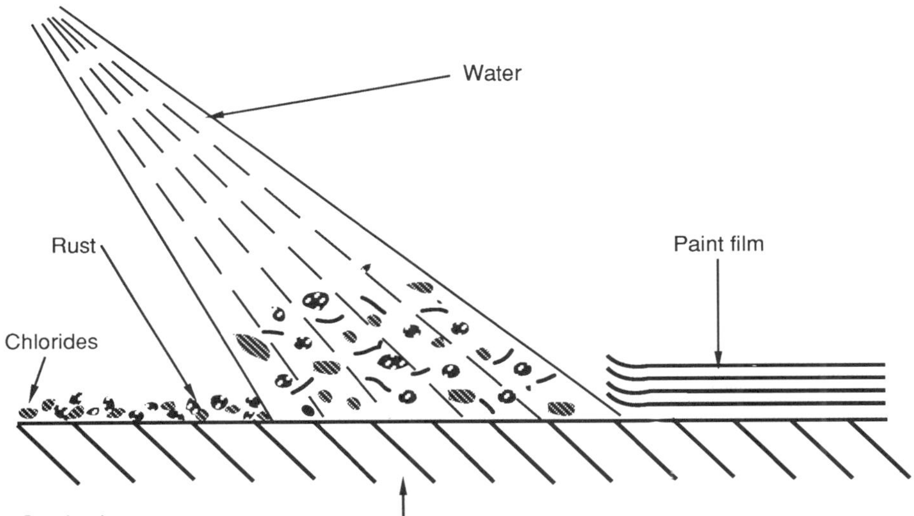

Fig. 3-19. High-Pressure Water Blasting. Steel is not scarified by high-pressure water blast. Paint, rust, and ionic materials are successfully removed.

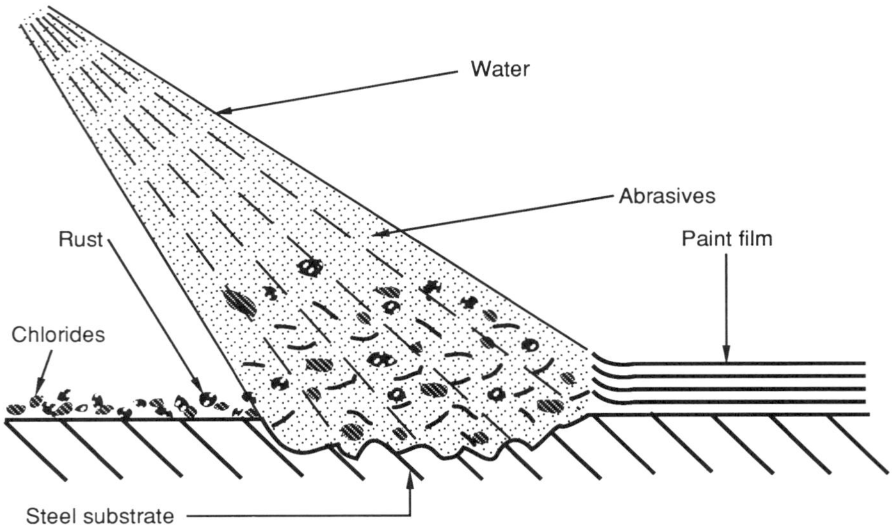

Fig. 3-20. Wet Abrasive Blasting. The combination of water, abrasive, and air removes steel, rust, paint, and all ionic materials effectively. The steel may rust before it can dry off unless care is taken. The procedure creates little air pollution, but does produce a messy sludge residue that poses a safety hazard to operators.

While it is often possible to satisfactorily dry the steel before corrosion sets in either by eliminating water or water and abrasive immediately after the wet blast (by dry blasting or blowing the newly blasted steel dry as a second step), the wet over-blast contamination of dried areas already blasted often cannot be avoided. This problem results in a need for constant reblasting and makes the technique less cost effective. The problem is particularly difficult in such areas as overhangs, open boxes, and the like. A sequenced dry-blast–wet-blast procedure for ferrous sulfate removal is described by Dasgupta and Ross.[62] An excellent, more general treatise on wet abrasive blasting and the newer surface preparation techniques may be found in the *SSPC Manual*, Vol. 1, 2d edition.[63]

Further refinement of the wet-blast method involves the use of inhibitive solutions introduced into the water (see Figure 3-21). This technique results in passivated surfaces that resist rust-back for several days after blasting. Inhibitors employed may be chromates, phosphates, nitrates, and so on, as well as water-soluble amines. Although the theoretical possibility of adhesion problems or osmotic blistering of barrier systems applied to such substrates exists in areas of high humidity, tests have shown few problems with adhesion of self-curing, water-based, inorganic zincs to this type of substrate.[64] The process has been developed with excellent results in Great Britain by K.U.E. Engineering of Bradford[65] and is said to provide coating systems with much longer service life. The British unit is self-contained and mounted on a truck from which two blasting nozzles can work.[61] The operation is controlled by intercom from a controller at the truck to the blasters in order to provide rapid adjustment of the mixed volumes of air, water, and abrasive. Using air volumes of 3 to 4 cubic feet per minute into which up to two gallons of water per minute are entrained, abrasive is added up to 7 pounds per minute and the mixture is delivered through wide-mouthed, cone-shaped nozzles of up to one inch in diameter at pressures up to 100 psi. Depending on the surface being cleaned, production rates of 50 to 300 square feet per hour are claimed, and control is so accurate that the system is as equally adaptable to scarification of single coats of paint as it is to the complete removal of multicoat paint systems. Because the abrasive is continually wet with water, the methodology allows the use of silica sand, whereas this abrasive is prohibited in dry-blast applications.

Several U.S. units are also available and are discussed in a comprehensive report on wet-blasting equipment and techniques issued in 1985 by the SSPC.[66] The report describes units in which water is added to the abrasive upstream, downstream, and just behind the nozzle. Some considerable diversity is noted in design, abrasive and water consumption, efficiency, and versatility of use. The report subclassifies the various units as follows: (1) air abrasive wet-blast units, (2) air–water abrasive slurry units (both of which use more abrasive than water), (3) high-, and (4) low-pressure water abrasive blast units that use higher volumes of water than abrasive.

The air abrasive wet blasters[67,68] are described as very efficient (80 to 90% of dry blasting) with little loss of visibility and good control of airborne dust. They are messy, however, and somewhat cumbersome, requiring two hoses to mix water and abrasive at the nozzle. Slurry abrasive blasting units,[69] such as the K.U.E. system described above, are for large production and are more expensive. In these units, the water and abrasive are mixed in a control unit that must be separately manned and that feeds the slurry to multiple-operating nozzles. Advantages are high efficiency, low water-flow rates, automatic inhibitor monitoring, and greatly reduced dust.

High-pressure water abrasive blasting[70] uses much less abrasive but high water rates. This allows the use of longer hoses, but efficiency is reduced (30 to 50% of dry blasting). The units are expensive and are probably best employed in the removal of old paint and rust. Visibility during operation is poor and high thrust (6,000 to 15,000 psi) results in operator fatigue. Less expensive and less exhausting are the low-pressure

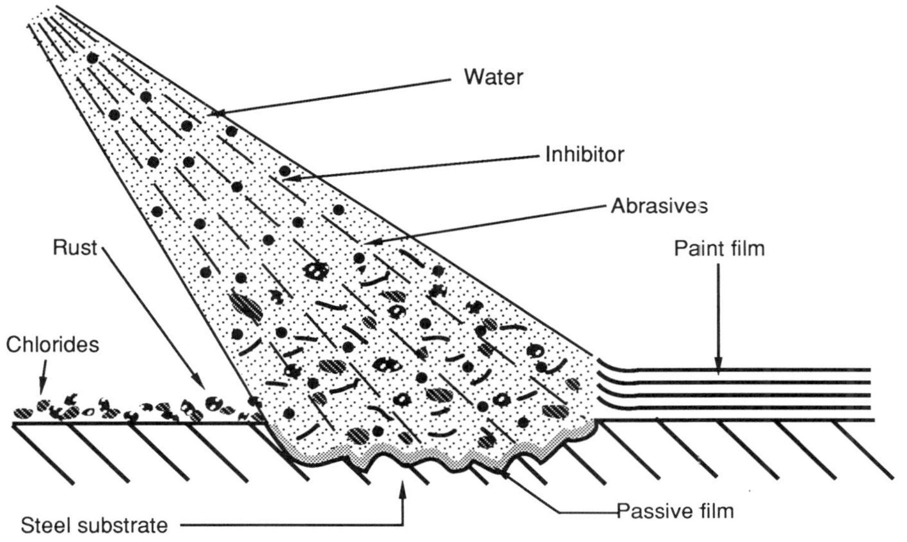

Fig. 3-21. Wet Abrasive Blasting with Inhibitors. The combination of water, abrasive, and air scarifies the steel and removes rust, paint, and all ionic material effectively. An inhibitor in the water stream, passivates the steel surface and prevents corrosion for several days. The procedure produces little air pollution, but its messy sludge residue creates safety hazards to operators and traffic.

(2,000 to 4,000 psi) water abrasive blasting units.[70] These units have much poorer efficiency and are again best used for rust and paint removal rather than for surface scarification. They operate at lower water-flow rates and use short hoses only, but have the advantage of better portability.

Experimental Preparatory Procedures

In recent years, environmental concerns and a greater awareness of the possible toxicological problems that may be associated with the removal of old lead-paint systems from bridge structures in urban areas have probably done as much as basic engineering concerns to motivate the search for alternative preparatory techniques. Local control of toxic aerosols from open-air blasting by means of vacuum-blasting techniques, although efficient on flat, uninterrupted surfaces, is not as effective on bridges where a profusion of rivets, angles, and corners reduce vacuum efficiency. The technique is also slow and tiring for operators and, again, reduces efficiency. Aside from wet abrasive blasting, little success has been realized in producing a viable, cost-effective alternative to dry abrasive blasting, and environment-protecting developments involving containment of the dry-blast debris have been more successful.[71,72,73]

Many exciting new concepts are being currently explored in the surface preparation of metal. Many of these are reviewed in a recent SSPC report.[43] They range from such space-age variants of flame-cleaning techniques as laser[74,75] and zenon lamp,[76] to high-velocity ice pellets,[77] and from ultrasonic[78] and bacterial cleaning[79] to citric-acid cleaning,[80] coating removal by means of cryogenic techniques, and high-shrinkage strip coatings, such as those based on polyvinyl pyrrolidone.[81] At this time, however, none of these techniques appears practical for the surface preparation of steel bridges.

Recent experimentation has involved an evaluation of cavitation blasting.[82,83,84] In this process, air bubbles introduced into a high-velocity water stream collapse against the surface being cleaned and produce a severe localized pressure reversal that literally pulls contamination and old coatings from the steel surface. Difficulties have been experienced in achieving practical production rates with this concept and, to date, a satisfactory, cost-effective, practical adaptation remains elusive. Stand-off distances are critical to cavitation effectiveness and the procedure does not scarify the steel surface.

PREPARATION OF CONTAMINATED WEATHERED A-588 STEEL SURFACES

A growing concern to the industry, particularly in the snow belt and along the coast, is the deterioration of unpainted weathered steel such as A-588. Studies[85,86] have shown that in high chloride-ion environments weathering steel is subject to severe corrosion in those areas that are well sheltered from the washing effects of rainfall. The mechanism of attack is particularly dangerous because the steel develops a honeycombed deterioration, with chloride solutions becoming deeply embedded into the grain structure of the steel. This effect causes considerable difficulty in cleaning. The cleaning and painting of these surfaces has been the subject of considerable study by Tinklenberg at the Michigan Department of Transportation (DOT) and Raska at the Texas Department of Transportation.[87] Experiments have been conducted with various water-washing and abrasive-blasting combinations prior to coating with a variety of different coating systems. Unfortunately, the entrainment of chlorides in such steel in these honeycombed deteriorated areas is so deep that the removal of chloride by this technique is never adequate. Once these surfaces have corroded under such circumstances, total rectification of the steel appears to be considerably more difficult (if not impossible) than does the rectification of A-36 steel. Tinklenberg concludes that simple abrasive blasting is, in this case, as appropriate a technique as the various blast–water-wash combinations examined and produces results that are quite comparable. Other studies indicate the benefits of wet blasting and/or washing, however. Many of these are discussed in an SSPC state of the art review.[88]

CLEANING OF EXISTING PAINT SURFACES

In the maintenance painting of existing bridges, where funding restrictions or the extent of deterioration is insufficient to justify the complete removal of the existing paint system, it remains a common practice to spot clean and prime deteriorated areas of the structure and then to apply full coats of conventional oil/alkyd intermediate and/or finish coats over both spot repairs and existing films in good condition. The practice is widely continued throughout the United States, and in most states lead-based systems are reapplied. While not so common and less advisable, such spot techniques have also been employed with high-performance lacquers and chemical-converting coatings such as epoxies.[3] In this technique, the preblast cleaning of the existing paint film becomes more critical because the existing film becomes, in effect, the new substrate, and attention must be paid to ensure that this film is free of all contaminants (see Figure 3-22).

In some cases where existing films are hard, old, and highly crosslinked, such as might be the case where old epoxies were used, it may be necessary to provide mechanical scarification by water or sweep blasting in order to obtain a reasonable key for adhesion of the new finish. In the majority of cases where such older films are oils and alkyds, such radical techniques are unnecessary, although the removal of oil, grease, road dirt, chalk, and ionic materials from the films to be recoated remains important. In the snow belt after winter or on the coast throughout the year, the cleaning of accumulated chlorides from such structures before blasting or power-tool cleaning becomes even more important. So, too, does the removal of sulphates from bridge structures in industrial environments. As in painting bare steel, the degree of care necessary in preblast preparation will to some extent depend upon the type of new coating system to be used (oil paints and alkyds being more forgiving than high-energy systems, which require better surfaces for good adhesion).

The several techniques discussed and grouped under SSPC-SP-1[89] provide methodologies for such cleaning. Additional information may be found in Volume 1 of the *SSPC Manual*.[90] A more detailed discussion of soil removal may be found in Spring's text.[91] Unfortunately, not all techniques discussed under SSPC-SP-1 are universally applicable to the removal of all contaminant species (see Table 3-3) and those that are are usually the most expensive. Effectiveness depends on the selection and mastery of the correct technique. Water, for example, will remove ionic contamination, especially chlorides, but will not touch oils or road dirt. Solvent cleaning will remove oils, greases, and road dirt but not ionic species. Neither method, without some mechanical assistance, will remove chalk from existing films. Detergent solutions and steam cleaning (with or without the use of such chemical agents as metasilicates) are more effective, although in all cases where chemicals and/or detergents are employed, a two-stage process in which paint films are rinsed after cleaning to remove chemical contamination is necessary. In detergent-solution cleaning, it is often necessary to use mechanical assistance, such as brush scrubbing or brooming of the surface before rinsing. In steam cleaning, it is important to allow the hot detergent solutions to work on the oils, grease,

Fig. 3-22. Cleaning of Existing Painted Surfaces for Repainting

and road dirt before a second clear-water pass is used to rinse down the coating film. Water blasting may also be employed and, with high pressure and good pressure control, may remove rust and loose scale from the spot-deteriorated areas in addition to old, poorly adherent paint. This may eliminate the need for subsequent blasting or power-tool cleaning where suitable low-energy coatings are to be employed in the new system. In many cases, as in subsequent spot blasting, it may be expedient to use hammers and chisels before water blasting to remove highly adherent, heavy rust scale. Like water blasting, steam cleaning is also useful for removing old oxidized paint systems. In water blasting, as in both steam cleaning and high-pressure washing techniques, control of the time the process dwells in one particular area is important because all techniques are capable of removing still-viable films of old paint.

Table 3-3
SSPC-SP-1 Methodologies and Their Effectiveness in the Removal of Contaminants

Cleaning Method	Road Soils	Deicing Materials	Chalk	Bird Droppings	Organic Chemicals	Inorganic Chemicals
Water washing	Fair	Good	Poor	Fair	Poor	Good
High-pressure water wash	Good	Very good	Fair	Very good	Fair	Very good
Detergent wash/water rinse	Good	Very good	Fair	Good	Good	Very good
Alkaline cleaning/water rinse	Good	Very good	Good	Very good	Very good	Very good
Emulsion cleaning/water rinse	Good	Very good	Fair	Very good	Good	Very good
High-pressure detergent wash/water rinse	Very good	Excellent	Very good	Excellent	Very good	Excellent
Detergent wash with mechanical scrubbing/water rinse	Very good	Very good	Very good	Very good	Very good	Very good
Solvent cleaning	Good	Poor	Fair	Fair	Very good	Poor
Steam cleaning with water	Good	Very good	Good	Very good	Good	Very good
Steam cleaning with detergent/water rinse	Excellent	Excellent	Very good	Excellent	Excellent	Excellent

Chapter 4

Coating Systems for Steel Bridges

Practical paint systems for bridge structures are designed to employ one or more of the fundamental mechanisms of corrosion control outlined in Chapter 2. The possible permutations of primer/finish coat systems that might be employed to produce viable coating systems for bridge structures are very large and increase as each new adaptation of polymer type to coating function is added. Many combinations are, of course, not possible because of the incompatibility of polymer properties with coating characteristics (e.g., the sensitivity of vegetable-oil-based systems to saponification when employed as vehicles for zinc rich systems. In other cases (acrylics or aliphatic urethane primers and midcoats), economic considerations discourage adoption, while in still others (vinyls, acrylics, and chlorinated rubbers as thermoplastic finish coats instead of phenoxy-resin-based topcoat vehicles), the market opportunities have already been more satisfactorily fulfilled by other coatings. Consequently, the vast array of potential candidates is trimmed by a sort of natural selection to a more manageable list, at least from the generic standpoint.

Each new system brings its own particular requirements for surface preparation and application as well as its own peculiarities that relate not only to its film-formation methodology and its mechanism of protection but also to its resistance to moisture, sunlight, corrodents, and physical abuse. None is as universal, nor as tolerant of variations from the ideals of preparation, application, and use as were the old lead-based systems, and most are less understood by specification writers and applicators alike. On the other hand, there is little doubt that if carefully selected and properly applied in any given environment (particularly more demanding environments), the best of the new systems will produce substantially better and more cost-effective protection than will the systems they replace. Each year, the protection of more and more bridges is consequently being converted from the lead- and chromate-based inhibitive approach to zinc and barrier systems.

Unlike the classical red lead/linseed oil systems (formulations of which are well known and have changed little over the past 200 years) or the BLSC systems developed and marketed so well in the early 1960s, the majority of the newer coating systems have been adopted from proprietary products. Individual paint manufacturers have introduced these proprietary products directly to the states and bridge authorities. The formulations even of such relatively commoditylike items as ethyl-silicate-based zinc rich primers have, in consequence, been closely held and, even now, for the most part remain inadequately appreciated by the states that specify them. Furthermore, since one product may radically differ from a generically similar product of another manufacturer, the older compositional forms of specification have given way in many cases to performance specifications that may or may not have a quasi-compositional outline (weight per gallon, minimum zinc content, etc.) but that give little real guidance as to the details of the composition.

In consequence, while "experts" bandy about terms like "high-build epoxies" and "aliphatic polyurethanes," they are inevitably generalizing about large families of products that may not only vary considerably in formulation (pigmentation, PVC, CPVC, volume solids, solvent systems, etc.) but may actually be polymerically very dissimilar. High-build epoxies, for example, may employ different types of curing agents, including such families of resins as polyamides, amido-amines, aliphatic and cycloaliphatic amines, and aromatic amines with or without such catalysts as nonylphenol or tertiary amino-phenolic materials, epoxy resins of various molecular weights, and modifications utilizing a variety of mono- and di-functional diluents, acrylic monomers, hydrocarbon resins, butylated urea formaldehyde, and silicone resins.[1] Families such as polyurethanes, with their rich diversity of available polyols and choice (if more limited) of di- and poly-isocyanates, are still more difficult to categorically specify, despite qualifications of polyols as polyesters or hydroxy-functional acrylics. The American Society for Testing and Materials (ASTM) five-family categorization for urethanes does little more than classify the polyurethane family in terms of curing mechanisms. A similar plethora of formulations is probable for many other systems and, in some cases (such as vinyls), an incomplete or erroneous specification of polymer type can lead to severe performance difficulties in service.

While all vinyl chloride/acetate resins will, for example, adhere to films of other vinyl chloride/acetate lacquers, only the maleic-acid-modified terpolymer will adhere to bare steel. Unfortunately, this carboxylated vinyl will not adhere to the wash primer (DOD-P-15328D),[2] which is required pretreatment for steel to ensure the adhesion of the hydroxylated vinyls. These and the unmodified copolymers will not adhere well to bare steel.[3] On the other hand, certain butyl-maleate-modified vinyls will adhere satisfactorily to both wash primer and to bare steel.

Equally destructive are the misrepresentations of coating-system failure that generalize about system type. Because one coating system is found sensitive to a particular failure mode, it does not necessarily follow that all coatings of the same generic type will be similarly sensitive. Formulation effects, pigmentation differences, solvent systems, and differences in the basic polymer (even differences in the molecular weight of a given thermoplastic) can have significant effects on the susceptibility of a product to a specific failure mode.

As a result, it is important to keep in mind the dangers of such generalizations when reviewing the extended treatment of coating systems for steel bridges that follows. Where possible, attempts are made to discuss elements of the formulation and to direct the reader to more detailed sources. Since so many proprietary materials are now in use, however, this is by no means possible universally.

OVERALL TRENDS IN PAINT SYSTEM USE ON BRIDGES

The most recent revolution in bridge-paint systems may be summarized best as an abandonment of the inhibitive systems for the zinc rich and barrier systems. This change was fostered not only by the bridge authorities' desire for better paint-system performance, but also by environmental pressures to (1) initially curtail the use of photochemically reactive solvents (e.g., aromatics, olefinics, branch-chained ketones, etc.) and then to curtail the general use of solvents once other solvent types were also found to be photochemically reactive and (2) to eliminate lead and hexavalent chromium. At the same time, the drive for faster-drying systems that could be applied in fewer coats at thicker film thicknesses was not conducive to oxidizing-cure mechanisms. In addition, the consistent push by the larger protective-coatings manufacturers for new markets that were secure from cutthroat competition also led to the swing away from the older oil and alkyd systems. These older paint systems of established formulations were readily available to the many smaller firms that for years had made a living from intense competition in the bridge-paint market. They were, therefore, of little interest to the industry giants that required larger profit margins. On the other hand, the bridge-paint market itself was a valuable source of new revenue for the larger companies, and it was one to which they addressed themselves by selling the idea of the "high-performance protective-coating system" based on zinc and barrier coatings.

During the 1970s, an increasing number of states and bridge authorities first experimented with and then adopted many of these newer systems, initially for new construction but then, with increasing frequency, for radical field maintenance. While some have experimented with one or more of the many new inhibitive pigments, few if any will concede that even the best of these new pigments is a match for the classical lead and chromate materials.

Perhaps the major factor that has retarded the complete abandonment of the inhibitive system has been the number of bridges across the country that still have these older systems and require maintenance. Funds are simply not available for the complete removal of existing paint from these bridges all at once. Therefore, the majority are simply spot repaired and repainted with lead- and chromate-based systems as they have been for years. To date, removal of lead for bona fide environmental reasons has been a practice only in local urban pockets of the country, although this is changing rapidly. The impact of such removal on job cost and difficulty is gradually increasing and is becoming a grave concern to contractors. Usually, only when a bridge has deteriorated to the point at which complete paint removal is justified is the inhibitive system abandoned.

In drier and more rural climates, a second factor discourages the change. Here, red lead- and zinc-chromate-based inhibitive systems are found to give long-term protection that reduces the cost-effectiveness claims of the higher-performance systems. Also, where the new coatings cure by moisture-activated mechanisms (e.g., alkyl-silicate-based zinc rich primers), there are problems with both drying and recoating in very dry climates. The next factor that appears most persuasively to discourage some states and authorities from abandoning the inhibitive systems has been the frequent failures that have attended the use of the new high-performance systems.[3] While "teething" problems will inevitably accompany any new technology, paint failures are expensive and embarrassing. The bridge engineer with a million-dollar paint failure on his hands finds little solace in such system peculiarities. More recently, concerns over the difficulties and expense of disposing of the spent abrasive from the removal of old lead-based coatings have also influenced decisions regarding the abandonment of traditional finishes.

Our review of current paint systems for steel bridges begins with a review of those old and new inhibitive pigment-based systems that remain in use.

INHIBITIVE COATING SYSTEMS

Inhibitive systems now in use by state highway departments may be categorized by five types:

1. Red lead-based systems
2. Zinc-chromate-based systems
3. Systems based on BLSC
4. Nontoxic systems
5. Latex systems.

With the exception of the latex systems, all are oxidizing primers and employ oxidizing intermediate and finish coats using either conventional pigmentation (titanium dioxide, chromium green oxide, phthalocyanine blues and greens, etc.) or aluminum. Vehicle systems are either unmodified linseed oil, alkyds (long- and medium-oil alkyds) with and without linseed oil modification, phenolic varnishes, or modified alkyds, such as silicone alkyds and uralkyds. In California, inhibitive latex primers based on zinc chromate/zinc phosphate[4] combinations or zinc hydroxy phosphite are employed in four-coat systems using, in some areas, latex aluminum topcoats. The five basic systems are shown in Table 4-1. Each is reviewed with its various options below.

System Group I: Red Lead/Linseed Oil/Alkyd-Based Systems

In its classical form (red lead/linseed oil primer, red lead/iron oxide/linseed oil intermediate coat, and white lead or blue lead/linseed oil finish), this system has changed little over the century and a half that preceded the 1950s. The complete system is now hardly employed at all, although versions of the primer and intermediate coat remain the primary maintenance systems in several states.

Red lead/linseed oil primers are generally of the TT-P-86E (AASHTO M-72-74) type, employing either straight red lead and linseed oil (Type I) or pigmentary modifications with red iron oxide and alkyd modifications of the vehicle (Types II, III, or IV). The classical 24 to 25 pound-per-gallon Type I red leads are still the best paints for the poorest surfaces, and there are few benefits obtained from better surface preparation.[5] Alkyd modification reduces such tolerances slightly, although all these systems are good for hand- and power-tool cleaned substrates. Dry-times are very slow, and two- to three-day recoat periods are most usual over the soft films. Increased alkyd shortens these recoat intervals considerably. Film thicknesses of these red lead primers usually vary from 1.5 to 3.0 dry mils, although film thicknesses are rarely specified. At very high film thicknesses, skinning and wrinkling become problematic, although these defects seem not to have serious effects on performance, particularly in the short term. On certain jobs, this author has seen Type I red lead films applied as high as 8.0 dry mils.

Intermediate coats employ more iron oxide at the expense of red lead and generally introduce more alkyd modification.[6] These systems may also employ extenders such as magnesium silicate and even graphite.[7] Despite the heaviness of the pigmentation, settling is not often the problem it could be, because of the shortness (thixotropy) of these systems (through lead-soap formation). Aluminum stearate, once widely used as a thixotrope, has now been replaced by the organo montmorillonites and hydrogenated castor-oil derivatives.

Most frequently, red lead systems today utilize BLSC finish coats[8] (and even intermediate coats[9]) of much the same composition as is used over the TT-P-615 Type I and V BLSC-based primer systems. Alkyd or phenolic-bound leafing aluminum finishes are also popular.[10] Nonaluminum finishes are quite variable in color and range from grays, blues, and greens to beiges, reds, and maroons. Until quite recently, Arizona used a blue lead topcoat,[11] although the use of white lead appears now to have been totally

Table 4-1
Utility and Application Comparison of Inhibitive Coating Systems for Steel Bridges

System Group	System Description	Volume Solids (%)	Typical Film Thickness (Dry Mils)	Surface Preparation (Minimum SSPC-SP)	Adaptability to Poor Surfaces	Probability of Early Failure from Deviations in Application	Recoat Times (Days)	Solvent Type	Usual Mode of Application	Compatibility with Old Coatings	Ease of Spot Repair	Ease of Cleaning (Salts and Soil)
I	*Primer:* red lead/linseed oil	95	2.0	2	Excellent	Very low	3	Mineral spirits	Brush	Excellent	Excellent	Good
	Intermediate: red lead-iron oxide/oil-alkyd	80	1.5						Brush, roller, or spray			
	Finish: aluminum/alkyd or phenolic	48	1.0				2					
	Finish: (alternative) BLSC[a]/oil-alkyd	75	1.0									
II	*Primer:* zinc chromate/alkyd	48	2.0	3 or 7	Good	Low	1	Mineral spirits	Brush	Excellent	Excellent	Good
	Intermediate: zinc chromate/alkyd	48	1.5						Brush, roller, or spray			
	Finish: aluminum/alkyd or phenolic	48	1.0									
	Finish: (alternative) BLSC/oil-alkyd	75	1.0									

III	*Primer:* BLSC/linseed oil-alkyd	77	2.0	3 or 7	Very good	Low	2	Mineral spirits	Brush	Excellent	Excellent	Good
	Intermediate: BLSC/linseed oil-alkyd	75	1.5						Brush, roller, or spray			
	Finish: BLSC/linseed oil-alkyd	75	1.5									
IV	*Primer:* nontoxic inhibitive/oil-alkyd	63	2.5	6	Fair	Moderate	1	Mineral spirits	Brush, roller, or spray	Good	Very good	Good
	Intermediate: nontoxic inhibitive/oil-alkyd	51	2.5									
	Finish: nontoxic inhibitive/oil-alkyd	45	2.0									
	Finish: (alternative) aluminum/alkyd or phenolic	48	2.0									
V	*Primer:* inhibitive latex	45	2.0	6	Fair	Moderate	1	Water	Brush or air spray	Fair	Excellent	Poor
	Intermediate: inhibitive latex	45	2.0									
	Finish: noninhibitive latex (2 coats)	39	3.5									
	Finish: (alternative) noninhibitive latex aluminum/latex	37	1.5									

[a]Basic lead silico chromate.

supplanted. Although certain states still retain old lead and oil topcoats in their specifications, the amount of usage is probably small. Without lead pigmentations, it is virtually impossible to get unmodified linseed oil to dry without severe wrinkling, and most lead and oil topcoats have now been replaced with 100% alkyds or heavily alkyd-modified vehicle systems pigmented with titanium dioxide, chromium green oxide,[12] and the like. Longer oil-length systems are possible when BLSC is employed in the topcoat pigmentation,[13] although, in these finishes, too, the trend is generally toward slightly shorter oil-length vehicles. These give better flow and leveling and provide a better gloss with some sacrifice of brushability. The lead pigmentation of TT-P-102 and AASHTO M-70-74(1982) (still on the books of certain agencies) has now been replaced with titanium dioxide, and the original linseed oil/bodied oil vehicles have been replaced with very long oil alkyds.[14]

It is the fashion today to view these older lead systems as passé. Yet they still have a place and probably will continue to do so (unless environmental restrictions block their use) until all existing bridges are totally stripped of old coatings and upgraded. Some states (Colorado, Montana, Wyoming, Arizona, North Dakota, and South Dakota) even retain these systems for new construction, despite the disruption of shop scheduling caused by the exceptionally long dry-times. In these states, it is inevitable that lead-painted bridges will be extant for a long time.

Performance of the lead-based systems, moreover, is hardly marginal: three-coat, 4.5-total-dry-mils systems are reported to last twenty years in rural Alabama (S. Cauthen, personal communication, 1985); in the Dakotas, twenty to thirty years are expected;[11] and in the valleys of Oregon, forty- and fifty-year service lives are claimed with five- to seven-coat, red lead-primer/alkyd topcoat systems.[12,15] Given the low surface preparation requirements, the minimal demands on contractors and inspectors, and the relative freedom from the risks of failure that haunt the newer systems, many state bridge engineers who are still using these systems are justifiably not motivated to switch to the newer systems.

Yet, as noted in Chapter 2, linseed oil and alkyds are subject to alkaline saponification and, in aggressive high-chloride-ion environments, the corrosion process itself may cause degeneration of these oxidizing binders, with susceptible ester groups in the backbone of the polymer. These limitations have led to inevitably shorter service lives in the snow-belt states and coastal locations, especially in the Southeast. In industrial areas, the service lives of red lead systems are also severely foreshortened. Bridge structures in these Zone-2 (particularly Zone 2B) and Zone-3 environments might well benefit from the more modern paint systems.

The zoning categorization of bridge environments is discussed by Keane[16] and is described in detail in chapter 1, volume 2 of the *SSPC Manuals*. Zone 1 refers to mild, normally dry interior (1A) and exterior (1B) environments. Zone 2 describes wet exterior environments (Zone 2A is frequently wet with fresh water, and Zone 2B is frequently wet with salt water). Zones 2C and 2D are fresh- and salt-water immersion environments. Zones 3A through 3E environments are chemical environments: Zone 3A, acidic pH of 2.0 to 5.0; Zone 3B, neutral pH of 5.0 to 10.0; Zone 3C, alkaline pH of 10.0 to 12.0; Zone 3D, solvent exposure; Zone 3E, severe exposure to oxidizing chemicals, extreme pH ranges, strong solvents, and/or high temperatures.

The cost of the Type I primers on the open market today is high because of the amount of red lead they contain. While no longer the commodity that red lead/linseed oil primer once was, in certain parts of the country intense competition still depresses the market price. The price per gallon is somewhat misleading because red lead/linseed oil primers were the first high-solids coatings, and many have volume-solids levels as high as 95%.

System Group II: Zinc-Chromate/Alkyd-Based Systems

The direct inhibitor zinc potassium chromate (ASTM-D-478), also known as zinc yellow, zinc chromate, and, in Europe, as basic zinc chromate, has been employed in alkyd and phenolic vehicles for many years. Its use on bridge structures has not been widespread, however, but seems to have increased as lead pigmentation has diminished. The concern over toxicity that has restricted lead pigments has not spread to the chromates, despite concern over the carcinogenic toxicity of hexavalent chromium. In fact, Alabama, which allows the use of either red lead or zinc-chromate systems, reports that, for the most part, only the zinc-chromate primers are now being used because of workers' concerns over lead's toxicity and because of the extremely slow drying rates of red lead primers (S. Cauthen, personal communication, 1985).

Zinc-chromate systems appear to be used more in the South than elsewhere in the United States. Alabama, Arkansas, Florida, and Texas use the systems, in one form or another, with alkyd topcoats. In Alabama[17] and Arkansas (proposed specification),[18] the first field coat is identical to the shop coat, although Alabama adds black to the intermediate coat to differentiate it from the red-oxide-colored primer. Alabama uses either an AASHTO M-69 Type 1 aluminum as a finish or a green BLSC topcoat with a predominantly alkyd-vehicle system. The Alabama primer, like the SSPC Paint 11[19] that it resembles, employs a binder that is one-half linseed oil. The use of linseed oil as a binder for zinc chromate has been discouraged elsewhere.[20]

Alabama reports good service with their system with an estimated ten-year minimum service life for the three coats over an SSPC-SP-6 blast. Alabama does not specify film thickness but relies on opacity to achieve 1.5 mils per coat. Arkansas, using a pure alkyd primer over the same blast quality, specified wet (not dry!) film thickness of 3.5 mils per coat. Green or blue finish coats are detailed in the Arkansas specification, or the M-69 Type 1 aluminum is used (3 wet mils per coat). Both states give detailed compositional specifications for all coats and variants.

Florida also relies on the zinc-chromate alkyd primer when zinc rich primers are not considered appropriate, and Texas uses a zinc-chromate primer in rural areas only. Alabama, Texas, and, especially, Florida, use zinc rich systems in Zones 2A, 2B, and 3.[16]

System Group III: Basic Lead Silico Chromate/Oil-and Alkyd-Based Systems

System Group III, originally introduced in several forms by The National Lead Company in the late 1950s, replaced the red lead systems in many states. Primers are, for the most part, heavily pigmented with BLSC (iron oxide is added for opacity) and based on linseed oil and long-oil alkyd vehicles. They are typified by federal specifications TT-P-615 Type I through Type V (AASHTO M-229-80, Types I, II, III, and IV). Very-long-oil systems (TT-P-615, Types I and V) are usually employed as primers in most states. As with other lead-based systems, the more alkyd that is introduced into the vehicle, the faster the drying properties but the poorer the tolerance for poorly prepared surfaces. In general, BLSC systems have somewhat more alkyd modification in their vehicles than do straight red lead systems.

Midcoats are generally shorter in oil length than the primers, with lower levels of BLSC and higher levels of other pigments, such as iron oxide.[21] Typical finish coats[22] may have much higher percentages of alkyd, and range from all BLSC to levels in the 1- to 2-pound-per-gallon range. Certain states employ no BLSC in certain finish coats when colors are light or when aluminum topcoats are employed. Generally, however, finish coats are pigmented with BLSC, which is over-tinted to a variety of colors, including greens and grays, with titanium dioxide and chromium green oxide. In Pennsylvania, silicone-alkyd finish coats are employed as part of the BLSC systems.[23]

Formulations and specifications for most states were originally developed as part of an aggressive marketing campaign.[24] The resultant specifications were extremely well documented and have remained essentially unchanged since the early 1960s. Like the red lead systems, the low-energy, oil-modified systems were originally applied to new mill scale (later to brush-off and commercially blasted surfaces), but have given satisfactory performance in rural areas for over fifteen years. Like the red lead systems in the Dakotas, thirty-year performances in the rural dry climates of Colorado and Wyoming have been the rule, and final service lives of thirty to forty years are projected. Shorter service lives are seen in the industrialized East, and such states as New Jersey and Massachusetts may realize no more than seven- to twelve-year service lives in marine or industrial environments.

Again, like the red lead systems, the low-energy vehicles make BLSC paint systems relatively invulnerable to catastrophic early failures due to compromised surface preparation, while their relatively high-solids content and normal three-coat defense-in-depth[24] form provide some insurance against very low film builds. Some states specify film thicknesses, although this is rare. Tennessee, for example, calls for 2.0 dry mils of primer and 1.5 dry mils each of the two field coats.

Unfortunately, manufacturers have been known to take advantage of the low specific gravity of BLSC (approximately equal to barytes) and to replace the BLSC with barytes (with or without red lead to maintain the lead content). The practice has been particularly common in finish coats where serious color floatation problems from the BLSC can cause field difficulties. This connivance has often caused problems in drying among the long-oil primers and midcoats, particularly in cold, damp climates at the beginning and end of seasons, because of reduced-through-drying catalysis of the low-lead systems. Recent analytical procedures designed by NL Industries are aimed at forestalling such replacement.[25]

Currently, about twenty states still employ BLSC systems for new construction, although only one-half of them regard it as their preferred system for new work. Twenty-five states use the system for spot maintenance (although not necessarily for radical maintenance) in the field, and most of these regard this as their preferred system.

A second variant of the BLSC system utilizes an aluminum topcoat instead of the colored finish. It is presumed that the decision to employ this alternative finish coat reflects the somewhat better performance of this type of finish (better UV resistance, increased impermeability). States employing this type of system include Virginia, Wisconsin, and Montana (which employs the system in both new construction and maintenance and considers its results excellent) and, to a lesser extent, Kentucky and Idaho. New Jersey uses a graphite-based finish coat over BLSC primer and an intermediate coat on the underside steel of bridge decks.[26]

An apparent advantage of these systems over red lead, and one highly touted by National Lead (lower coating weight by virtue of lower specific gravity of the pigment) was recently cited as a major reason for the adoption of the BLSC system in the repainting of the Brooklyn Bridge. The system change saved 100 pounds in weight per linear foot on the old structure.[27] Paint weight can be a significant consideration, especially on older bridges or on bascule structures where paint buildup must be removed periodically to maintain the counterweight balance.

System Group IV: Nontoxic Inhibitive Pigment-Based Systems

Despite the number of candidates ready to replace lead and hexavalent chromium pigments, comparably few have found their way into bridge-paint specifications or even onto trial bridges. Where they have, the change has been forced on highway departments by political and public concern over the environment. In response to such pressures, Massachusetts in 1979 began to test various new pigments in bridge-paint systems put together by the pigment manufacturers. In 1980, the state opted to phase out the BLSC system and to use instead a system based on the nontoxic calcium borosilicate.[28] After five years of exposure, the results were mixed; the same system was doing well on some structures and poorly on others. Postpainting inspection has found that the 7.0 dry mils called for in the three-coat specification was rarely if ever met and that surface preparation was generally poorer than specified. The state concluded that for such nontoxic systems to do well, the same sort of care in application and inspection must be given that is called for with the zinc and barrier systems. To counteract the deficiencies in surface preparation and application, changes were made. A fourth coat, a white intermediate between the red primer and the maroon intermediate, was added. (This also provided a better distinction between coats.) At the same time, the blast requirement was upgraded from a SSPC-SP-7 to SSPC-SP-6. After five years, bridges bearing this system are said to be giving very satisfactory performance. The borosilicate primer is an oil-modified alkyd and two pure long-oil alkyd intermediate coats are employed. The finish coat is a silicone alkyd, and calcium borosilicate is employed in all coats.

The U.S. Army Corps of Engineers, using a borosilicate primer and intermediate system with two aluminum topcoats on three bridges over the Cape Cod Canal, has seen the same sort of variation in performance associated with the quality of work; that is, structures now seven years old look better than structures similarly painted more recently.

The New York and New Jersey Port Authorities have employed a borosilicate-pigmented silicone-alkyd finish coat over lead primers in the maintenance of the Goethals Bridge and the outercrossing between Staten Island and the New Jersey shore.[29]

A nontoxic system based on barium metaborate has been authorized as an alternative to the borosilicate system by Massachusetts, although, to date, few bridges have been painted with this system. Other states, such as California and Idaho (a single instance), have also experimented with nontoxic pigments (zinc phosphate and zinc hydroxy phosphite) in latex paints. California initially settled on a mixture of zinc phosphate fortified with zinc chromate in the first two coats of their four-coat latex system.[4] Later the state adopted a similar primer based on zinc hydroxy phosphite without the chromate.

North Carolina is currently evaluating two nontoxic inhibitive oil/alkyd systems: one based on zinc hydroxy phosphite, the other based on calcium borosilicate.

Several zinc-phosphate-based paint systems have been approved by the British Department of Transportation for use on bridge structures. These include systems based on vinyl/alkyds, epoxy esters, chlorinated rubber/alkyds and epoxies.[30] Use of zinc phosphate is not as prevalent in the United States, but is increasing.

Much test data on the performance comparison of many of these newer nontoxic inhibitors has been reported by the SSPC in their Project PACE Study. Unfortunately, in this study, pigments and paint systems are treated as proprietary materials and are not generally identified. Nevertheless, this study provides good general data on the performance of this class of system against the more traditional lead and chromate pigments.[31]

System Group V: Inhibitive Latex Systems

Apart from the oleoresinous oxidizing system, the only other bona fide inhibitive systems currently adopted on any large scale are California's latex-based inhibitive systems. These systems[4,32] are based on zinc phosphate/zinc chromate (more recently, zinc hydroxy phosphite) and a styrene acrylic resin for the primer, with acrylic latex finish coats. They were introduced in 1979 in response to the increasing limitations on solvent emissions in that state. They are employed as both spot- and full-coat systems. Later variations on these same systems eliminated the zinc chromate from the primer and first coat, probably because of concern over the toxicity of the hexavalent chromium. Chromate pigmentation also tends to promote destabilization of the latex polymer,[33] however, and it is felt that this, as well as the environmental restrictions, limited the levels of zinc chromate in the original systems.

By 1984, these latex systems had largely replaced other maintenance coatings in California, despite an apparent relaxation of the original regulatory demands: the twenty-four individual district bodies legislating volatile organic content (VOC) in California became, for a time, more lenient. As a result, the original goals of 250 grams per liter VOC by 1984 were by no means universally adopted. Nevertheless, apart from some severe coastal applications that still employ the older phenoxy/wash primer/vinyl system that the latex system replaced, the latex system is being used throughout the state. More recently, the same authorities have resumed their original intentions, and the depression of VOC is becoming even more intensive.

In its present form, the primer is applied over a commercial blast in two dry mil coats (a red and a gray) over which a latex finish system is applied in two coats of 1.5 to 2 dry mils each. In most applications, the finish is a simple colored latex paint without inhibitor, although on the San Francisco–Oakland Bay Bridge (also bearing the latex system), and on certain draw bridges, the finish coat is pigmented with leafing aluminum. Interestingly, the use of aluminum pigmentation on the draw bridges has been associated with some freedom from mechanical seize-ups caused by the expansion of these structures when nonaluminum pigmentation is employed (R. Warness, personal communication, 1985). It is theorized that reflection of solar radiation by this type of finish reduces metal temperatures and, therefore, expansion.

In some early formulations, the third coat was based on nonleafing aluminum, but now this coat employs an ordinary gray pigmentation. The aluminized-latex finish is a relatively simple formulation devoid of surfactants and the like, but, because of stability problems (hydrogen generation), the aluminum and latex are packaged in separate components.

The four-coat Caltrans system is applied over a commercial blast and seems to give service as good as earlier solvent-based systems, although some deterioration at the edges, seams, and lower flanges is seen on some structures. These premature deteriorations have been associated with such problems as insufficient film build and applications in low-temperature and high-humidity conditions. High chloride-ion concentration in the environment may also play a part, because it is in coastal areas that the inappropriate temperatures and humidities have been experienced. In an effort to control the film thickness, California has now restricted the use of airless spray application because of application control problems. Some say it is difficult to get contractors to realize that first-coat film thickness is to be measured from the profile peaks and that additional material must be used to fill the profile before the film thickness count can begin.

While it remains too early to make a final evaluation of the effectiveness of the California latex system, it is interesting to note that no higher incidence of failure occurs on bridges using this system as compared to such other high-performance systems as vinyls and zincs. A service life of twenty years in the interior valleys may be feasible if the material is applied successfully. In the more demanding coastal applications, five to seven years seems more probable. California observers conclude that location and overall quality of job execution (film thickness, surface preparation, and inspection) are just as important as the specific nature of the coating system to the successful outcome of bridge painting.

Some questions remain as to whether the California experience with latex paints will spread. Experimental systems based on latex primers have been employed elsewhere (as early as the late 1960s in Massachusetts), but resistance remains in the Northeast because of the effect of cold weather and its limitations on the painting season. High levels of deicing salts on most bridges in the Northeast may also preclude adoption of this system. It is interesting to note that in the experimental evaluations at Georgia Tech, the latex system's performance in salt-fog tests was poorer than that of solvent-based systems, although less difference was apparent in actual marine exposure. This may say as much for the inappropriateness of the salt-fog test itself as anything else.[34]

Until quite recently, few latex resins were considered suitable even for the maintenance paint system. Therefore, it is interesting that many formulators would, at this stage, concur with California's selection of the particular resin type (styrenated acrylic) and the specific vehicle.[33] New advances in latex technology are being introduced each year, however, and show some promise. The vinylidene chloride terpolymers[35] are examples of such innovations that employ a new acid-buffered formulating technology. While early results are ostensibly interesting, the propensity of this type of polymer to undergo dehydrochlorination, thereby attacking the steel substrate, gives some cause for concern. California used some of this type of primer on experimental sections of the San Francisco–Oakland Bay Bridge in 1984.

ZINC RICH COATING SYSTEMS

The zinc rich primer/barrier-finish concept (see Table 4-2) has, to a considerable extent, assumed the mantle once held by the inhibitive system for bridge painting. In its many forms, this composite system serves as the vanguard of new system design for many states and bridge authorities. At least thirty states now utilize the system for new construction, and, despite the exacting surface preparation and application requirements, over twenty-five have employed this type of system in maintenance applications. Over a dozen generic variants of the system have been employed, ranging from alkaline silicate zinc rich/vinyls (with a WP-1 tiecoat) to uralkyd zinc-primer systems with aluminum/styrene acrylic topcoats. As with so many of the new systems, because of the proprietary control of so many formulations, the generic classification is expanded dramatically by the numerous individual systems that are being or have been employed on bridges.

In all zinc rich based systems, the zinc coats are the primary defense against corrosion and, in several instances, these (particularly the alkaline silicates) have gone untopcoated in environments of suitable pH and achieved service lives of ten to twenty years or longer. More usually, however, both to prolong the coating life and to provide color and the required aesthetics, zinc rich primers are recoated with such simple barrier systems as vinyls, epoxies, and urethanes as one- or two-coat finishes. In these cases, results give a more finished appearance to the structure and eliminate the aesthetic drawbacks caused by white rust (i.e., zinc corrosion). In some cases, coloring pigments have been added to the zinc in an attempt to provide colored zincs in a single coat, but these attempts have been frustrated by color-uniformity problems and the inevitable color changes caused by corrosive deterioration of the zinc surface.

Table 4-2
Utility and Application Comparison of Zinc-Based Coating Systems for Steel Bridges

System Group	System Description	Volume Solids (%)	Typical Film Thickness (Dry Mils)	Surface Preparation (Minimum SSPC-SP)	Adaptability to Poor Surfaces	Probability of Early Failure from Deviations in Application	Recoat Times (Days)	Solvent Type	Usual Mode of Application	Compatibility with Old Coatings	Ease of Spot Repair	Ease of Cleaning (Salts and Soil)
	Inorganic Vehicles											
VIA	*Primer:* alkyl silicate zinc	65	3.0	6	Poor	High	1–7[a]	Alcohol/glycol esters	Airless spray	Poor	Poor to fair	Very good (Zinc—poor)
	Finish: high-build vinyl	33	5.0					Ketones/ esters/ aromatic hydrocarbons				
VIB	*Primer:* alkyl silicate zinc	65	3.0	6	Poor	High	1–4[a]	Alcohol/glycol esters	Airless spray	Poor	Poor to fair	Very good (Zinc—poor)
	Intermediate: WP-1 vinyl wash primer	9	0.3				0.02–1.00	Alcohol				
	Finish: high-build vinyl	28	5.0					Ketones/ esters/ aromatic hydrocarbons				
VIC	*Primer:* alkyl silicate zinc	65	3.0	6	Poor	High	1–7[a]	Alcohol/glycol esters	Airless spray	Poor	Poor	Excellent (Zinc—poor)
	Intermediate: epoxy/ polyamide (high build)	60	4.0				1	Alcohol				

	Finish: aliphatic urethane	52	2.0					Ketones/ esters/ aromatic hydrocarbons				
VID	*Primer:* alkyl silicate zinc	65	3.0	6	Poor	High	1–7[a]	Alcohol/glycol esters	Airless spray	Poor	Poor	Excellent (Zinc—poor)
	Intermediate: epoxy/ polyamide (high build)	60	5.0					Ketones/esters aromatic hydrocarbons				
VIE	*Primer:* alkyl silicate zinc	65	3.0	6	Poor	High	1–7[a]	Alcohol/glycol esters	Airless spray	Poor	Poor	Excellent (Zinc—poor)
	Intermediate: high-build aliphatic urethane	60	4.0					Ketones/esters				
VIIA	*Primer:* alkaline silicate zinc	60	3.0	5	Poor	Very high	1+	Water	Airless spray	Very poor	Very poor	Very good Zinc
	Intermediate: WP-1 vinyl wash primer	9	0.3					Alcohol				
	Finish: vinyl	17	2.0				0.25–1.00	Ketones/esters aromatic hydrocarbons				
VIIB	*Primer:* alkaline silicate zinc	60	3.0	5	None	Very high	1+	Water	Airless spray	Very poor	Very poor	Very good (Zinc—poor)
	Intermediate: vinyl	20	1.0				0.1	Ketones/esters				
	Finish: vinyl	25	1.5					Aromatic hydrocarbons				

(continued)

Table 4-2
(Continued)

System Group	System Description	Volume Solids (%)	Typical Film Thickness (Dry Mils)	Surface Preparation (Minimum SSPC-SP)	Adaptability to Poor Surfaces	Probability of Early Failure from Deviations in Application	Recoat Times (Days)	Solvent Type	Usual Mode of Application	Compatibility with Old Coatings	Ease of Spot Repair	Ease of Cleaning (Salts and Soil)
	Organic Vehicles											
VIIIA	*Primer:* phenoxy zinc	38	3.0	6	None	Moderate to high	1+[a]	Esters/ketones	Airless spray	Fair	Fair	Very good (Zinc—fair)
	Intermediate: WP-1 vinyl wash primer	9	0.3				0.02–1.00	Alcohol				
	Finish: vinyl	28	5.0					Ketones/esters/aromatic hydrocarbons				
VIIIB	*Primer:* phenoxy zinc	38	3.0	6	Fair	Moderate to high	1+[a]	Ketones/esters	Airless spray	Fair	Fair	Excellent (Zinc—poor)
	Intermediate: epoxy/polyamide (high build)	60	4.0				1	Aromatic hydrocarbons				
	Finish: aliphatic urethane	52	1.5									
IX	*Primer:* chlorinated rubber zinc	68	3.0	6	Fair	Moderate	1+[a]	Aromatic/aliphatic hydrocarbons	Airless spray	Fair	Good	Good (Zinc—fair)
	Intermediate: chlorinated rubber (high build)	35	3.5				3	Ketones/esters				
	Finish: chlorinated rubber	36	2.0									

X	*Primer:* vinyl zinc rich	20	2.0	10	Poor	Very high	0.2–1.0	Ketones/esters/aromatic hydrocarbons	Airless spray	Poor	Good	Very good (Zinc—fair)
	Intermediate: vinyl	28	3.0				0.2					
	Finish: vinyl	28	3.0									
XIA	*Primer:* epoxy/polyamide zinc rich	47	3.0	6	Good	High	1.0	Alcohol/glycol esters	Airless spray	Fair	Good	Excellent (Zinc—fair)
	Intermediate: epoxy/polyamide	60	4.0				1.0	Ketones/esters				
	Finish: epoxy/polyamide	60	2.0					Aromatic hydrocarbons				
XIB	*Primer:* epoxy/polyamide zinc rich	47	3.0	6	Good	High	1.0	Alcohol/glycol esters	Airless spray	Fair	Very good	Very good (Zinc—fair)
	Intermediate: epoxy/polyamide zinc rich	47	1.5				1.0	Ketones/esters				
	Finish: vinyl aluminum	14	3.0					Aromatic hydrocarbons				
XIC	*Primer:* epoxy/polyamide zinc rich	47	3.5	6	Good	High	1.0	Alcohol/glycol esters	Airless spray	Fair	Very good	Good (Zinc—fair)
	Intermediate: epoxy/polyamide red lead (marine environments only)	49	1.5				1.0	Ketones/esters				
	Finish: vinyl toluene/acrylic	50	1.5					Aromatic hydrocarbons				
XIIA	*Primer:* uralkyd zinc rich (2 coats)	55	3.5	6	Very good	Moderate to high	1.0	Aromatic hydrocarbons/aliphatic hydrocarbons	Brush/alrless spray	Fair to good	Very good	Good (Zinc—fair)
	Finish: vinyl toluene/acrylic	50	1.5									

(continued)

Table 4-2
(Continued)

System Group	System Description	Volume Solids (%)	Typical Film Thickness (Dry Mils)	Surface Preparation (Minimum SSPC-SP)	Adaptability to Poor Surfaces	Probability of Early Failure from Deviations in Application	Recoat Times (Days)	Solvent Type	Usual Mode of Application	Compatibility with Old Coatings	Ease of Spot Repair	Ease of Cleaning (Salts and Soil)
XIIB	*Primer:* moisture-curing urethane zinc	63	2.0	6	Fair	High	1.0	Ketones/esters	Brush/ airless spray	Fair	Fair	Excellent (Zinc—fair)
	Intermediate: epoxy/ polyamide	60	2.0				1.0	Aromatic hydrocarbons				
	Finish: aliphatic urethane	52	2.0									
	Nonpaint Vehicles											
XIII	*Primer:* galvanizing	100	5.0	8	Poor	Moderate to high		None	Dip	—	Fair	Excellent
XIIIA	*Primer:* galvanizing	100	5.0	8	Poor	High	0.25	None	Dip	—	Good	Very good
	Intermediate: WP-1 vinyl wash primer	9	0.3				0.02–1.0	Alcohol	Airless spray			
	Finish: vinyl	28	5.0					Ketones/ aromatic hydrocarbons				

XIIIB	*Primer:* zinc metallizing	100	5.0	5	Poor	High	0.25	None	Metallized spray	None	Good	Very good (Zinc—poor)
	Finish: vinyl (carboxylated)	20	2.0					Ketones/ aromatic hydrocarbons	Airless spray			
XIIIC	*Primer:* zinc metallizing	100	5.0	5	Poor	High	0.25	None	Metallized spray	None	Good	Very good (Zinc—poor)
	Intermediate: WP-1 vinyl wash primer	9	0.3				0.02–1.00	Alcohol	Airless spray			
	Finish: vinyl (hydroxylated)	17	2.0					Ketones/ aromatic hydrocarbons				

[a]Critically dependent on attainment of full cure and solvent release.

Because of the numerous individual formulations, attempts at a truly comprehensive treatment of the many possible zinc systems is quite beyond the scope of this volume, although the most common systems are considered in terms of their broad generic descriptions. Where necessary, additional comments on the more obvious aspects of the formulation design are offered.

The most reasonable authority for categorization of these systems is the type of primer vehicle, which may be divided into organic and inorganic vehicles. Subclassification within these two categories is again made by virtue of specific vehicle type (see Chapter 2). Such classifications have been adopted by the SSPC[36] and are discussed by Berger[37] and Munger.[38]

Inorganic Vehicles

System Group VI: Alkyl-Silicate Zinc Rich Systems

The bulk of present inorganic zinc rich systems in use on bridges are two-pack alkyl-silicate-based materials formulated with partially hydrolized ethyl silicate. This type of product is typified by Carboline's "Carbo Zinc II" and Mobil's "Mobil Zinc 7", upon which some state specifications are based. Most manufacturers now make this type of product, which is fast becoming a commodity. Commercial offerings of the prehydrolized-silicate vehicle have been available for several years,[39,40] and the basic formulations are hardly complex, although thixotropes, extenders, and other pigments (i.e., coloring pigments and inhibitors) are not as well documented.

Variations based on zinc content by weight are offered by several manufacturers, but most market an 82% zinc version (percent by weight of zinc on the dry film weight) that has served well on numerous bridges and other steel structures across the country.

Single-pack variants are also available, and these are now being adopted by several states that formerly adopted the two-component ethyl-silicate material. Some authorities think that some sacrifice in performance and certainly in stability has been exacted for the convenience of no field mixing, although such studies as the Mann's Harbor Paint Evaluation have shown some single-pack systems to give performance ratings on a par or, in some cases, better than the two-pack systems after thirty-three months of exposure.[41] Some authorities, in response to questions concerning the relative merits of one- and two-pack zinc systems, thought that while the single-pack zincs were not as good as the two-pack systems, bridge service was not so severe that such differences were critical. On the other hand, at least one major manufacturer will not supply single-pack zincs and advocates that no single-pack zinc should be topcoated, citing poor cure response and the dangers of lateral cohesive failure under vinyls, epoxies, and urethanes.[42] With some commercial single-pack systems, difficulties with the untopcoated films (mudcracking and delamination) have been noted (see Figure 4-1), while with others difficulties with protracted curing rates have indeed been found. Some of these difficulties have been experienced on actual bridge structures. Commercial single-pack resins, although not necessarily as uniform in composition as the prehydrolized resins for two-pack systems, are also available from several resin manufacturers for the preparation of these primers.[38,43]

An additional advantage in the use of single-pack inorganic zinc rich primers over two-pack systems is somewhat common to all two-pack systems. At the end of any painting day, especially on small jobs where primer systems of limited pot life are employed, there is a strong temptation among contractors to avoid mixing a new pail of primer to finish a small area, because much of this material inevitably will be wasted. There is an equal aversion to leaving new blasted steel to turn overnight so that it

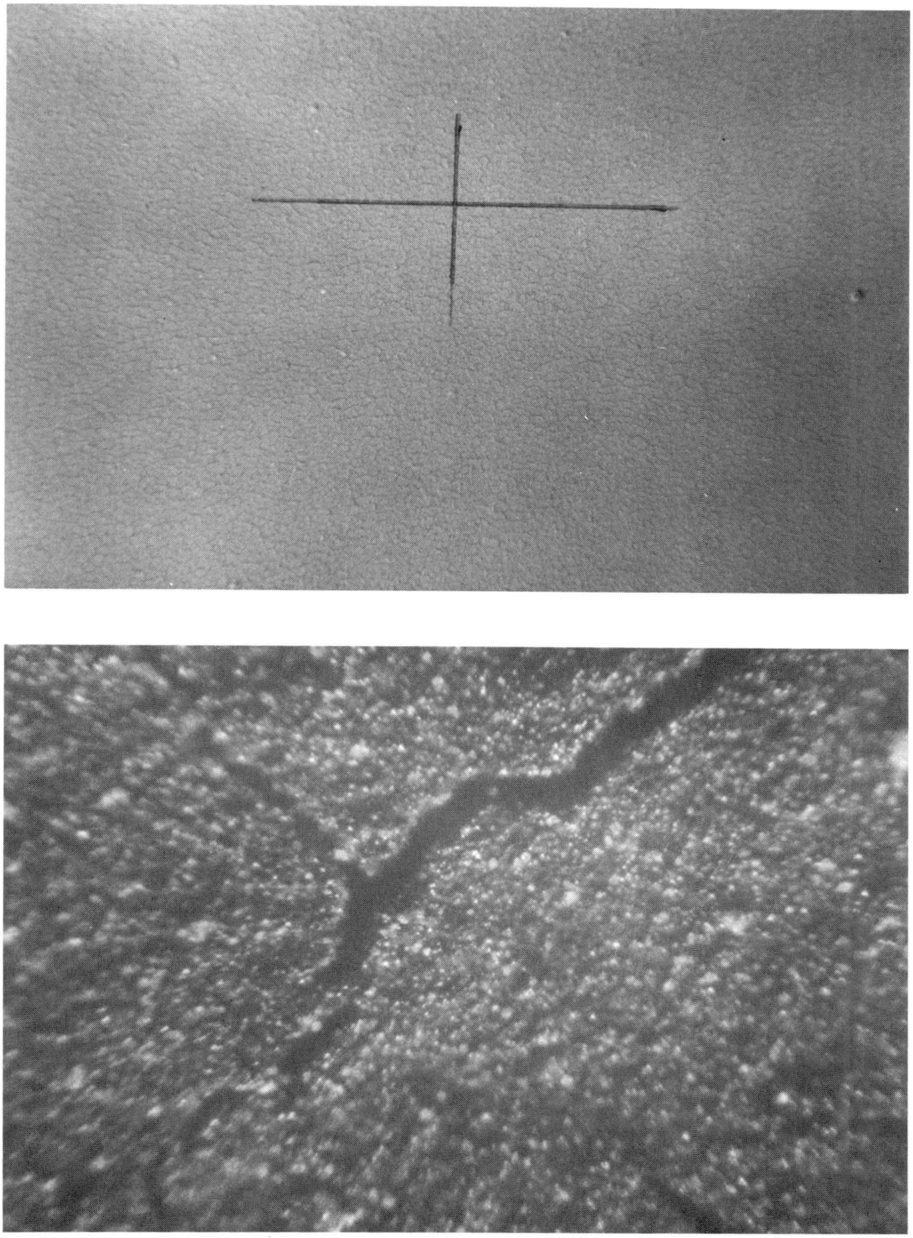

Fig. 4-1. Mudcracking of Inorganic Zinc Primer Films in Areas of Excessive Film Thickness

requires reblasting in the morning. As a result, the practice of thinning out inorganic zincs and applying no more than just enough paint to hold the steel is not uncommon. If this is missed by the inspector, it may be a couple of days or so before such thinned areas are discovered and these then have to be touched up or given a make-up coat. At the very least, this statistically increases the likelihood of problems (because of an additional interface), and, in the recoating of inorganic zincs, this can be a particularly difficult problem. Where inspection is minimal, large areas painted with primers of insufficient film thickness may exist, and performance will be reduced.

Some disagreement exists among manufacturers concerning the procedures for touch up of inorganic zincs. Their advice varies from simply thinning the second inorganic zinc coat to sweep blasting the original coat before applying the make-up coat. Very often, organic zinc make-up coats are employed and applied over hand- or power-scarified surfaces. Epoxy/polyamides and phenoxies have been used in this way.

The widespread use of organic zincs of high PVC over inorganic zinc films is to be avoided because inevitably some of the vehicle binder from the organic zincs will be absorbed into the porous inorganic matrix, thus effectively raising the PVC/CPVC ratio of the organic primer and decreasing its cohesive strength. This increases the risk of later system failure by lateral cohesive problems in the organic zinc continuum.

As with the two-pack alkyl-silicate specifications, however, the single-pack zinc specifications remain vague and reflect the proprietary interests of the various manufacturers of the products from which the specifications are derived. Perhaps most seriously lacking is a more complete specification for vehicle systems (including such additives as silanes, borates, etc.), but the manufacturers of the silicate vehicles show little interest in formalizing specifications as was done with alkyds, phenolics, and the older vehicle types. Even in the latest specifications, such as those of the state of Virginia,[44] the vehicle is often incompletely defined as an organo silicate in appropriate solvents and additives.

In-depth compositional specifications for two-pack alkyl silicates are beginning to appear from some bridge authorities, such as the State of Massachusetts Public Works Department and the Massachusetts Turnpike Authority.[45] Some specifications, like those of the state of Ohio, refer to the SSPC Paint System Guide PS 12.00,[46] although this does not spell out specific formulations either.

As stated in Chapter 2, the use of inorganic zinc dictates a minimum surface-preparation quality of SSPC-SP-6, and some specifications prefer SSPC-SP-10. Mil profiles on new steel are preferably 1 to 2 mils, although in maintenance applications involving the removal of old existing paint, higher profiles of 3 to 4 mils are often necessary. Film thicknesses for the inorganic zinc primer usually range from 2.5 dry mils to 5.0 dry mils, although too high a film thickness (particularly in the single-pack systems) would produce serious film defects, such as mudcracking and delayed-cure response. These concerns restrict the use of inorganic zinc on structures of complex design with surfaces that are difficult to reach or clean or on which applications of controlled film thicknesses are difficult. Inorganic zinc-based systems are most ideally suited to the modern, simple rolled-beam designs with large areas of uninterrupted steel surface (see Figure 4-2). Major advantages of the alkyl silicates are common to all inorganic zinc systems: long service life, reduced application temperature requirements, and a high coefficient of friction (higher, in fact, than that of sandblasted steel). This high coefficient of friction allows application to faying surfaces of steel members in the shop and eliminates the need to mask these areas (an advantage not universally appreciated by bridge design engineers). The coefficient of friction is dependent on carefully controlled application and, for faying surface applications, film thicknesses that remain in the 1–2 dry mil range per mating surface area necessary.[47] Applied over wet abrasive blasted surfaces, inorganic zincs are also being evaluated for weathering steel surfaces.[48]

System Groups VIA and VIB: Alkyl-Silicate Zinc/Vinyl Systems

The most common barrier finish coat for alkyl-silicate inorganic zinc primers is the high-build vinyl system in films ranging from 3 to 6 mils. The propensity of these vinyl finish coats to bubble during application (see Figure 4-3) over the inorganic zinc[49,50,51]

Fig. 4-2. Simple Rolled-Beam Bridge Design Well Suited to Application of Inorganic Zinc Rich-Type Coating Systems

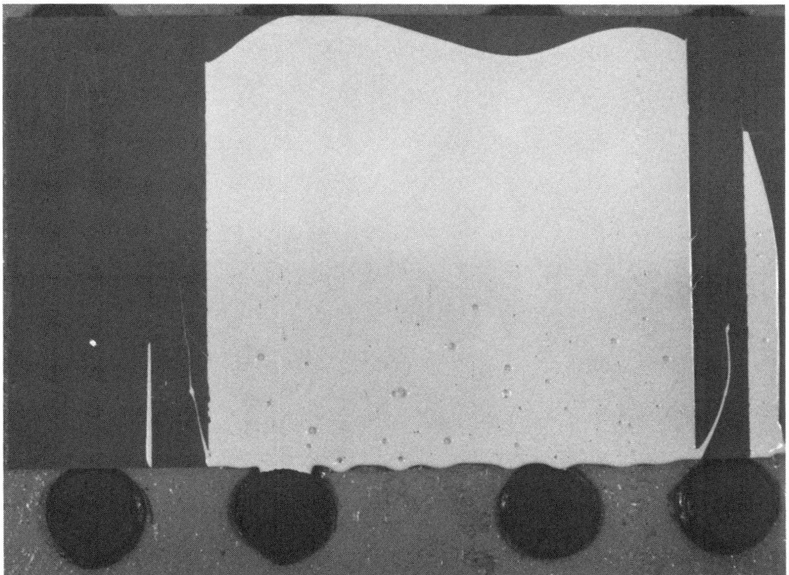

Fig. 4-3. Bubbling and Cratering of Finish Coat Applied over Inorganic Zinc Rich Primer

has in many states encouraged the use of vinyl tiecoats or the vinyl wash primer WP-1 (DOD-C-15328D),[2] which, in films of 0.3 to 0.5 dry mils, is probably the most effective solution to the problem (System VIB). Although the wash primer adds to the adhesion of the vinyl topcoats over the wash primer and may improve the curing of the primer, the SSPC in a large study of the topcoating of inorganic zinc rich primers concludes that the wash primer neither improves nor detracts from the performance of any topcoated zinc rich system.[50]

In some cases, in lieu of the wash primer, carboxylated vinyls are also used at low solids as tiecoats (System VIA). Mist coats have also been used, as have zot coats (i.e., topcoats that are thinned radically with solvent and applied prior to a full unreduced coat), to eliminate the bubbling phenomenon. However, mist coats are not as universally reliable with vinyls and other high-molecular-weight thermoplastic lacquers as they are with other finishes, such as epoxies. At least in the short term, bubbling and the consequent crater formation as the bubbles burst, although aesthetically displeasing, do not appear to compromise the performance properties of the system.

A second and potentially more serious problem with inorganic zinc/vinyl systems is the lateral cohesive failure discussed in Chapter 2. This failure occurs when the vinyl topcoat is applied before the inorganic zinc has completely hydrolized (see Figure 4-4). North Carolina has reported numerous failures of this type on bridge structures treated in 1976 and 1977. It is thought that low shop humidity during application of the zinc, rapid topcoating, and a relatively tight (low PVC) vinyl film may have added to the failure by first delaying hydrolysis and then preventing further cure with the impermeable topcoat (W. M. Medford, North Carolina Department of Transportation, personal communication, 1984). As with all moisture-curing systems, humidity and temperature

Fig. 4-4. Lateral Cohesive Failure (Intracoat Splitting) of Topcoated Inorganic Zinc Rich Primer

will affect both the curing of the alkyl-silicate primer and the necessary recoating interval. Job schedules, particularly in maintenance operations, are often in conflict with delayed topcoating requirements, and these differences can lead to serious problems. One manufacturer allows only rapid (one-day) recoating of such inorganic silicate zincs if the WP-1 (phosphoric-acid-cured) wash primer is used as a tiecoat. This manufacturer believes this will increase the rate of cure. It is interesting to note, however, that in the North Carolina experience, a wash primer was, in fact, employed.

Some claim that vinyl-butyral-modified alkyl-silicate primers with fine particle-sized zinc greatly diminish both topcoat bubbling and lateral cohesive problems by producing films with more rapid cure and reduced porosity.[52,53] The organic polymer is said actually to combine chemically with the ethyl silicate, producing improved adhesion and a denser product. These claims have been disputed elsewhere.[54]

By the mid-1980s, many of those states that retained oxidizing systems for spot maintenance in the field were employing inorganic zinc rich/vinyl systems in new construction. The inorganic zinc is applied in the shop, touched up in the field, and the finish coat system is then applied. In some states, where bridges in the field need complete removal of the old paint, the inorganic zinc rich approach has also been adopted. Such systems are not generally advocated for older structures with such inherently unsuitable design features as heavy lattice work, back-to-back angles, and the like. Even use on heavily rivetted structures should not be attempted unless excellent job control (inspection) can be maintained. Some problems with delayed recoating on shop-applied zinc surfaces (e.g., salting, contamination with shop soils, dirt, and oils) have proven to give significant field painting problems, and some states have abandoned the approach in favor of complete in-shop painting.[55]

Like most vinyl systems, either of two formulation approaches may be employed in high-build-vinyl finish coats. The first utilizes a carboxylated system for direct-to-zinc application; the second uses a hydroxylated vinyl that necessitates the use of the wash primer as a tiecoat. Either vinyl may be used over carboxylated tiecoats. In some cases, mixed resin types are employed. Many paint manufacturers would prefer the carboxylated vinyl approach because this not only reduces the number of coats required but makes for easier formulating. Molecular weight for molecular weight, the carboxylated vinyls give lower solution viscosities than do their hydroxylated counterparts. This allows for higher volume solids (with easier high-build capabilities) or lower cost solvent systems. On the other hand, many feel that the wash primer/hydroxylated vinyl system produces added engineering benefits in improved interfaces and adhesion, and they rely on thixotrope/solvent systems for high-build capability. Some manufacturers also use the lower viscosity, butyl-maleate-modified vinyl (which displays adhesion to both metal and wash primer alike) as an auxiliary resin.

In most cases, the service life of these inorganic zinc/vinyl topcoat systems is most excellent; even in relatively demanding Zones 2A, 2B, and 3 environments, claims are made of fifteen years and more of service. Earliest systems were applied in the late 1950s and, in many instances, are still in good shape. The experience of the state of West Virginia is typical. That state, which first adopted a two-coat system in the early 1970s (only later was a tiecoat employed) expects thirty years of service from the system. Apart from some early delamination problems before the tiecoat adoption, West Virginia has so far not needed to repaint any structures. Borrowed almost directly from the proprietary products of a large manufacturer, West Virginia's specifications do not delineate the type of topcoat or the finish coat vinyl formulation, instead leaving this up to the manufacturer.[56]

One problem that has haunted the West Virginia system as well as that of several other states, such as Maine, Tennessee, and Missouri (using similar vinyl topcoats), has been topcoat chalking and fading (see Figure 4-5). The phenomenon is reported to depend on color and may well be related to specific pigments or pigmentation levels. Other manufacturers note some chalking and fading of the vinyl system after five years or so, but in Maine, West Virginia, and other states, dramatic color changes have occurred in less than two years. Curiously, Maine, using the same topcoats from the same manufacturer in the same color, has had some bridges that fade badly, while others, with equally exposed paint films, have retained excellent color stability. Like other performance specifications, the Maine specification does not control composition, and failure analyses have not been performed.

Review of the high-build vinyl specifications of many states[57,58,59] reveals the same numbers (27% to 29% minimums for pigments, 19% to 21% minimums for vehicle solids, 48% total solids, etc.) and similar language, betraying perhaps some uniformity of the source. Few provide real details of composition beyond references to aromatic/ketonic solvent blends, light stable plasticizers, and so on. Most high-build vinyl specifications of state transportation departments do not describe either pigment or vehicle makeup in detail, calling for a vinyl chloride/vinyl acetate copolymer or, in some cases, even a hydroxylated vinyl chloride/vinyl acetate terpolymer suitable for direct use over an inorganic zinc! It is presumed, in this case, that they rely on the manufacturer to correct the apparent discrepancy and modify the copolymer by using either a butyl maleated vinyl or by substituting the carboxylated vinyl, in spite of the specification. This type of specification generally does not use the wash primer tiecoat and is applied directly to the zinc. The danger of this type of quasi-compositional specification derived from a single source is exemplified in the chalking problems described above, although, in fairness, different interpretations of the same specification from other manufacturers have reported similar problems. Other high-build vinyl specifications[60] are even less compositional but generally call for the vinyl to be made by the manufacturer of the zinc

Fig. 4-5. Severe Chalking Failure on a High-Build Vinyl-Finished Bridge Structure

primer. The West Virginia specification allows great latitude in composition, but, oddly, specifies weight per gallon and viscosities as well as resistance to sulphuric acid and sodium hydroxide. South Carolina specifies that the vinyl shall have been used successfully over an inorganic zinc rich primer for at least two years in a similar environment.[61] Several states employ lower build systems for topcoating inorganic zinc, and, in general, these are more definitively specified. Maryland calls for SSPC Paint 9;[62] Massachusetts calls for a hydroxylated vinyl of specified composition;[63] and Minnesota provides complete formulations for eight separate vinyl paints for use over a wash-primed phenoxy zinc.[64] High-build compositional specifications have recently been specified by the Massachusetts Turnpike Authority.[45] Martell, in a review of vinyl technology,[65] describes a Massport compositional high-build vinyl specification[66] for use over a wash primer. A later variation of the same specification is even more detailed.

In the Massachusetts Turnpike specification, the vinyl is applied directly to the zinc and a butyl-maleate-based vinyl resin is used to modify the normal carboxylated terpolymer. Despite claims of increased moisture sensitivity and high methanol extractives (T. Ginsberg, Union Carbide Corp., personal communication, 1984), there is probably considerable use of this resin in other high-build vinyls because of its reduced cost. The Massachusetts Turnpike specification employs a low-build acrylic (with vinyl modification) as a two-mil topcoat for the four-mil high-build intermediate vinyl coat. The modification of vinyls with acrylic thermoplastics is not an uncommon practice to upgrade UV resistance where necessary.

The susceptibility of vinyls to fading and chalking, evidenced by experiences in Ohio, Missouri, Tennessee, and Florida, is by no means categorically proven to be a defect in the generic system. Many individuals blame the fading not so much on resin type but on pigmentation selection and specific formulations. The position is given much credence by the significantly better UV resistance of the low-build systems, such as those described in the Minnesota vinyl specification.[64] As noted in the case of Maine, some batches (and/or products) give better performance than others. In an unpublished study,[67] severe color deterioration, such as has been seen in Maine, has been shown to result from formulation practices not in keeping with the general guidelines originally published by Union Carbide Corp. many years ago (see Figure 4-6). These formulation guidelines are extremely comprehensive, delineating acceptable types and levels of pigmentations, plasticizers, stabilizers, and solvent systems. Experience would show that formulations within these guidelines do not undergo the severe nonuniform color degeneration that has been seen in many cases. Gary Tinklenberg of the Michigan Department of Transportation reports that color deterioration in exposed vinyl films seems to be related more to individual companies' products than to generic characteristics (personal communication, 1985). In 1983, Union Carbide Corp., at the request of the Florida Department of Transportation, formulated a group of high-build vinyls of medium PVC in a variety of colors in order that the specificity of chalking to individual formulations and colors might be determined. As of late 1985 all of these high-build systems are said to show good color retention in the field.

System Groups VIC, VID, and VIE: Inorganic Zinc/Epoxy/Urethane Systems

Somewhat newer and less widely employed on bridge structures at present are epoxy and urethane topcoats over an inorganic zinc. In general, high-build epoxies (3.5 to 5.0 dry mils) are used as intermediate coats directly over the zinc primer (if necessary using a mist coat technique) and then 1–2 dry mils of aliphatic urethane are applied as finishes (System Group VIC).

It is reported that in 1985 New York State officials used this type of system on the

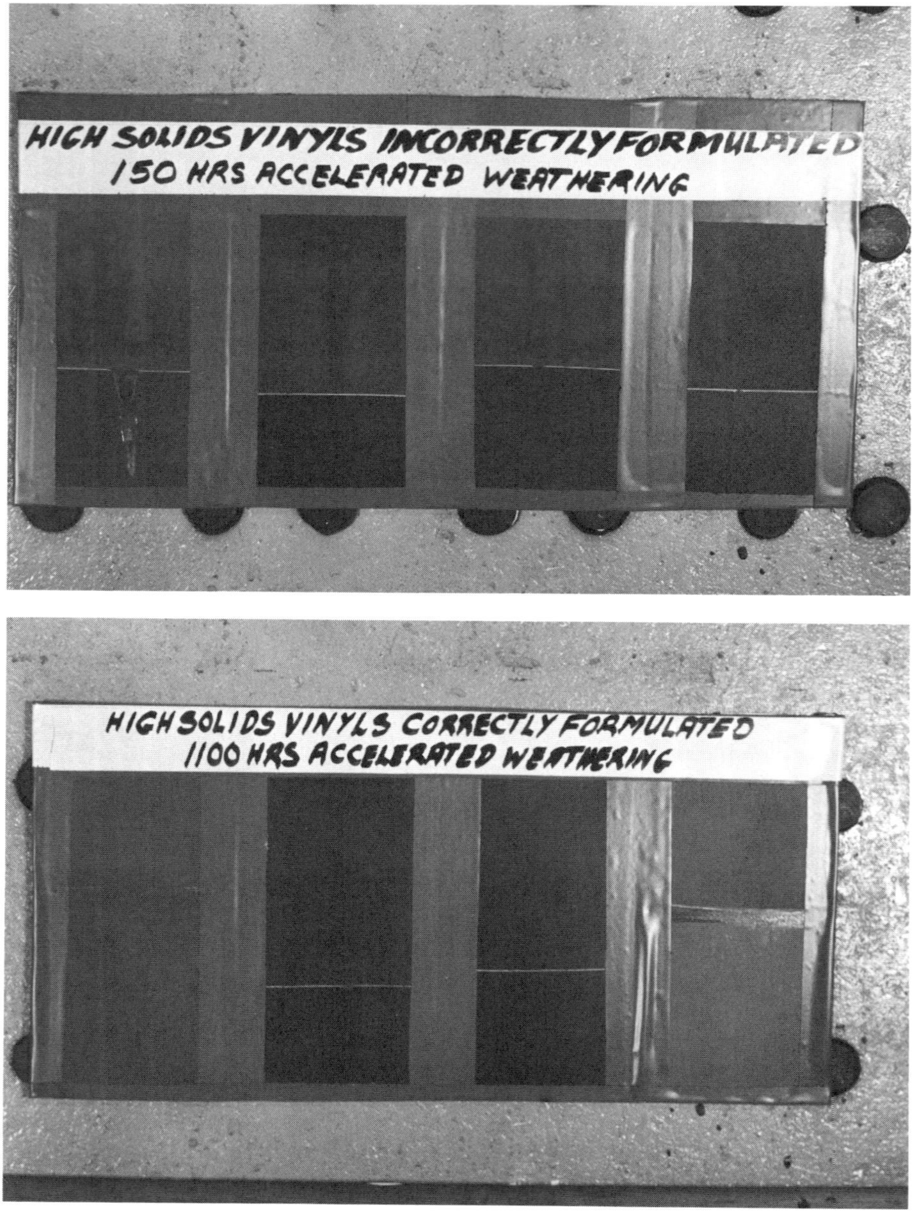

Fig. 4-6. Effect of Improper Pigmentation on the Accelerated Weathering Resistance of a High-Build Vinyl Finish. The upper part of each panel has been exposed, the lower part unexposed.

Green Point Bridge. Pennsylvania has painted several bridges with the system and state officials report good early data. In early 1983, the state of Michigan adopted this type of system for a total shop painting program,[55] and has since painted a number of bridges that show good early performance. The Michigan specification[68] is a performance specification without compositional additions. Several other states are either testing or considering the system (Ohio, Massachusetts, etc.), and the state of Wisconsin is now using it. The system was also selected for the new painting of the suspended span of the Newport Bridge in Rhode Island in 1984. Material costs are slightly less than the vinyl (epoxy intermediate at $0.030 per mil square foot, acrylic urethane at $0.055 per mil square foot), and the possibility exists for achieving further cost reductions by limiting the use of urethane to some exposed areas of the structure and using an additional coat of epoxy in those areas with reduced UV exposure. The Michigan specification at one time eliminated urethane from areas other than the facia beams and simply retained a two-coat inorganic zinc/epoxy system in nonexposed areas (System Group VID). More recently, the state has begun to adopt the three-coat urethane finish system across the entire structure.

High-build epoxy midcoats are usually applied directly over the zinc. They are somewhat less sensitive to topcoat bubbling than are the vinyls, probably because of their initially lower molecular weight, which tends to result in reduced early surface skins. Bubbling is somewhat dependent on formulation, but in most cases can be reduced by mist coating techniques. High-build epoxy midcoats are subsequently recoated with urethane or another coat of epoxy, although once formed, bubbling and its resultant cratering tend to telegraph from coat to coat. Epoxy midcoat solids normally range between 55% and 70% by volume. Resin systems at this time are most often based on medium molecular weight Bisphenol-A type epoxy resins with polyamide cures, although the use of lower molecular weight epoxies is increasing. PVCs are higher than the urethane finish coats and gloss is often reduced to semigloss or eggshell to improve intercoat adhesion and disguise gloss reductions in applications where urethane finish coats are not used.

Two principal types of polyols are used in the urethane systems: a hydroxylated acrylic or a hydroxylated polyester. The acrylic is typified by the products of Rohm & Haas Co.[69] and the Henkel Corp.[70] The polyesters are produced by the Mobay Chemical Company, which also makes a range of diisocyanates.[71] The only isocyanate of any real consequence is the adduct of hexamethylene diisocyanate. All companies produce good formulative support, although only Mobay has specifically directed its marketing thrust at public works and highway departments and paint companies alike. The acrylic polyols, however, seem to have at least as large a share of the young market at this time as do the polyesters. While the polyester-based products have the edge in chemical, solvent, and abrasion resistance, the acrylics have slightly better UV resistance and faster initial drying profiles. They are also less costly. Conventional volume-solids ranges for the urethane finish coats are between 50% and 60%.

Despite the improved dissemination of formulating data (particularly on the polyester urethane system), coating systems of this type are most usually purchased from prequalified manufacturers' lists (without compositional requirements). This technique is employed in Michigan and Florida.

In-depth compositional data on the epoxies, like that supplied on the urethanes by Mobay, is less readily available to the specifying agencies, despite the greater maturity of the epoxy coatings. The Maine Department of Transportation[72] provides a relatively detailed specification that calls for definitive parameters for PVC, pigment weight, solids by weight and volume, and weight per gallon of both components and the mixed material. This provides much assistance to a would-be formulator. Similar requirements are also listed in the same specification for a polyester urethane. Curiously, neither document specifies a required mixing ratio.

Unfortunately, neither acrylic- nor polyester-based systems are sufficiently mature to make specific comment on the service life at this time, although claims are made that they will be as good or better than the more established vinyl-topcoated system. Early data is promising, particularly in UV resistance and salt-fog resistance (for the total system). Numerous advances in urethane (particularly polyol) chemistry are being made, and higher-solids systems are now becoming available.

A further advance on the system above is the use of a single high-build urethane directly over the inorganic zinc (3 mils of inorganic zinc, 4 dry mils of urethane). This type of system (System VIE) has been employed in the 1985 reparations made to the Neil Underwood Bridge in Hampton, New Hampshire.[73] While individual material costs of this system can be somewhat higher than either the vinyl or the epoxy/urethane finish systems, the advantages in reduced labor costs of a two-coat system more than adequately compensate for any apparent disadvantage.

System Groups VIIA and VIIB: Alkaline-Silicate Zinc/Vinyl Systems

Bridge-paint systems based on alkaline-silicate zinc rich primers, although not as widespread as the self-curing ethyl silicates, are not at all unknown. Florida, for example, has employed postcured sodium-silicate-based zinc rich primers in both two-coat (inland, nonchemical areas) and three-coat (coastal and high-corrosion areas) vinyl-finished systems.[74] These specifications are detailed, essentially compositional documents, and call for 2.5 to 3.5 mils of primer and 0.75 to 1.0 mils of vinyl-aluminum topcoats with or without an intermediate tiecoat of 1.0 to 1.5 dry mils of vinyl (pigmented with BLSC). Surface preparation requirements are for a near-white blast (SSPC-SP-10) with a 1.0 to 1.5 mil profile.

The high levels of performance that can be achieved with sodium-silicate zinc rich systems are exemplified in the performance of an untopcoated 3-mil system on the Port Orange Drawbridge near Daytona Beach. This system gave ten years of corrosion-free protection before a topcoat was applied.[75]

A self-curing alkaline zinc rich based system has been employed on the Golden Gate Bridge in San Francisco since the early 1960s. In this case, a WP-1 wash primer is employed as a tiecoat with the vinyl finish. Alkyd finishes have also been used on the structure, with the WP-1 tiecoat serving as an effective insulation between the zinc and the alkyd, thereby preserving alkyd adhesion.[76]

Other examples of alkaline-silicate-based systems are reported by Munger and include the 3-mil sodium-silicate-based system on the Batman Bridge in Tasmania.[38]

Organic Vehicles

System Group VIIIA: Phenoxy Zinc Rich/Vinyl Systems

Perhaps the most widely used organic zinc rich based system, the phenoxy zinc/vinyl system was one of the many coating systems developed in California's transportation department laboratories (Caltrans).[77] Like all of Caltrans's specifications, the documents are comprehensively compositional and involve no difficulty in interpretation for any manufacturer. Early versions of the system were problematic because of intercoat adhesion difficulties. While the phenoxy resin is basically a high molecular weight linear epoxy thermoplastic, the solvent resistance is such that the normal thermoplastic solvent weld is unreliable. Addressing this, some manufacturers have deliberately modified phenoxy primers with phenolics and polyketone resins with some success. In other applications, including Caltrans's, vinyl finish coats based on hydroxylated-vinyl-chloride/vinyl-acetate resins are applied over a tiecoat wash primer (DOD-P-15328D) in order to guarantee adhesion to the zinc. Generally, commercial blast or near-white blasts are specified for this system, 2.5 to 3.0 mils of primer being applied. Wash primer and vinyl topcoat film thicknesses are 0.25 to 0.5 mils and 3 to 5 mils, respectively, although in many systems two vinyl topcoats are employed of 2 mils each. Service lives of ten to fifteen years have been achieved in California with a four-coat system, although earlier breakdown has been encountered, especially near the coast.

Several other states have either adopted, modified, or copied this specification and have had good service with it. These states include New Jersey, North Carolina, Massachusetts, and Minnesota. New Jersey uses two coats of the phenoxy primer. A series of colored, compositional, and in-depth specifications, including formulations for the two low-build vinyl topcoats of this type of paint system are included in the Minnesota 3529 specification.[64] Both Louisiana and Florida[74] report reduced success with phenoxies as compared to the epoxy organic zinc rich primers.

The phenoxy vinyl system is often used (as is the inorganic zinc/vinyl counterpart) as an alternative to conventional lead systems for Zones 2A, 2B, and 3 service in such states as New Jersey, where environmental zoning is utilized for the selection of bridge-coating systems. Phenoxy zinc rich primers are sometimes used also for touch-up of inorganic zinc rich systems.

System Group VIIIB: Phenoxy Zinc Rich Systems with Other Topcoats

The recoating difficulties with phenoxies may be reduced by modification with certain other resins, such as phenolics. Using the modified phenoxy to improve recoatability, one major manufacturer has employed phenoxy zinc rich primers as the base coat for a high-build epoxy (4 mils) with a urethane finish coat (1.5 mils). This system has given almost fifteen years of service on many structures, but its use on bridges is limited; two to three years of service life, albeit reasonably successful, is all that has been recorded to date.

Intercoat adhesion problems are most common with vinyl systems, however, and other finish coats have been used without modification. Such applications include phenoxy zinc rich primers, which have been used on joints and bearings and topcoated with 16 mils of a C-200-type coal tar epoxy.[78] As with many systems, service life is quite variable from bridge to bridge, but the phenoxy/coal tar system has given good service for almost ten years in these applications on some bridges. There is, however, some question as to whether use of the zinc rich primer beneath high builds of a coal tar epoxy is necessary or even advisable. The high-build barrier properties of the coal tar coating, in eliminating the passage of electrolyte to the zinc, severely reduces the cathode/anode area ratios at scratches and breaks and leads to reduced protection of the bared steel at the scratch.

The state of California has also used chlorinated-rubber/aluminum finish coats over the phenoxy/washprimer systems (R. Warness, Caltrans, personal communication, 1984). This system is said to give at least seven to ten years of performance without problems. Finally, in spot-blast applications over old oleoresinous systems, some states have employed the phenoxy/vinyl-washprimer/alkyd-finish-coat combination. Some problems with this system have been experienced with active solvents from the phenoxy attacking the old alkyd in areas peripheral to the blast.

System Group IX: Chlorinated Rubber Zinc Rich/Chlorinated Rubber Systems

The chlorinated-rubber system[79] was selected for use on the five-year painting program on the Tobin Memorial Bridge in Boston and, despite the many reservations that were voiced, was giving excellent service after ten years. The system consists of a 3.0-mil application of a two-pack chlorinated-rubber-based zinc rich primer, a 3.5 mil chlorinated-rubber midcoat, and a 1.5- to 2.0-mil finish of a low-build chlorinated-rubber enamel. Surface preparation is a near-white blast. A similar system that was used as a spot repair of a red lead/chromium green-oxide alkyd-oil system on the legs of the same bridge is doing better than expected, although some corrosion was occurring after five years around the periphery of the spot blast as the old alkyd continued to break down. Interestingly, while this system performs well in the field, it does not do well in salt spray and attests to the limitations of this test[80] as a true measure of field performance.

Early difficulties were experienced with the chlorinated-rubber primer, which was originally proposed as a one-pack system (TT-P-1046).[81] Unfortunately, in metropolitan Boston, the solvent systems of all paints must conform to Rule 66/3 (reduced levels of certain photochemically reactive solvents) and the early use of acetate-type solvents with lower-than-specified levels of oxirane-acid acceptor produced severe dehydrochlorination in this composition, leading not only to reduced stability and performance but also to fires and explosions both in the manufacturing plant and on the job. In consequence, a two-pack system[82] was implemented, wherein the zinc was introduced to a highly stabilized resin system on the job site. This points out the complexity of coating formulations and the need to test when changes are made.

System Group X: Vinyl Zinc Rich/Vinyl Systems

A third thermoplastic organic zinc rich system is based on the three-component vinyl zinc rich primer developed by the U.S. Army Corps of Engineers.[83] This rather cumbersome composition is being used in Oregon where early success during the first three years has been reported. Because of the necessity of zinc-to-steel contact (which precludes the use of the WP-1 wash primer as a steel pretreatment and, therefore, of the hydroxylated-vinyl resin) and the reactivity of the carboxylated-vinyl resins with zinc pigment, the vinyl zinc rich primer must utilize the unmodified vinyl-chloride/vinyl-acetate copolymer, which has inherently poor adhesion to metal. This deficiency is rectified by the addition of a small amount of suitable organo-functional silane, which is introduced as one of the three components (the zinc also being mixed separately) as the paint is prepared on the bridge. Two relatively high-build vinyl coats of 3 mils dry film thickness each are applied over 2 mils of this primer in rapid succession.

The state of Oregon specifications for this system are compositionally thorough[84] and are a substantial improvement over many vinyl specifications. Interestingly, Oregon posits that no vinyl shall be accepted that displays any tendency to chalk. This would again seem to intimate that, in this case, chalking is not a peculiarity of vehicle type but of the formulation. Unlike Oregon's many disappointing experiences with high-performance systems, this same vinyl system initially appears to be giving good service in both coastal and inland areas of the state. Fewer insurance claims from passing motorists during the application of this fast-drying system were also considered an advantage.

System Group XI: Epoxy/Polyamide Zinc Rich Systems

Epoxy zinc rich primers are usually polyamide-cured, medium molecular weight systems, although other curing agent types may be employed and epoxies of different molecular weight used. They are always two- or three-pack systems because of the chemically reacting cure mechanism, concerns over pigment settlement, and in-can gassing. Compositional specifications for the zinc rich primer are typified by the Corps of Engineers's E-303 specification.[85] In general, epoxy/polyamide zincs have been found to provide some of the best performance of all the organics, but their use on bridge structures in the United States does not seem as widespread as is the use of inorganic zincs or even phenoxies. Nevertheless, the Florida Department of Transportation describes the epoxy/polyamide zincs as outperforming both the chlorinated-rubber and the phenoxy thermoplastics.[74] Michigan has adopted this type of primer and is said to prefer it to inorganic zincs over previously corroded surfaces; officials there advocate its use over weathered and corroded A-588 surfaces after blasting.

Like most organic zinc rich primers, the epoxy/polyamide zinc rich primer takes on many of the characteristics of the vehicle. The film is hard, tough, and solvent resistant, as well as highly adherent. Tolerance for poorer surfaces is rather better than the inorganics, although poorer surfaces may confound cathodic protection. The dilution of resin components produced by the high PVC/CPVC ratio of the organic zinc rich primer tends to reduce cure kinetics to some extent and give longer pot lives.

System Group XIA: Epoxy/Polyamide Zinc Rich Primers with Epoxy/Polyamide Finish Coats

System Group XIA, while very sensible, appears to have been used less often than the analogous systems described below that utilize thermoplastic topcoats. Film thicknesses recorded for various applications are 2–4 dry mils for the primer, 2–3 dry mils for the midcoat, and 2–3 dry mils for the finish. A similar three-coat system (utilizing 2–3 mils of zinc primer, 3–5 mils of red lead primer, and 2–3 mils of green finish) was applied by the painting contractor Oliver B. Cannon & Sons as a test section against a three-coat red lead primer/midcoat/green finish (4–6/2–3/2–3 dry mils, respectively) on extensive sections of the Newport Bridge in Jamestown, Rhode Island, in the mid-1970s. In fact, after ten years, the all red lead systems appear to have a very slight edge over the zinc. Although both systems are in quite good shape, some slight pinholing corrosion is seen on the epoxy zinc rich-primed area.

Michigan now reports excellent early success with a variation of this system in field maintenance. The Michigan system uses high-build epoxy intermediate coats with urethane finishes and is being used in the field for complete renewal of old bridges. The Michigan philosophy is that by replacing the old mill scale beneath existing lead-based systems with a good blasted substrate they not only break the lead cycle but also eliminate the need for future radical blasting. While they foresee necessary maintenance of the new zinc-based system, this future maintenance involves attention to finish-coat renewal only and not to the renewal of the entire system (Gary Tinklenberg, Michigan Department of Transportation, personal communication, 1986).

System Groups XIB and XIC: Epoxy/Polyamide Zinc Rich Primer with Thermoplastic (Vinyl Type) Finish Coat

System Group XIB is typified by the specification for Louisiana. Originally a two-coat system in which the tinted epoxy/polyamide zinc rich primer was recoated with a vinyl aluminum, this system proved to have an inadequate service life. Deterioration occurred in six to eight years at edges and seams, and the system has now been replaced with two coats of epoxy/polyamide zinc. The first coat is tinted red,[86] while the second is tinted green for contrast.[87] The epoxy zinc coats are applied in 3.0 and 1.5 dry mils, respectively, over a near-white blast. The system is finally recoated with a vinyl aluminum topcoat[88] and now appears to be giving good service (ten years minimum), although problems are encountered with the vinyl-aluminum (leafing) specification, a low-build system from which 3 dry mils are required in a single coat. Despite the use of double passes, the build requirements are reportedly impractical. On new jobs, the topcoat is applied in the field after touch up of the zinc (where necessary) and overall cleaning of the primer. There is a maximum time of four months between the application of the second zinc coat and the vinyl (a lacquer that is based on a mixture of both hydroxylated and carboxylated resins). Any delay in topcoating beyond four months results in a mandatory application of a WP-1-type wash primer (MIL-P-15328A) before application of the vinyl. While the specification calls for the elimination of lap marks in topcoat application, the fast-solvent system (toluene and butyl acetate) has given problems with this.

The Louisiana specifications are compositional in nature with full documentation in pigment, vehicle, and solvent makeup, and products are prequalified. In addition, there is a performance specification with requirements for salt fog (1500 hours), sunshine and weatherometer exposure (1500 hours), and fluorescent UV condensation exposure (QUV) with four-hour cycling 16°C (60°F) for 1500 hours. The test is performed on the total system, which is evaluated (albeit qualitatively) for blistering, chalking, checking, cracking, delamination, discoloration, rusting, and undercutting.[88] Interestingly, in this test, paints are applied at dry-film thicknesses and over surface preparation qualities required by the paint company, not by the state of Louisiana, although the specification is quite specific on blast quality, mil profile, and film thicknesses required by the state.

A second, rather similar system is a new primer-topcoat system being introduced in Texas (J. Raska, Texas Department of Transportation, personal communication, 1985). The system (System XIC), a three-component epoxy/polyamide zinc rich primer with a vinyl-toluene-based finish, was formulated in the laboratory at the Texas Department of Transportation in the early 1960s. By 1984, it had been employed only once. Its use has been discouraged since then because of the three-component mix. Data would now show the epoxy/polyamide system to be slightly better than the uralkyd zinc that was originally adopted by the state instead of the epoxy/polyamide, and plans are now to change primers. (Service lives, indicated from the early experimental applications of the epoxy/polyamide system, are in the range of fifteen to twenty years.) In the years since its inception, the vinyl-toluene/acrylic resin of the topcoat has been discontinued by the resin manufacturer, and the new system employs a styrene/acrylic-solution lacquer that has been used more recently over the uralkyd zinc. Film thicknesses for the new system will be 3.5 dry mils minimum for the primer and 1.5 dry mils of finish. In salt-water environments, the system will be fortified with an additional midcoat of epoxy/polyamide red lead (also 1.5 dry mils). Like the original system, the new specification will not only be compositional in nature but will specify formulas.

System Groups XIIA and XIIB: Urethane Zinc Rich Primer Systems

Zinc systems based on both oil-modified urethane (uralkyd) and moisture-curing urethane vehicles have been developed. The oil-modified urethane primer is a rather unique specification (System XIIA) that was developed by formulators at the Texas State Department of Highways[89] at the same time as the three-component epoxy system. The primer is a zinc rich product based on a linseed oil modified polyurethane.[90] The zinc powder (89% minimum by weight of product) is mixed into the vehicle in the field. Two coats of primer are recoated with a vinyl-toluene/acrylic finish[91] (more recently a styrene/acrylic) and, in some cases, an aluminum-pigmented, bodied linseed oil modified with a hydrocarbon resin.[92]

Confounding the experts who warn not only of the inadvisability of such oleoresinous vehicle blends for zinc but also against the use of oil-based systems as topcoats, these systems have provided good protection in Texas over the past twenty years or so and are now replaced only because of the greater acceptance of the three-component epoxy primer (favored by the formulators).

All specifications are quite detailed and are presented in formulation format, with individual raw materials carefully specified and duplicates allowed only after prior laboratory testing. Careful formulation is said to be essential for good chalk resistance and color control in the vinyl-type finish. Grays and greens based on titanium dioxide and chromium green oxide are used. Phthalocyanine blues and greens and other organic colorants are not allowed.

Originally, the primer system was applied at 5.0 dry mils with 1.0 dry mil of finish, but since 1972 the required film thickness has been adjusted downward to 3.5 dry mils minimum with 1.5 dry mils of finish. The only difficulties that have been reported with the system are application related, primarily a tendency to blister at high film thicknesses of primer, which should run no higher than 10 mils. In one case of serious failure of the system, 80 dry mils of primer were discovered in some areas.

Cost of the system, including the blast (SSPC-SP-10), are said to range widely from job to job (from $1.10 per square foot on low structures having both easy access and nonintricate design to $2.75 per square foot on higher, more complex structures that have significant traffic impact on job schedules). More reasonable cost ranges are discussed in Chapter 9.

New to the market in 1984 were single-pack, moisture-curing, zinc rich polyurethane systems (System Group XIIB), based on the same diphenylmethane diisocyanate that is the binder for the urethane-aluminum barrier system (System Group XVIII). While Michigan reports some favorable data from evaluations of these systems, they are essentially too new for definitive comment. Typical of this type of system is that being specified for two new bridges in Allegheny County, Pennsylvania. In this system, the urethane-zinc primer is recoated with an epoxy/polyamide intermediate coat and an aliphatic-urethane finish.

Other oleoresinous zinc rich systems (epoxy-ester based) have been used in Japan in bridge systems,[93] although little use has been made of these materials in the United States. Proprietary products[92] of this type were used experimentally under chlorinated-rubber finish coats that were applied over power-tool-cleaned surfaces in local areas (some of the legs and sails) of the Mystic Tobin Bridge, Boston, Massachusetts, to avoid difficult blast containment problems. After five years, these showed no sign of corrosion breakdown. The same system has also been used extensively in industry, but only once on a bridge structure (the Peace Bridge in Niagara Falls) (S. Willox, Norfolk Corp., personal communication, 1984) for galvanizing touch-up. After ten years, this system is said to have given satisfactory service.

Nonpaint Vehicles

System Group XIII: Other Zinc Systems

Cathodic protection techniques by means other than paints are also increasingly being employed on bridges. While formal cathodic protection by means of impressed current or solid anodes is not possible because of the noncontinuity of the electrolyte on bridges, such techniques as galvanizing and thermal spray (metallizing) are being employed.

For the most part, galvanizing is employed only on small sections of bridge structures because of size limitations of the hot-dip kettles, although several bridge structures have been erected entirely from galvanized stock.[94,95,96] Galvanized sections of bridges are quite common, however, and galvanizing is a most cost-effective technique for the protection of pans, railings, and so on. The cables of suspension bridges, such as the Golden Gate Bridge, are often galvanized before erection.

Taking the process one step further, duplex systems[97] involving color-coated galvanizing have also been employed on overpasses and railings. In this process, the galvanizing, pretreatment, and coating are applied in the shop and the structures are shipped to the field, where no more than touch-up is required. By this method, the new "suicide-control" railings of the Bourne and Sagamore Bridges in Massachusetts were protected with a galvanized/vinyl system (System XIIIA).[98] Newer systems utilizing urethane topcoats are also under evaluation. A duplex system consisting of a galvanized-steel base with 2 mils of a zinc-dust/zinc-oxide primer and 2 mils of a phenolic finish has been used on the Hood Canal Floating Bridge in Puget Sound.[99]

Because galvanizing costs are levied on a weight basis, cost comparisons with normal coating techniques are difficult. Cost per square foot for the galvanizing approximates $1.00 to $1.50 per square foot (depending on member size), and $1.75 to $3.00 per square foot for the total galvanizing (0.3 mils of WP-1 wash primer/5 mils of hydroxylated vinyl) (duplex) system. Common thicknesses of galvanizing coatings range from 3.4 mils upward. More detailed cost data may be found in Chapter 9.

The painting of galvanized surfaces can be problematic, particularly where oleoresinous topcoats are employed.[100] As with zinc rich systems, acidity from the applied coatings forms zinc salts at the interface, and, in time, this invariably produces adhesion loss. Preweathering the galvanizing, acid treatments and zinc dust and zinc-dust/zinc-oxide primers, as well as primers based on calcium plumbate have been used with variable success. The most universal design approach would seem to be the use of a WP-1-type washprimer pretreatment (DOD-C-15328D).

Somewhat more applicable to the wholesale protection of bridges is the use of the "metallizing" or thermal-spray approach.[101] In this process, metallic zinc metal (or alloys of zinc and aluminum) are applied from gas or electric arc guns directly onto the white-blasted steel (see Figure 4-7). Thicknesses are somewhat higher than hot-dipped galvanizing films and average about 10 mils. Because the surface of the metallized film (unlike galvanizing) is porous, a sealer—usually a carboxylated vinyl or, in marine or chloride exposure, a wash primer with a hydroxylated-vinyl finish—is normally required to reduce field contamination and for aesthetics. Applications have been made in both field and shop and the versatility of the process has also been useful on galvanized structures where sections are too large for hot dipping.

Thermal spray has not been used as much in the United States as in Europe, where several large and many smaller bridges have been coated. These include the Forth Road Bridge,[102] protected with 3 mils of zinc, wash primer, two coats of zinc chromate phenolic, and two coats of micaceous-iron-oxide finish (one applied in the shop and one in the field). On this bridge, the lower surfaces of the girders, which are subject to salt spray and slow dry-off conditions, have begun to break down after fourteen years, but the upper areas of the structure remain in excellent condition.

During the late 1970s, the Pierre-Laporte Bridge in Quebec, Canada, was thermally sprayed in the field (after abandonment of a BLSC oil/alkyd system) with 5 mils of zinc, washprimer pretreatment, and two 1-mil coats of vinyl (System XIIIC).[103] Up to twenty years of service are predicted for this structure. New York state has recently experimented with an 8-mil zinc/aluminum (85/15) thermal-sprayed primer using a urethane finish over a wash primer on the Robert E. Moses Bridge on Long Island, New York. Extensive use of thermal spray is reported in Scandinavia, France, and Great Britain. In rural areas, a sealed thermal-sprayed zinc coat of 6 dry mils is predicted to last well over twenty years; in industrial and sea-water-splash zones, about 10 dry mils is required for the same sort of protection.[104]

Fig. 4-7. Thermal Spray Application (Metalizing)

BARRIER COATING SYSTEMS

Although, in its most extreme form, the barrier coating system (Table 4-3) is perhaps the heaviest duty coating system yet devised, it has not been employed to the same extent as the zinc rich systems in the corrosion protection of bridges. It is exemplified in both thermosetting systems (epoxies) and all-thermoplastic systems (e.g., vinyls) and, occasionally, in mixtures. Definitions are complicated somewhat by the inclusion of inhibitive pigments in some formulations, but their inclusion is ignored in our categorization below because it is unlikely that the inhibitors themselves play any major role in the protective mechanism of these relatively impermeable systems.

System Group XIV: All-Vinyl Barrier Systems

The all-vinyl-chloride/vinyl-acetate lacquer barrier system is one of the oldest high-performance coating systems for steel structures. It was pioneered by the U.S. Army Corps of Engineers in the early 1950s, and since then has been used widely on diverse structures.[63] Vinyl coatings are tough, flexible, adherent systems that are highly resistant to moisture-vapor transmission, particularly when pigmented with aluminum. They have good resistance to saponification and chemicals. They are permanently solvent sensitive (ketones, esters, etc.), but this allows intercoat adhesion by means of solvent weld. A more complete review of their properties may be found in the National

Table 4-3
Utility and Application Comparison of Barrier Coating Systems for Steel Bridges

System Group	System Description	Volume Solids (%)	Typical Film Thickness (Dry Mils)	Surface Preparation (Minimum SSPC-SP)	Adaptability to Poor Surfaces	Probability of Early Failure from Deviations in Application	Recoat Times (Days)	Solvent Type	Usual Mode of Application	Compatibility with Old Coatings	Ease of Spot Repair	Ease of Cleaning (Salts and Soil)
XIVA	*Primer:* vinyl (carboxylated)	20	1.5	10	Poor	Moderate	0.1	Ketones/ esters/ aromatic hydrocarbons	Airless spray	Poor	Excellent	Very good
	Intermediate: vinyl (2–5 coats)	20	1.5(×3)				0.1					
	Finish: vinyl	20	1.5									
XIVB	*Primer:* WP-1 vinyl wash primer	9	0.3	6	Fair	Moderate to high	0.02–1.0	Alcohol	Airless spray	Poor	Very good	Very good
	Intermediate: vinyl (hydroxylated) (plus 2–4 additional coats)	17	1.5(×3)				0.1	Ketones/esters				
	Finish: vinyl	17	1.5					Aromatic hydrocarbons				
XV	*Primer:* chlorinated rubber	32	1.5	6	Fair to good	Moderate	1.0	Aromatic hydrocarbons/ aliphatic hydrocarbons	Airless spray	Good	Excellent	Good
	Intermediate: chlorinated rubber	35	3.0				1.0	Ketones/ esters/glycol esters				

	Finish: chlorinated rubber	33	1.5									
XVIA	*Primer:* epoxy/polyamide	55	2.5	6	Good	Moderate to high	1.0	Alcohol/glycol esters	Airless spray	Good	Good	Excellent
	Intermediate: epoxy/ polyamide	60	2.5				1.0	Ketones/esters				
	Finish: epoxy/polyamide	60	2.0					Aromatic hydrocarbons				
XVIIA	*Primer:* aluminized epoxy mastic	85	5.0	2 or 7	Very good	Moderate		Aromatic hydrocarbons	Brush spray	Very good	Very good	Excellent
XVIIB	*Primer:* aluminized epoxy mastic	85	5.0	2 or 7	Very good	Moderate	1.0		Brush/ airless spray	Very good	Very good	Excellent
	Finish: aluminized epoxy mastic	85	5.0									
XVIIC	*Primer:* aluminized epoxy mastic	85	5.0	2 or 7	Very good	Moderate	1.0	Alcohol Ketones/esters	Brush spray	Very good	Very good	Excellent
	Finish: epoxy/polyamide color coat	80	5.0					Aromatic hydrocarbons				
XVIIIA	*Primer:* moisture curing urethane aluminum	50	2.5	6	Fair to good	Moderate to high	1.0	Aromatic hydrocarbons	Brush/roller	Good	Fair	Excellent
	Intermediate: moisture curing urethane aluminum	50	2.5				1.0					
	Finish: moisture curing urethane aluminum	50	2.5									

(continued)

Table 4-3
(Continued)

System Group	System Description	Volume Solids (%)	Typical Film Thickness (Dry Mils)	Surface Preparation (Minimum SSPC-SP)	Adaptability to Poor Surfaces	Probability of Early Failure from Deviations in Application	Recoat Times (Days)	Solvent Type	Usual Mode of Application	Compatibility with Old Coatings	Ease of Spot Repair	Ease of Cleaning (Salts and Soil)
XVIIIB	*Primer:* moisture curing urethane aluminum	50	2.5	6	Fair to good	Moderate to high	1.0	Ketones/esters	Brush/ airless spray	Good	Fair	Excellent
	Intermediate: moisture curing urethane aluminum	50	2.5				1.0	Aromatic hydrocarbons				
	Finish: aliphatic urethane color coat	52	2.0									
XIX	*Primer:* coal tar epoxy	71	8.0	10	Poor	High	0.2–1.0	Aromatic hydrocarbons	Airless spray	Poor	Poor	Very good
	Finish: coal tar epoxy	71	8.0									

Association of Corrosion Engineers (NACE) Publication 6H177. Two basic types are employed. The first employs a carboxylated vinyl in a noninhibitive primer coat applied directly to the steel. Over that, multiple coats of vinyl intermediates and finishes are built up to the required film thickness. These systems are typified by the Bureau of Reclamations VR-3 and VR-6 systems.[105] Because intermediate and finish coats do not require the 1% maleic-acid-modified polymer for adhesion to steel, some systems may utilize the unmodified copolymer in coats other than the primer, the solvents of which partially dissolve the vinyl vehicle in the coat already applied and attain intercoat adhesion by the solvent weld typical of the thermoplastic system. Other vinyls, such as the butyl-maleate-modified materials and even the vinyl chloride/vinyl ether systems may be employed in these intermediate and finish coats. As in the high-build systems, thermoplastic acrylics and polyketone resins may also be employed. All of these vehicle modifications are more rare in the barrier vinyl systems, however.

Inhibitive pigmentation is rarely built into the carboxylated primer because of reactivity, although some relatively insoluble inhibitive species (e.g., zinc phosphate) may be added with care.

The common alternative approach to the carboxylated-vinyl barrier systems involves the use of a WP-1 (DOD-P-15328D) wash primer for adhesion to steel and a hydroxylated vinyl (a vinyl-chloride/vinyl-acetate/vinyl-alcohol terpolymer) applied over that. Because the hydroxylated vinyl is not necessary for adhesion in further coats and is a more expensive resin, intermediate and finish coats may again utilize the straight copolymer. While the hydroxylated-vinyl primer has been pigmented with both red lead and zinc-chromate pigmentation, there appears to be little difference in service life between vinyl systems based on this approach and on the VR-3 approach that utilizes no inhibitor. Much confusion can arise from the specificity of vehicle type to the surface being coated, and in some cases (such as the SSPC Paint 8) both the carboxylated and hydroxylated vehicles are employed in the same primer, giving the product versatility over both bare steel and the wash primer with only a slight compromise in adhesion. The approach is also used in the SSPC Paint 9 specification utilized by the state of Maryland.

The hydroxylated vinyl is more compatible with other resin systems than is either the carboxylated resin or the straight copolymer. This characteristic has been used in vinyl/alkyds, which have been used widely by the U.S. Navy.[106] Newer types of hybrid coatings include vinyl/epoxies[107] and vinyl/urethanes.[108] In the latter case, the hydroxylated vinyl acts as a polyol for the isocyanate curing agent. Both vinyl/urethanes and vinyl/epoxies are two-package systems and are quite new. These systems have been applied to bridges in both Maine and Florida with encouraging results after more than three years of service.

The normal all-vinyl barrier system, however, is a lacquer, and its high molecular weight (of necessity) reduces volume solids because of viscosity and application considerations. As a result, high-build paint systems must be applied in multiple coats. Usually four to seven coats are necessary in order to build up 7- or 10-mil systems. Vinyls are high-energy systems and require careful surface preparation (near-white or white blast) for best adhesion and good barrier properties.

The use of vinyl barrier systems on bridges has apparently not been either as widespread or as uniformly successful as it has on other structures, such as potable-water storage tanks, locks, ships, and dams, where service lives of twenty or more years are not at all unusual. This reduced usage on bridges is probably related to the number of coats required to achieve the desired system film thickness and to the difficulties in achieving proper application and inspection in the field.

In California and Oregon, seven-coat vinyl systems have been used on a substantial number of bridges, although with mixed success, particularly near the ocean. In Oregon, a seven-coat all-vinyl system on the Columbia River Bridge at Astoria showed severe deterioration within the first six months and required extensive repainting after three years. The state complained about the low film builds, and data would indicate that this was the primary reason for such poor performance. Only 1 and 2 dry mils were realized in some areas from seven coats of paint! Also, one must suspect poor inspection. California has had better success with the same seven-coat system in coastal areas where service lives of over twenty-five years have been achieved. Even better performance is recorded in the interior valleys. California, too, has had mixed results and attributes much of the variability to application effects and the different environments.

It is possible that despite the number of coats employed, the geometric complexities, angles, and edges on many bridges make the application of adequate thicknesses of such low-build lacquers difficult without constant and careful inspection. High-build intermediate and finish coats (like those used as topcoats with zinc rich systems) are also useful in reducing the number of coats required to attain an overall film thickness, although the high-build vinyl does not appear to have the same degree of impermeability (mil for mil) as do the low-build lacquers.

Other hybridized systems utilize combinations of hydroxylated-vinyl resins and alkyds in shop primers upon which full-vinyl barrier systems are built (New Jersey). The Connecticut vinyl primer is recoated with an alkyd intermediate- and finish-coat system. Both the New Jersey and Connecticut systems use inhibitive pigmentations in the vinyl primers.

System Group XV: Chlorinated-Rubber Barrier Systems

Other thermoplastics used as barrier systems include the chlorinated-rubber resins. Chlorinated-rubber barrier systems have a similarly high degree of chemical resistance to vinyls. Somewhat more susceptible to dehydrochlorination, they need more careful stabilization than do vinyl systems, and lead compounds and epoxies are employed to this end. Solvent resistance is poorer than that of vinyls, but moisture-vapor transmission resistance is probably greater. A lack of ester groups on the chlorinated-rubber molecule lends this polymer excellent resistance to saponification and chemicals. Chlorinated-rubber resins are usually modified with straight chlorinated-hydrocarbon resins and chlorinated-hydrocarbon plasticizers. Properties are more completely considered in NACE Publication 6H275. As barrier systems, chlorinated-rubber coatings are unmodified and require good surface preparation. Alkyd-modified hybrids are more forgiving as primers but are more susceptible to saponification.

Like the chlorinated-rubber zinc rich system used on the Tobin Memorial Bridge in Boston,[79] however, chlorinated-rubber-based barrier systems are more rare in the United States than they are in Europe. In Great Britain, they are much more common and have been used in combination with micaceous iron oxide for some time by British Railways. Bishop[109] reports that chlorinated-rubber paints work particularly well when primers are pigmented with zinc phosphate; and red lead, BLSC, and dibasic-lead-phosphite pigmentations have been advertised for use by Hercules Inc. and Imperial Chemical Industries (ICI) in the United States.

A chlorinated-rubber barrier system was selected for use on the Hambell Bridge in Great Britain and is described by Ferguson[110] as consisting of a wash primer and a chlorinated-rubber metallic-lead primer, followed by a micaceous-iron-oxide-pigmented, chlorinated-rubber midcoat, all applied in the field. These were followed by two gloss field coats of chlorinated-rubber enamel. Metallic lead primers have been used before in Great Britain. It has been suggested that lead (electrochemically more "noble" than steel in its macro form) has a more active potential when finally divided.[111] There seems to have been no confirmation of this, however, and, if metallic lead pigmentation works, some alternative mechanism seems more likely.

In the United States, Minnesota,[112] California, and Massachusetts have developed chlorinated-rubber systems. SSPC Paints 17, 18, and 19 are compositional specifications for a chlorinated-rubber system, although SSPC Paint 17 contains inhibitive pigments. Aluminum finish coats have been advocated by California. Minnesota has specified chlorinated-rubber finish coats for use over a variety of primer types, including red lead, BLSC alkyds, and phenoxy zinc rich primers pretreated with a wash primer. These systems, however, appear to have been used only occasionally.

System Group XVIA: All-Epoxy/Polyamide Barrier Systems

Dunn,[1] in an excellent review of epoxy curing agent combinations for coatings, lists over two dozen possible curing agents for use with epoxies. In bridge painting, relatively few of these materials are applicable because of curing difficulties at ambient temperatures, pot-life requirements, applicable viscosities, color, and other physical and chemical properties. The principal curing agents for use at room temperature are polyamide, amido-amines, and aliphatic and aromatic amines. The polyamides have the larger share of the market for bridge work, although, as the drive for higher solids increases, the other materials are being used more frequently. The chemistry and properties of epoxy resins for protective coatings have recently been discussed by Brady.[113] A more comprehensive, if less specific, discussion of epoxies may be found in the Lee and Neville handbook.[114]

Epoxy barrier systems are rigid thermosetting systems that have less flexibility than many coating systems. They are, however, tough, hard, and afford excellent protection, despite a tendency to chalk badly in some exposed areas. NACE Publication 6H182 describes in detail the properties of this type of coating system in atmospheric exposure situations. Application is not difficult if such flow agents as butylated urea formaldehyde or silicone resins are used to offset a slight tendency to crater. Because of the excellent wetting properties of the polyamide resin, the primers are relatively tolerant of application over poorer surfaces, even those contaminated with traces of moisture, but, as a rule, SSPC-SP-6 blasts are required. Application seasons may be limited because of the severely reduced kinetics of the normal epoxy/amine (or amide) reaction below 10°C (50°F). More recently, lower-temperature coating systems have been introduced. These are said to cure at temperatures down to 2°C (35°F). Pot lives of these two-component systems depend on curing agent, catalyst, solids, molecular weight of the resin, and atmospheric conditions, as well as the amount of pigmentation. Medium molecular weight epoxy/polyamides are still the most common combinations for this type of service, with pot lives generally in the range of eight to sixteen hours. Newer systems based on lower molecular weight epoxy resins, modified amines, amido-amines, or polyamide curing agents, with higher solids (75%–95%) have much shorter pot lives, which may be as low as four hours unless carefully modified.

The classical low molecular weight epoxy barrier system is typified by the three-coat Navy specification MIL-P-24441, although this system is not overly high in solids.

One of the earliest applications of epoxy systems on bridges was the 1966 application of a three-coat epoxy system (bearing red lead in primer and midcoat and utilizing the green finish coat) used on the steel of the Newport Bridge in Rhode Island. The original system was shop applied with only 5% field painting. A twenty-five-year service life was predicted in spite of the live loading and stress reversals common on large bridges and their effect on the highly crosslinked film. The catastrophic adhesion failure that subsequently occurred on this structure even before the steel was erected is well documented, but seems to be related more to blast quality, profile difficulties, contamination, film thickness, and the like than it is to stress-related effects. Following the original failure, this same system, applied over newly blasted surfaces in the field in the early 1970s has done well to date. The epoxy system used was a relatively simple medium molecular weight polymer crosslinked with a polyamide resin. The same system, with some changes in PVC, was used in all coats.

As is the case with vinyl coatings, the epoxy coatings are so impervious to moisture at high builds that it seems doubtful that the inclusion of an inhibitive pigment does as much for the system as does the impermeability of the vehicle. Strontium chromate is a common inhibitive pigment for epoxies, and this type of pigmentation is included in one of the MIL-P-24441 primers and in U.S. Naval Air Development Center primer MIL-P-23377D. In most MIL-P-24441 systems, however, there are no inhibitive pigments, and the specification remains a true barrier system.

An epoxy system utilizing glass flake was also employed on one of the towers of the Newport Bridge in a more exposed zone. This system failed early in the extremely aggressive salt-water environment. The towers are now painted with a single-pack inorganic zinc, over which is applied an epoxy/polyamide finish. The article written by Antonino[115] that describes the system before the failure provides interesting reading today, especially when compared with later articles on the failure, such as that by Berger.[116]

Like all thermosetting vehicles, epoxies are far more easily adapted to the high-solids technologies because of their initially lower molecular weight at the time of application than their thermoplastic counterparts from which films of high molecular weight are cast solely by solvent evaporation. Still lower molecular weight bisphenol A/ epichlorohydrin-based epoxy variants, with alternative curing agents formulated at high solids to reduce solvent emissions, have received wider use on bridges in more recent times. For the most part, these materials are discussed under the aluminized-epoxy barrier systems (System Group XVII). These materials have short pot lives and give less flexible films unless suitably modified (as many of the mastics are). While such properties as chemical resistance and hardness are improved, they seem less important in bridge use than flexibility.

All epoxies are potentially subject to amine blush, although many curing agents, such as the polyamides, are more resistant than others. Because of the required induction time (a thirty- to forty-five-minute period after mixing and prior to application, during which the individual reactants are allowed to "sweat in" or begin to form longer chained molecules that do not migrate to the surface of the film), the usable pot life may be further shortened in some systems. Amine blush is a troublesome phenomenon, however, and can result in the formation of greasy films of amine carbonate (the reaction product of the stratified amine and carbon dioxide from the air) that easily can go undetected. Amine carbonate films also can subsequently nullify the adhesion of any additional coat. As application temperatures are lowered, lower reactivities extend required induction times and compound the amine blush problem.

Intercoat adhesion difficulties can also occur when old, hard epoxies are recoated after several years of exposure. For this reason, it is often necessary to scarify mechanically old epoxy films and to remove chalking deposits with a sweep blast or a water blast before the application of the repaint systems.

Cost of the epoxy system will depend on the polymer type, the particular formulation, the pigment content, volume solids, and so on. Epoxy primers may range from \$0.020 to \$0.030 per mil square foot.

System Group XVIB: Epoxy/Polyamide Primer with Urethane Topcoat Barrier Systems

Because of the chalking difficulties of the epoxy, updated systems of this type, like updated epoxies over zinc rich systems, often employ urethane topcoats that have better UV resistance, and, therefore, score high in color and gloss retention. Two-pack urethane systems, like epoxies, are thermosetting materials and are strong, hard, and almost as chemically resistant as the epoxies. Final properties, again like the epoxy systems, will depend very much on the individual formulation and particularly on the choice and level of polyol. Two primary polyol types are in use on bridges: one is a hydroxylated polyester that gives excellent chemical resistance, hardness, and abrasion resistance but is somewhat more sluggish in dry-to-touch times and more expensive than the second type, a hydroxylated acrylic polyol. The latter materials seem chemically resistant enough for bridge application and dry faster than do the polyesters. They seem easier to use and are somewhat less expensive. Both the polyester and the polyhydroxy-acrylic aliphatic urethanes have excellent weathering resistance.

Lower-cost aromatic urethanes based on toluene diisocyanate are possible for use as midcoats, but are not used extensively. They are prone to recoating difficulties and their color and gloss retention is far poorer than epoxies. The moisture-curing aromatic urethane is based on diphenylmethane diisocyanate and is discussed under system group XVIII.

System Group XVII: Aluminized-Epoxy-Mastics Barrier Systems

Aluminized epoxy mastics are two-pack, very-high-solids, low-molecular-weight epoxy systems that are pigmented with aluminum and designed for application over hand- and power-tool-cleaned substrates as well as brush-off and commercially blasted substrates. They are primarily designed as two-coat maintenance systems for spot-blast applications over old alkyd systems in order to upgrade the old alkyd inexpensively. Other applications, however, have included use as finish coats over zinc rich primers and as primers under vinyl, epoxy, and urethane finish coatings. Some have also advocated the use of this type of coating for the repair of old inorganic zinc systems.

This type of coating was first introduced by a single manufacturer in the late 1960s, and was advocated for application over rusty substrates with a minimum (SSPC-SP-2) of surface preparation. The idea of applying barrier coatings over a rusted substrate is hardly in keeping with theory, and it seems likely that only high film-thickness applications, preferably with aluminum-flake pigmentation (both of which effectively prevent or slow down water absorption), will give reliable performance. Epoxy mastics, with their high-build capabilities, are perhaps best suited to fulfill these requirements, and indeed they have been found superior to other systems in a controlled study performed by Rensselaer Polytechnic Institute and Avondale Shipyard, New Orleans, Louisiana.[117] On the other hand, better surface preparation should produce theoretically more viable coating systems at lower builds.

The initial epoxy mastic, which is still the most widely used and probably the most successful, was based on a low molecular weight epoxy cured with a proprietary curing agent modified with what appears to be a bituminous phenolic derivative. Specific details remain undisclosed. The product is thick (100–140 Krebs Units [K.U.]) and 90% in volume solids. Application is impossible without thinning, and the manufacturer recommends a 10% dilution for spray application, while some applicators use higher dilutions. Like many of the lower molecular weight, epoxy-resin-based systems, the material has a short pot life (four hours at 24°C (75°F) when thinned 25%). The modified curing agent is said to add significantly to its wetting properties but is probably responsible for a marked tendency of the product to undergo a patchy bronzing, particularly when brush applied. This is an aesthetic drawback that is less obvious when the product is sprayed on and that is said to become more diffuse on aging. While essentially a barrier system, this particular product once contained inhibitive lead pigmentation. Newer formulations, however, are reported to be lead free.

Despite the low molecular weight resin, most aluminized epoxy mastics are relatively flexible compared to other epoxies, probably from resin modification, which serves well in bridge applications. Use of aluminum pigment not only reduces moisture penetration, but reflects UV radiation and heat and protects the epoxy resin.

The success of the original aluminized-epoxy mastic undoubtedly was influential in the general development of this kind of product by other manufacturers. Other houses, however, instead of duplicating the original bituminous phenolic-modified mastic, made use of nonmodified systems based on amide and amido-amine curing agents, and, in the case of at least one manufacturer, developed the technology to include nonaluminized versions of similar high-build systems. Many of these systems have lower viscosity than the original product, although solids are still in the same range and thinning for spraying is recommended, especially for the first coat. Pot lives remain relatively short (2.5 to 3.0 hours at 21°C (70°F), but the nonmodified systems do not bronze. The nonaluminized variants are normally formulated at a lower solids level than the aluminized systems (80% to 85% volume solids). To date, the newer variants of this type of coating have not received the same field use on bridges as has the original material, and the field history is not as securely established.

Aluminized-epoxy mastics are usually applied as two-coat systems in 5+5-mil coats, although some manufacturers recommend one coat (5–8 mils) only. The consensus is that two-coat applications of the aluminized epoxies are necessary for viable long-term performance. Where they are applied in spot applications, the first spot coat is used over the power-tool-cleaned (or blasted) spots, and the second coat is applied as a total topcoat. It is reported that in Iowa, this type of system has been used to spot repair older zinc/vinyl systems.

As in the aluminum moisture-curing urethanes, recent developments utilize similar vehicle systems with zinc dust. This system, while new to bridges, has been used in some off-shore oil rig applications and is said to provide good performance over salt-contaminated surfaces.

The epoxy mastics have been adapted for maintenance applications in several states, including New York (New York State Thruway), Ohio, Wyoming, South Carolina, Massachusetts (on railroad bridges maintained by the public works department), and Virginia. The product is typified by the Virginia specification,[118] which appears to be based on the original product (possibly in its original form). The state of Wyoming uses the aluminized-epoxy mastic on overhead structures and for painting bridge railings.[119] South Carolina uses the same product as a finish coat over inorganic zinc.[120] The Waldvogel Memorial Viaduct in Cincinnati, Ohio, was painted with a single 5-mil coat of this system over a wire-brushed surface. The Fort Steuben Bridge in Steubenville, Ohio, and the Grand Island Bridge in Niagara Falls, New York, also were painted with this type of paint system, and the coating is also being employed for added protection on the expansion bay steel of the Newport Bridge (1984 painting). The aluminized-epoxy mastics seem most widely used in bridge maintenance as an alternative to inorganic zinc systems on those bridges where, because of design difficulties (back-to-back angles, latticed features, and heavily riveted or bolted construction), neither blast quality nor film thickness control can be maintained without continuous, costly inspection (see Figure 4-8).

Specifications in most cases are hardly definitive, particularly those based on the initial product, and may best be described as quasicompositional with performance criteria. The Virginia specification contains a prequalification clause that calls for at least two years of service as a one-coat, 5-dry-mil film in a similar service and environment to that being considered. The Virginia specification does not appear to utilize the epoxy mastic alone but as a 5-mil primer over a commercial blast and beneath 3 mils of high-build vinyl, or as a 5-mil finish over 3 mils of inorganic zinc applied over a near-white surface.

The amount of diversity between products in this class creates great difficulty in categorizing these products. Some contain inhibitive pigments, some are straight barrier systems, some employ leafing aluminum, and others are based on nonleafing alumi-

Fig. 4-8. Structure of Complex Design Suited for Use of an Aluminized Epoxy Mastic System in Lieu of an Inorganic Zinc System. The zinc system is precluded because of unreliability of surface preparation and primer film thickness control.

num. Still others have abandoned aluminum pigmentation altogether, with what sacrifice in performance is yet unknown.

Service-life claims by the manufacturer are as much as fifteen to twenty years for the original aluminized product, although other reports indicate more variable success. This variability may reflect either poor inspection, poor application practices, or both. Early applications on the Route 290 viaduct in Worcester, Massachusetts, were disappointing, although thin films were cited, and later, more controlled application did better. One Ohio contractor reported poor results with a similar system leading to extensive system rectification. Test data would show that single-coat systems are less dependable than two-coat systems, and it is possible that variable performance is directly related to film-thickness. Ohio now regards its single-coat system as a failure (Gary Tinklenberg, Michigan Department of Transportation, personal communication, 1986). Difficulties also have been related by the manufacturer to the degree of pitting on steel surfaces, and it is now recommended that highly pitted surfaces receive blasting preparation only. Various caveats are found in many data sheets concerning the effects of variability in the interpretation of the SSPC-SP-2–3 specifications and the effects of surface on performance. Many of the test coupons utilized newly rusted, white-blasted steel as substrates for the coating. This, unfortunately, is not typical of the type of rusted steel found on many existing bridge structures.

System Group XVIII: Urethane/Aluminum Barrier Systems

Few recent paint systems have been developed directly from raw-material manufacturers. One notable example is the aluminized barrier system based on a diphenylmethane diisocyanate moisture-curing resin and nonleafing aluminum pigmentation that was introduced in the early 1970s.[121] Like the aluminized-epoxy mastics, these moisture-curing, urethane-type primers are advocated as an alternative to zinc rich systems on structures where design considerations might restrict their application. Film thickness specifications would, in many cases, seem at least as critical as zinc systems (as far as the upper limit is concerned) and certainly more so than the epoxy mastics.

Two variants have been used: (1) System Group XVIIIA, a three-coat, aluminized moisture-curing system with all coats based on the same diphenylmethane diisocyanate, but bearing slightly different nonleafing aluminum levels in successive coats and using small amounts of tinting pigments to differentiate between coats; and (2) System Group XVIIIB, wherein the final coat of the moisture-curing urethane is replaced with a two-component aliphatic urethane based on the reaction of the hexamethylene diisocyanate adduct Desmodur N and the polyesters Desmophen 650A and Multron R221-75.[71]

These systems (mainly XVIIIB) have been used on several bridges (Patton Street Bridge and Glenwood Bridge) in the Pittsburgh area, and test areas on the Bessemer and Lake Erie Railroad Bridge are now thirteen years old and reported to be in excellent condition. While the two-coat system applied to the railroad bridge was applied over a hand-tool-cleaned surface, newer recommendations suggest commercial blasting of these surfaces and a three-coat application. In 1978 on Homestead Bridge in Pittsburgh, a spot blast was adopted in which a SSPC-SP-6 blast was used on rusted areas with an SSPC-SP-7 on tightly adhering old oil paint. The bridge was spotted with the diphenylmethane diisocyanate primer system and then given a full coat of the same product. Finally, a full coat of the polyester/aliphatic urethane two-pack system was applied. In a paper by Bracco,[122] ten individual coating systems designed for bridge structures are described.

The manufacturer advocates application of the systems above by brush and roller only at film thicknesses of 2.0 to 2.5 mils dry film thickness (DFT) each (1.5–2.0 mils for the aliphatic urethane topcoat). The manufacturer's stipulation of brush and roller for the system may be related to a need to optimize wetting over pitted surfaces on old bridges. It may also be related to difficulties with the wind-carried aluminum overspray, which is a severe problem with all aluminized finishes. Additionally, it appears that higher builds of the aluminum primer tend to develop severe bubbling tendencies on application (probably carbon dioxide generation) and brush and roller application may control against excessive film thickness better. Based on early field data, the resin manufacturer expects ten to twelve years of service from this three-coat system and twenty-five to thirty years if a commercial blast is substituted for the power-cleaned surface preparation.

Systems of this type are available from many manufacturers, including one major one that is evaluating new variations, including zinc rich primers based on the same moisture-curing urethane resin and high-build urethane topcoats based on hexamethylene diisocyanate. Other manufacturers think that the 2 dry-mil-per-coat film thickness may be too low for a viable protection system, especially over highly pitted steel. Less bronzing has been recorded with diphenylmethane diisocyanate primer systems than with some of the epoxy mastics described above. One manufacturer describes slight bronzing developing after initial application of the urethane system, but this is mitigated as the system ages and defuses to a more uniform gray color. When the aliphatic urethane is used over the moisture-curing primer, these problems are not pertinent.

A further variation of the systems above was qualified by Michigan for use on riveted, plated girders and trusses in the field. In this system, two coats of moisture-curing primer are first recoated with a high-build epoxy and then finished with an aliphatic two-component urethane.[123] By 1985, however, the state of Michigan had reversed itself and had abandoned this system as a result of extensive delamination failures.

While the literature of the Mobay Chemical Company concentrates on the polyester/urethane finish only, it is important to bear in mind that the company does not market acrylic polyols, and these materials might reasonably be examined as less expensive polyols for topcoat applications in lieu of the polyester-based polyol systems.

Costs for the aluminized primer are said to approximate $0.045 per mil square foot. Costs for the finish coat may be relatively variable and somewhat high. They range from $0.050 per mil square foot for the hydroxylated-acrylic-based system to $0.070 per mil square foot for the polyester-based type.

System Group XIX: Coal Tar Epoxy Systems

One of the most common corrosion-resistant barrier coating systems, the multicoat coal tar epoxy system finds only limited use on bridge structures because of its color and poor UV resistance.

Coal tar epoxy systems exhibit both the very low moisture-vapor transmission and excellent acid and alkali resistance of the coal tar component and the general chemical and abrasion resistance of the epoxies. Resistance properties, moreover, seem synergistically improved by the blending. A more complete outline of the physical and chemical properties of this coating system may be reviewed elsewhere.[124,125]

So excellent are the performance properties of this system that it has found much success in protecting some of the most demanding areas of bridge structures, such as the expansion bay steel that tends to be more susceptible to deterioration because of leaky joints and so forth. On the Massachusetts Turnpike, the coatings have also been employed on the beam ends and bearings, although here the two normal 8-mil applications are applied over a 3-dry-mil film of zinc rich phenoxy (see Figure 4-9).

The state of New York, citing a need to protect both A36 and A588 steel in highly vulnerable areas beneath bridge decks that are repeatedly wet and show slow dry-off characteristics, has adopted the use of a two-coat coal tar epoxy system (one coat is tinted).[126]

The major drawback to this heavy-duty system is its sensitivity to sunlight. The product chalks and may become brittle on exposure to the actinic rays of the sun. For this reason, and because of the product's ugly black color, its use on bridges is almost universally limited to areas that are hidden from view. Some state highway departments, such as that of New York, have used other systems as topcoats (even alkyds), but the solvents of the recoat system must be carefully selected in order to avoid bleeding of the coal tar. Also, in recoating coal tar epoxy, care should be taken to ensure intercoat adhesion; this may be difficult with generically different topcoats.

Even self-recoating is difficult if too long an interval is allowed between coats, and sweep blasting is often necessary after the first coat has aged more than twenty-four hours under summer conditions. The use of powerful solvent softeners, such as the slow-evaporating n-methyl-2-pyrrolidone, applied by brush and allowed to sit thirty minutes before the second coat is applied, have been used in some cases in lieu of the sweep blast, although this technique is not foolproof.

Coal tar epoxy systems cost between $0.12 and $0.15 per mil square foot for each 8-mil application, but service lives in even the most demanding areas are as long as any

Fig. 4-9. Use of Coal Tar Epoxy System for Added Protection to Beam Ends and Bearings

system now offered. Total materials for a three-coat, inorganic zinc/two-coat, coal tar epoxy system would probably cost about $0.35 per mil square foot.

Because of the tremendous cohesive strength of the super high build of coal tar epoxy, care must be taken to ensure a good balance between cohesion and adhesion strengths if delamination problems are to be avoided. Consequently, specifications normally call for a white or near-white blast with relatively deep profiles (4 mils or so).

Similar systems have also employed inorganic zinc, but it remains a point of some discussion as to whether any kind of zinc primer is really necessary. So impermeable to water is the coal tar epoxy that the sacrificial protection by the zinc at small holes or "missed spots" (holidays) or scars is afforded only by the sides of the scar (the depth of the film thickness if zinc film), and corrosion is inevitable at the scribe because of inadequate area effects.[50]

Specifications for coal tar epoxies have been well developed, and formulation examples, such as the C-200 specification of the U.S. Army Corps of Engineers, are available.[89] Actual formulations may also be found under SSPC Paint 16,[125] which approximate the Corps of Engineers's specification. Some manufacturers, while not offering the C-200 specification, have their own specific products that are essentially similar, although they may employ different curing agents or even different epoxy resins.

In certain amine-cured coal tar epoxies, extreme care must be taken to ensure sufficient induction time to prevent amine blush with consequent amine-carbonate formation. The slightest interfacial contamination of the coal tar-epoxy system can lead to severe intercoat adhesion problems.

System Group XX: Wax Barrier Coating Systems

One species of barrier coating that has been in use for many years is the wax coating. These are still used in certain states (e.g, the Maine Turnpike uses this type of coating on the beam ends and bearings of overpasses), and other states (e.g., Texas, Oklahoma, Georgia, and Ohio) are evaluating newer variations of the same type of materials.

In recent years, this type of system has seen some renewed interest in several applications, including bridge painting, because of the introduction of a family of metal salt dispersions.[127] These are blended and used with microcrystalline waxes and other additives to produce a family of coatings with excellent corrosion resistance, unusually high resistivity, and very low moisture-vapor transmission. Claims of corrosion resistance[127] have been confirmed in recent studies, and aluminized variants, like the epoxy mastics, have been tested successfully at high film thicknesses over rusty surfaces.[128] Lower film builds require better surface quality for the same degree of protection. It may be concluded that performance over poor surfaces is directly related to the product's ability to restrict water ingress through the film.

Unfortunately, like all wax coatings, these systems do not form hard, abrasion-resistant films but remain quite soft and even tacky. Their unique cohesive characteristics and self-healing tendencies, however, make them highly resistant to such corrosive failures as peeling, blistering, and cracking, which plague more conventional coating films.

With judicious utility on such nontraveled surfaces as pans and below-deck areas, application of these systems on bridge structures may well be justified because of their unique form.[129] Certain British applications, with similar calcium-sulphonate-modified wax-coating systems that employ a colored sealant as a topcoat, are claimed both to extend system life and provide a tougher, drier film.[130]

Wax barrier coatings are normally applied at high film thicknesses and costs are about $0.020 per mil square foot. Little data on service lives seems to be available at this time.

DISCUSSION

While every effort has been made to discuss the most common coating systems that have been employed in the protection of bridge structural steel at this time, the treatment above is certainly not exhaustive. New systems are being introduced every day, and many potential systems have yet to be fully evaluated for this kind of work. Thus, such diverse formulations as water-soluble or water-dispersible alkyds; polyurethane latexes, vinyl toluene, styrene butadiene, and styrene-acrylate systems; single-pack, colored, moisture-curing urethanes or uralkyds; polyesters; and so many others have not been considered. These omissions do not reflect an unsuitability for bridge service but merely the fact that bridge applications are as yet unknown.

Finally, as noted elsewhere, there is at this time a state of flux existing in the bridge painting discipline. In contrast to the 1960s revolution in bridge-paint systems, during which BLSC replaced red lead in many states, there are now many more candidates vying to displace the BLSC and other conventional systems. Furthermore, these systems are more fundamentally different not only from the oil/alkyd systems they replace but also from each other. For the most part, less real compositional data seems available on the systems, and claims and counterclaims abound concerning their relative merits.

Many of the newer coating systems have now been employed on bridge structures in various parts of the country for more than twenty years—first as experimental systems and then more generally, although structural casualties have occurred and there have been some disastrous failures. First among the many candidates now employed, however, would seem to be the inorganic zinc rich primer with the vinyl topcoat, although even here satisfaction is not complete and performance (primarily in color retention of the vinyl) is variable. Compositional vagueness has made it difficult to establish whether such differences are formulatory or inherent to the system. Different performances from product to product might indicate the former.

Despite this emerging trend, new systems are constantly being introduced, and it remains possible that still further improved coatings (e.g., inorganic zincs/epoxy/urethane systems) may supplant even the inorganic zinc/vinyls as experience with these systems develops (although the long-term "recoatability" of the hard solvent-resistant thermosets may, in the end, preclude this). One interesting advantage that has emerged from the plethora of systems that have been introduced is the versatility and choice that the systems have made possible. While it remains true that such increased choice favors increased adaptability in coating-system design, it is nonetheless true that any given high-performance system is less universal than was the lead system it replaced.

Despite the marked trend away from the inhibitive primer and toward the utilization of zinc rich and barrier coating systems, many states and bridge authorities still cling to the lead-based paint systems that they have used with considerable success for years. In many states, today's trends seem set as much by the marketing plans of the larger paint companies and the concerns of bridge authorities, applicators, public health and environmental agencies, and the general public over the toxicity of lead (and, to a lesser extent, hexavalent chromium-based pigments) as by the desire to rectify deficiencies in existing paint systems. In fact, in rural areas of the country, actual service lives in case histories with red lead and BLSC systems are of durations that exceed those now wishfully predicted for many of the newer high-performance systems.

Undoubtedly, the move to the newer systems can be justified on a performance basis, but only in those environments where established coating systems are not providing the performance required. Taking into account the additional initial costs of surface preparation and disposal, materials and application, the necessity for experienced control, and the inevitable risk involved in adopting something new, the advantage of a high-performance coating system is only of real value in those environments where the conventional systems cannot provide the extended levels of performance (or are for some other reason prohibited).

The importance of the environmental zoning concept is, in consequence, most relevant. Many states observe that the location and environment of a bridge is more critical to coating system performance than is the nature of the coating system employed. Reports of examples of major variability in performance of identical paint systems that have resulted from deficiencies in surface preparation, application, and film thickness also come from many states. These deficiencies are most apparent in the high-performance zinc and barrier systems than they were in the lead and chromate-based systems, which seemed more tolerant of deficient application and a lack of adequate inspection.

On the other hand, in coastal environments, in areas of the country where widespread use of deicing salts is the norm, where humidities are high, and where the atmosphere is contaminated with acidic oxides and other pollutants, the newer coating systems, when properly designed and applied, are providing a level of performance not before achieved with conventional paint systems.

Chapter 5

Coating System Selection and Specification Design

COATING SYSTEM SELECTION

The actual selection of a coating system for a given bridge introduces an entirely new set of complexities. In the past, such selection was a moot point. Little else was available but lead and oil, and steel was either new and uncoated or bore the same basic paint systems as those to be used for repair. Where alternative coating species were available, these differed in little more than pigmentation or, in the most extreme cases, in the particular modification of the basic vegetable oil (alkyds, phenolic varnishes, etc.).

In the present state of flux that exists within the industry, the selection and combination of individual components of a paint-system specification is a discipline unto itself, and it is one that is fraught with many uncertainties that arise not only from the complexities of coatings chemistry but also from the effects that numerous factors may have on the suitability of a given paint system for a given structure. Not only is there a statistically increased likelihood of system design error that is axiomatic with increased choice, but today's high-performance systems are higher energy coatings that do not easily tolerate incompetency in job execution and that are not nearly as adaptable to the variety of design features and surface conditions that may be found on any given structure as were the systems they replaced.

Furthermore, today's system-design engineers are faced with the repair of existing paint systems that were applied over the past twenty years. These systems may be equally as varied in type as the systems available for the repainting and are often incompletely documented. In addition, many of the newer systems are less easy to repair than were the old oil/alkyd types. Finally, the demands of the late twentieth century (increased traffic loads, high levels of deicing salts, etc.) on both bridges and paint systems are more exacting.

It is little wonder, then, that with the increased use of the high-performance coating systems, catastrophic failures, the likes of which had not been encountered in the lead-and-oil days, have occurred with greater frequency.

Such failures, however, cannot be laid entirely at the door of the system-design engineer. Too often, he is merely led by the paint manufacturer, who may be incompletely aware of all the nuances of the coating system himself, and particularly of the parameters of what constitutes acceptable application and utility and what does not. It is one thing for a coating system to function properly under ideal circumstances but another to maintain performance under practical field conditions where the ideal may be compromised either slightly or substantially.

Under these circumstances, it is becoming increasingly necessary for a paint specification to be designed to fit a particular structure than for the structure to be "adopted" by any given paint system. Only on large bridges, however, are paint systems designed for the specific structures. In reality, it is economically infeasible for every simple overpass to be evaluated individually and for a coating system to be tailored specifically for that structure. Nor is it necessary, for it is more usual to find a whole group of bridges of similar design in the same general location and environment and having identical existing paint systems. In this case, what works for one structure will most usually work for another. The adoption of a single specification for a group of overpasses is a viable methodology and one widely employed across the country.

On the other hand, the concept of zoning,[1] which is now used in several states, is a practical approach to specificity in coating system selection that is both economically feasible and engineeringly sound. Here, a state will employ a coating system in marine or industrial areas that is totally different from one it may use in rural, inland areas. While this strategy may not take into consideration all factors that may have an impact on the suitability of a given paint system for a given structure, it nicely deals with one of the most important factors: exposure. Many states in all parts of the country have observed the effects of the environment on the service life of coating systems and report that coating systems on coastal structures may give less than half the service life of the same system in the rural interior. Similarly, a paint system in metropolitan New Jersey will give no more than one-half or even one-quarter of the service that the identical paint system will give in the desert Southwest or in rural states, such as Wyoming or the Dakotas.

Moreover, states employing multiple types of coating systems, such as California and Oregon, report that in many cases they see fewer differences in the performance of different paint systems in any given area than they do in the performance of the same system in areas of different environmental severity. An all-vinyl barrier system has been seen to fail in about the same length of time as a conventional red lead system in coastal areas of the Northwest, but both will give much longer performance in the interior valleys. On the other hand, it is widely accepted that zinc/vinyl systems will outperform the BLSC and red lead systems in more aggressive environments, and such states as North Carolina, Texas, and Florida employ both coating systems, utilizing the zinc in marine or other severe environments where it is considered necessary.

A further example of increased specificity in coating system selection is the increased use of "specific utility." The concept takes advantage of an understanding of those areas of a particular structure that are most vulnerable to a specific type of deterioration and addresses these by specific utilization of pertinent coating systems, additional coats, or additional film thicknesses in these particular areas. In this way, money that is not needed in the overprotection of less vulnerable areas of the structure can be saved and spent better in areas where more serious deterioration is expected, based on studies of the deterioration patterns of previous paintings on the same or similarly designed and located structures. In its simplest form, one coat of a three-coat system may be saved on less vulnerable areas (e.g., interior webs), while an additional coat is applied to more critical areas (e.g., lower flanges, seams, edges, etc.). More complex examples include the use of UV-resistant finish coats (silicone alkyd, acrylics, aliphatic urethanes) on exposed surfaces, while less expensive systems (long-oil alkyds, vinyls, epoxies) are retained in areas having less UV exposure. Similarly, coal tar epoxies, hardly desirable or necessary on most sections of a bridge, are particularly valuable on expansion bay steel where some of the most severe deterioration occurs. The concept is discussed more fully elsewhere.[2]

COATING SYSTEM SPECIFICATION

Over the past fifteen years, the quality of the best system specifications (if not the specifications for the individual paints) has improved considerably, although, in contrast, the majority of specifications remains disappointingly inadequate. Modern bridge paint specifications still pay more attention to the individual elements than to the interaction of these elements and their effect on the overall system. In many ways, the typical high-performance specification betrays the considerable confusion and uncertainty with regard to composition (if not utility) that exists among many of the state specification writers. All too often, the blind dependence on the manufacturers shows from whence the high-performance systems of many states were derived.

It is a paradox that, while the heart of the specification must necessarily concern procedures and materials and their application and inspection—issues of considerable technical complexity—detail on these elements is invariably less exact and less extensive than are the legal provisions of the "boiler-plate" sections. In many cases today, specifications (especially the compositional specifications) have less definition than in the past. While the trend away from compositional control to performance control is by no means a bad thing (there is much justification for the assertion that composition is the prerogative of the coatings manufacturer not the specification writer), the trend conveniently obscures the ignorance of coatings' makeup that exists in many state departments of transportation.

Performance specifications could also be improved. Greater attention might be paid to those factors known from past service to be critical (i.e., postblast priming intervals, cure verification, recoat times, etc.). A better definition of acceptability and nonacceptability criteria could be spelled out. A response methodology for correction of common deficiencies, if delineated within the specification documents, would eliminate many legal hassles and claims for extra work orders and perhaps serve to prevent the deficiencies before they occur.

Where improvements in specifications are seen, they doubtlessly arise from the improved knowledge and experience with the use and limitations of the higher performance systems in the field. The fact that the improvement in specification writing is by no means universal seems to indicate one or both of two things: either there is poor communication of the use/performance data with respect to given products and specifications or that same unfamiliarity with coatings' makeup and use noted above is prevalent among many who prepare such specifications.

During the past fifteen years, a greater awareness of the importance of accurately specifying surface quality, film thickness, volume solids, sag resistance, climatic parameters under which painting can take place, and performance test procedures has developed. The use of control instrumentation is now specified in more detail as are, in some cases, those less quantifiable coating and coating-film defects that shall not be tolerated.[3]

Somewhat less widely defined seem to be issues concerning the type and purity of abrasive that shall be used (though not necessarily its size!), a mil profile range, both upper and lower film-thickness parameters for each single coat and for the total system, and dry adhesion tests in the field. It is hoped that the awareness of the importance of cure evaluation and recoat times (minimum and maximum) will continue to grow and be reflected in improved specifications. Similar improvements are hoped for in definition of climatic controls (substrate temperature vs. ambient temperature, more use of dew point instead of relative humidity, etc.). Also, with an increase in the necessity for containment of blasting residues will come additional controls on airborne emissions, tolerable wind speeds and directions, and the like.

What is perhaps more encouraging is that there seems to be increased emphasis on prequalification, not only for the paint manufacturer and the contracting firm but also for materials and individual operatives who do the work. There also is an increasing awareness that the paint manufacturer should be experienced not only with the product specification and the contractor on jobs that have used such systems in the past, but also that specifications should call for detailed evidence of a particular paint system's past performance on similar jobs. The new specification for the inorganic zinc/vinyl/vinyl-acrylic system for the Massachusetts Turnpike Authority demands that operatives have at least five years' experience using similar coatings and equipment, and requires documented résumés for each applicator.[4]

Modern specifications often call for continuous inspection and limit the number of work areas in which the contractor may work at any given period so that inspection will be adequate and cost effective. The specification for the Tobin Memorial Bridge near Boston not only required a team of experienced inspectors on the job but also called for field consultants in coating technology and environmental science.[3] Many specifications now call for technical field-service representatives from the paint manufacturer to be present for some period at the start of any job to school both contractor and inspector in the nuances of a particular system.

Shop and Field Specifications

Paint-system specifications for shop-applied new construction are inevitably less complex and technologically more definitive documents than are specifications for the repair of aged coating systems in the field. Apart from the provisions that must be made for field touch up, the uniformity and predictability of the new substrate, the ease of access for application and inspection, and the control of climate and contamination make for a cleaner, more certain specification. In effect, the specification writer is more assured that the actual product will more closely duplicate his conceived model.

Existing bridges are often less adaptable to newer paint systems than are newly constructed structures. Bridge design engineers who once designed their structures with little thought of paint or paint-system maintenance today show more care. Lattice members, back-to-back angles, intricate detail, and inaccessible box interiors (with which field maintenance is confounded) are now replaced with a greater emphasis on simple rolled beams and large uninterrupted areas, where surface preparation and application of today's more demanding coatings can be accomplished with much greater surety. Boxed members, once open, are now often sealed completely, particularly on cantilever or orthotropic spans. This practice has led to diminished corrosion inside the box members and eliminates many of the field-access difficulties that plague the painting of open boxes. It also reduces the area of steel requiring surface preparation and paint. Connections, once left bare in shop application, can now be protected with inorganic zinc rich films that have slippage coefficients higher than those of sandblasted surfaces.

Less clearly differentiated are the advantages and disadvantages of the various options that exist in new construction alone. The NACE Task Group T-6H-23 completed a study comparing the various possible sites and procedures for painting structural steel.[5] They cite four major options: job-site application before erection, after erection, painting at the applicator's yard site, or painting at the fabricator's plant. These options are multiplied when one considers that a three-coat system specification, as would be the case with each of the above alternatives except the first, applies in the application of the intermediate coat and the finish as well as the primer. There are, in effect, ten different options for a three-coat system. The specification writer must select which methodology most satisfactorily fulfills his requirements. It is not surprising, then, that in bridge construction, preparation and painting after erection is rarely adopted. It has few advantages, is the most costly procedure, and introduces many of the usual field-painting difficulties one experiences in maintenance. Similarly, job-site application has few advantages compared to work done either in the fabricator's yard or the applicator's shop. Handling damage may be minimized, especially if intermediate or finish coats are applied in the field, but costs are higher, and job quality will probably be poorer because of the more erratic application resulting from access difficulties and climatic effects.

Advantages and disadvantages of work performed in the fabricator's shop versus the applicator's yard involve cost savings and the quality of preparation afforded by centrifugal blasting. (Most fabricators use centrifugal-blasting methods, while most applicators use air-pressure blasting.) Also, application by the fabricator is often performed in-shop (under cover), while most paint applicators have historically worked without cover. On the other hand, a painting contractor will probably be more skilled in the use of coatings than will a fabricator, who, aware of the cost overruns and scheduling problems that haunt his painting operations, may actually refuse painting services, particularly those involving the application of coats other than the primer.

Various options as to which coats to apply in the shop and which coats to apply in the field remain. They revolve mainly around difficulties of field touch-up and the greater vulnerability of multicoat systems to handling distress. The experience of the Michigan Department of Transportation[6] would indicate that it is indeed possible to apply all coats in the shop with little difficulty, good control, and extreme cost effectiveness. While there are data to support cost savings in the use of inorganic zinc as a shop primer coupled with deferred (five-year) topcoating programs (including the treatment of topcoating as an expense instead of a capital cash outlay),[7] these data refer to plant construction and would not hold true for bridge work because of the high cost of access alone. For both aesthetic and engineering considerations, the field touch-up of a prime coat or even a midcoat may be more satisfactory than the touch-up of a finish coat (especially a thermosetting system), but field application is inevitably more costly and less secure than shop application. Tinklenberg[6] observes that a greater awareness of the critical nature of handling finished pieces in itself fosters greater care in their handling, which, in turn, reduces the amount of field touch-up of finish coats actually necessary in full shop-applied systems. The state of Michigan reports remarkably little field touch-up of its shop-finished system and is now firmly convinced of the approach. Bayliss has discussed the practicalities of lifting, transportation, and storage of shop-applied finished steel as well as the use of temporary protectives.[8] Single-site applications will also save money on inspection fees.

Interestingly, Montle,[9] in discussing the problem of lateral cohesive failure of inorganic zincs through premature topcoating, observes that jobs calling for in-shop primer and topcoat application are more suspect than field jobs because the fabricator, motivated by logistics, is often too eager to topcoat the newly applied primers and get them out. Gelfer and Vandorsten,[10] on the other hand, in an attempt to reduce the difficulties of long recoat times of alkyl silicate zinc rich films in the shop, have produced experimental data indicating that ultra-rapid recoating intervals are possible with conventional (vinyl, epoxy, chlorinated rubber, etc.) topcoats, hydrolysis and solvent release of the inorganic zinc primer apparently occurring in spite of the presence of the topcoat. These data still require field confirmation, however. Inorganic zincs based on the high $SiO_2 : K_2O$ ratio alkaline silicates are reported by Schaffer[11] to withstand rapid topcoating (two hours) without zinc-film splitting.

Design of paint systems for shop application in new construction are not restricted by the many factors that affect the design of repaint systems in the field. The factors that directly affect system selection in field maintenance are as follows:

1. Bridge design
2. Location and severity of exposure
3. Existing paint system—type, age, condition
4. Expertise of painting contractor
5. Availability and quality of inspection
6. Funding
7. Environmental restrictions on paint and painting process
8. Required service life
9. Field-performance record of candidate coating systems

Good field maintenance is more expensive than shop work, particularly in those instances where the jobs must comply with environmental regulations and so on. Despite the added cost, field application is all too often less successful than shop application. In the field, bridge design, access, location, and exposure heavily affect the selection of a paint system, as do the availability of inspection and the competency of the contractor.[12] Perhaps the most severely limiting factor in the design of field-maintenance specifications, however, is the existence of old paint systems.

While there is a gradually increasing field practice of radically blasting a bridge to remove all old coatings totally and renew the metal/primer interface, this is expensive, and most field maintenance must utilize coating systems over surfaces consisting of old paint that at best has been spot blasted to an SSPC-SP-6 or 10 and very often only power cleaned or brush-off blasted. Furthermore, the condition of the surface immediately presented to the new paint (which can be normalized to some extent by cleaning) is of less concern than the condition of the interfaces within the older, existing composite (which are inaccessible). With an old paint system that must, in part at least, perform as the substrate for the new coating system, the specification writer assumes the legacies of all the mistakes of those who went before him. Bad specifications, poor surface preparation, sloppy application, unduly long recoating intervals, and the sheer buildup of paint film thickness on many older structures produce a highly tenuous composite of tired, brittle, and weakly cohesive paint films, faulty interfaces, and other various malignancies that are usually built up over nothing better than adherent, hard, and unscarified mill scale (see Figure 5-1). Yet paint systems for many such older structures are still written blindly, without any careful analysis of the structure and the condition and type of existing paint.

In the past, when such structures were recoated with the older lead and oil systems, the universality of these coatings, their initial flexibility, and the mildness of their solvent systems caused few problems even under such conditions. With newer paint systems that have stronger solvents, stronger vehicles, and greater initial shrinkage on drying, these older composites are more severely strained with new applications, often

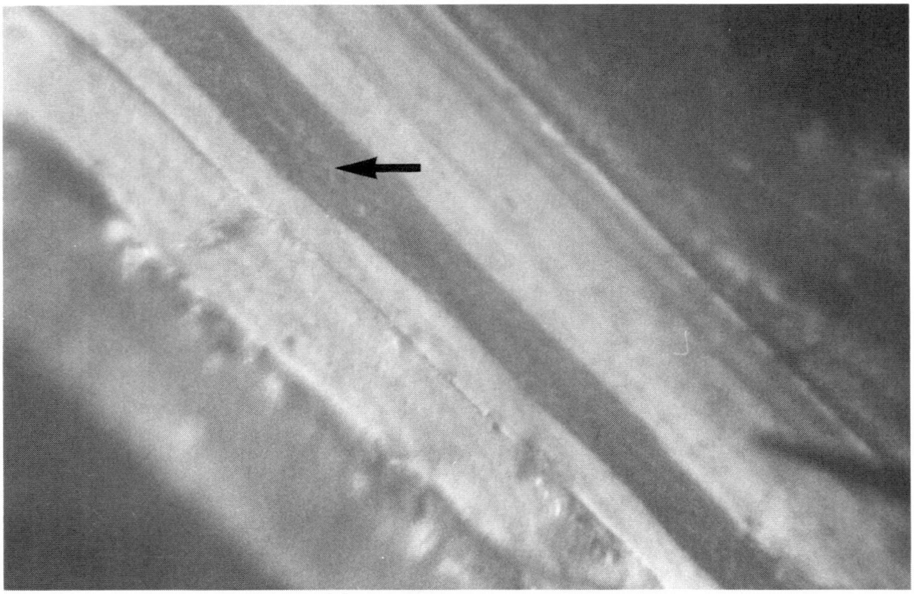

Fig. 5-1. Typical Cross Section of an Existing Paint Film Composite on an Old Structure That Must Serve as the Substrate for a New Application in a Spot-Cleaned Field Maintenance Painting. Note interfacial deficiency between two coats of finish hidden with the composite (*arrow*).

to the point of breakdown. Early adhesive and/or cohesive failures at one interface or another within the continuum of one film or another can result in system delamination, particularly in those areas peripheral to a spot blast where the old paint system was damaged almost to the point of failure during spot blasting. As old paint films gradually increase in film thickness with each additional application, as the more conventional lead and oil systems are used less and less due to increasing environmental restrictions, and as the mixing of new and old systems increases, it is probable that this type of coating failure will become more commonplace, at least until these older structures are totally stripped of their original paint composites.

In a limited number of cases, primarily in urban areas, environmental and community pressures have spurred the removal of old lead systems on bridges. The process is best exemplified by the partial repainting of the Tobin Memorial Bridge near Boston (see Figures 5-2 and 5-3) and by the containment of a truss structure in Burlington, Ontario.[13] Unfortunately, the premium exacted for the Boston work was substantial. Not only was the basic cost of the radical system removal high, as was that of the actual mechanism of containment (design of structures, use of scrubbers, disposal of grit, air monitoring, blood testing, etc.), but the impact of the press and the politics of environmental and community groups on job execution was even more inflationary. These influences were so great that unless a more realistic attitude is adopted on other jobs, it seems likely that bridge authorities (and certainly state transportation departments) will not be able to afford such radical deleading elsewhere. In California, Ohio, Louisiana, and Iowa, on the other hand, other less elaborate containment devices have been proposed and used. While less effective than those used on the Boston job, these devices have engendered less furor and have had less impact on job costs. Then again, even where such relatively simple containment procedures as hand-held vacuum blast nozzles are used, surface preparation costs are also substantially inflated.

In the vacuum-blasting process, the venturi nozzle itself is surrounded by a skirted vacuum line (see Figure 5-4) that theoretically sucks back the spent abrasive and debris from the steel as part of the blasting process. Although the unit works, production rates are protracted by some two to three times that of open-blasting rates. In addition, some units are heavy, and operator fatigue is a factor in production rates and containment efficiency. Vacuum blasting is only effective where nozzles can be held close to the steel surface. Their containment efficiency rapidly drops off when the unit must move over riveted or bolted areas, and the units have even less efficiency in corners and complex areas or on such open steel work as railings and gratings.

More often than not, while specifications call for the total removal of lead paint under suitable containment (even citing such methods as were used on the Tobin Memorial Bridge), the containment is not enforced, and the job proceeds with open blasting or token tarpaulins. Containment designs such as those used on the Tobin Memorial Bridge are also quite specific to bridge design and not easily adaptable to other structures. This frustrates those engineers who would seek to specify containment but are unwilling or unable (because of skill, cost, or time constraints) to design specific containment features for other structures. Vague phrases calling for containment "simi-

Fig. 5-2. Railing Area Containment Methodology on the Tobin Memorial Bridge, Chelsea, Massachusetts

Fig. 5-3. Leg Structure Containment Methodology on the Tobin Memorial Bridge, Chelsea, Massachusetts

lar to . . . " are inappropriate, unfair to the contractor, and invite disputes and even litigation. Much detail on containment practices can be found in the Snyder and Bendersky report.[13]

General Painting Specifications

A complete bridge-paint specification is a relatively complex document that is designed both to guide and to police the contractor's execution of the job within a set of acceptable parameters. The specification is usually divided into several sections, many of which are "boiler plate" in nature and apply to all contracts issued by the pertinent bridge authority. The following is the format of a typical bridge-painting specification:

General Provisions (Standard requirements for contracts with authority)
- Definitions and interpretations of contract documents
- Proposal requirements and conditions
- Award and execution dates
- Bonding and insurance requirements
- Legal requirements, permits, and responsibilities to public
- Penalties, incentives, and liquidated damages
- Indemnification
- Variances

Special Provisions (Detailed requirements of specific job)
- Scope and definitions of work
- Drawings
- Measurement of quantities and payment
- Minimum-wage requirements
- Utilities, sanitation, and cleanup
- Traffic control and interfacing with adjacent works and other contracts
- Police service
- Quality control, inspection, and sampling of materials
- Guarantees

Technical Specifications
- Cleaning, surface preparation, and painting specifications
- Paint-system specifications
- Individual paint specification
- Other materials specifications (sealants, etc.)
- Repairs
- Environmental provisions and disposal

Sample Contract Forms
- Corporate and individual contract forms
- Performance bond
- Payment bond
- Subcontractor forms
- Extra-work forms

Fig. 5-4. Vacuum Blast Nozzle Design

These general sections set definitions, proposal requirements and conditions, and give general information on the award and execution of the contract, including bond requirements and penalties, insurance requirements, permits, and other legal requirements. They may also describe the scope and control of the work, its progress and prosecution, and the procedures for the measurement of quantities and payments.

Separate sections pertain to the particular job. These sections are also normally nontechnical and usually itemize special job considerations. They provide more detailed requirements of the specific job and include the drawings, minimum-wage requirements, insurance, and necessary indemnification clauses. They may describe such provisions as traffic control, sanitation, utilities, debris removal, interfacing with adjacent work, interpretations of contract documents, and so on. They also may outline other control mechanisms, including inspection, sampling procedures, rigging requirements, pollution control, prequalification requirements for the contractor, and any extraordinary requirements that may pertain to the particular job.

The technical documents are contained in a separate section of the specification and are primarily concerned with requirements for cleaning, blasting, and painting, although quality-control provisions and environmental protection (if appropriate) may also be found here. The heart of this section, however, delineates the paint-system specification itself and the specifications for the individual paints making up that paint system.

Sample contract forms also are often enclosed in the specification, including those for contract (performance) and payment bonds, as well as the job contract itself and necessary forms for subcontractors.

The content of the technical specification will depend, of course, on the particular system used, not only in terms of the individual coating specifications but also in terms of the consequent specifications for cleaning, surface preparation, and job execution that are necessitated by the choice of such coatings. Once very simplistic, the better technical specifications for cleaning, surface preparation, and painting have, since the late 1960s, evolved into very detailed documents that continue to be refined as more and more is learned about the parameters of successful and unsuccessful application of each paint system. As experience with a particular paint system grows, ideally controls are continually added at each new letting of the specification to channel the job execution within the best painting practices for the application of the system. Where once control of thinning, brush technique, and number of coats were the important criteria, a multitude of quantifiable requirements typified by those items below are now often included.

Surface Preparation
- Allowable climatic conditions (temperature, humidity, dew point, wind speed, and, where applicable, wind direction)
- Surface preparation methodology
- Media type and size range (for blasting)
- Inhibitor type (wet blasting)
- Equipment type (for power-tool cleaning)
- Solution types and concentrations, equipment and rinsing schedules (for washing)
- Required surface quality
- Mil profile
- Blast-prime interval

Coating Preparation
- Homogenization
- Mixing (of components)
- Thinning (solvent type and amount)
- Induction time requirement

Coating Application
- Allowable climatic conditions (temperature, humidity, dew point, wind speed, and, where applicable, wind direction)
- Allowable application methodology
- Wet-film thickness (minimum and maximum allowable)
- Dry-film thickness (minimum and maximum allowable)
- Recoating intervals (minimum and maximum allowable)
- Pot-life limitations
- Description of required appearance
- Itemization of nonacceptable conditions and required rectification

Coating Materials (Compositional Specifications)
- Nonvolatile content
- Pigment content
- Nonvolatile vehicle content
- Weight per gallon
- Vehicle type as determined by infrared spectra
- Zinc content (where applicable)
- Quantitative determination of key pigmentary elements (lead, chromium, etc.)

Coating Materials (All Specifications)
- Viscosity
- Dry time—touch
- Dry time—hard
- Dry time—recoat (may include coin test and solvent rub tests for zinc primers)
- Pot life
- Color
- Gloss
- Flexibility
- Sag resistance
- Salt fog
- Weatherometer or fluorescent UV condensation exposure
- Immersion, salt water
- Immersion, fresh water
- Bullet hole

Coating Materials (Complete System)
- Adhesion and cohesion
- Salt fog
- Weatherometer or fluorescent UV condensation exposure
- Immersion, salt water
- Immersion, fresh water
- Hot/cold/UV/salt cycling environment

Field Evaluation of Finished System
- General appearance
- Dry-film thickness

Adhesion
Touch up procedures

The Specification Writer

The value of a tight specification that "locks in" these requirements is described by Kline,[14] who concludes that "the absence of specific requirements with respect to many of the factors means in effect that they are considered not important, and leave the applicator to his own devices to our collective detriment" (p. 169). It is, moreover, too late to implement such controls after the job begins, for to do so will be to modify the scope of the work as bid by the contractor and will cause difficulty between inspector and operative, encourage claims for extras or, in extreme cases, even spawn more serious disputes and litigation.

It is important, therefore, that those who produce such specifications are familiar not only with the paints and paint systems selected and with the typical day-to-day operations involved in the execution of the bridge painting process but also with the strengths, vulnerabilities, and field history of the coating system selected. It is also important that specification writers appreciate the demands that the design peculiarities of a particular bridge may have on these strengths and weaknesses.

It is rare that the specification for a new paint system will adequately cover all salient points and lead to satisfactory job execution in the first instance. As the job progresses, however, deficiencies can be addressed, if necessary by means of job extras and so forth, and these changes may be adopted in subsequent issues of the specification. Through this process of refinement, sound, workable specifications giving good performance can evolve.

Unfortunately, well-composed specifications are too often the exception rather than the rule in bridge painting, and many of the practices above describe today's best work rather than the norm. As so often happens throughout this discipline, the implementation of best-practice methodology lags sadly behind our realization of what should be done.

It is disturbing that many of those in state government and other bodies responsible for the generation of bridge-paint specifications have little practical experience with either paints, paint systems, or the painting process itself. Too much use of the "scissor-and-paste" approach to specification design still exists, and there is too little continuous review of emerging practices. Paint specification writers often will admit privately to their training inadequacies and confess to an imprecise knowledge of the materials and techniques that are the subject of their specifications. They play it safe and copy. The use of copying techniques under these circumstances is probably reasonably justified. But, while the awareness of self-limitation is commendable, it is inexcusable that there remain so many individuals with such limitations in such positions of responsibility. Also, as paint systems become more complex and specific in relationship to bridge design, size, location, and environment, and as the availability and quality of inspection remains distressingly deficient, there is an increasing danger in "play-it-safe-and-copy" thinking.

An additional complication is that in some cases those responsible for the preparation of paint specifications and those responsible for coating-system design and final specification production are not the same individual. In some cases, when individual paint specifications are prepared by those other than state employees, there may be little or no communication between such parties. Moreover, there seems to be too little communication between field engineers and inspectors and specification writers. It is in the field where specification deficiencies and required refinements are established. Formal channels wherein this type of data can be discussed directly between field representatives and specification writers (if not formulating chemists) are imperative to the refinement of paint-system specifications.

Sources of Instruction for the Preparation of Coating System Specifications

There is only slightly less awareness in the bridge-painting industry of the general deficiency in good specification writers than there is of the lack of painting inspectors. Unfortunately, there seems to be less effort directed toward a resolution of the former problem than to the latter problem. No courses are given that deal specifically with the preparation of bridge-paint specifications, and few books and papers seem to address the subject. Best are probably the *SSPC Manual* Volume 2, Weismantel's *Paint Handbook*[15] (particularly sections in chapters 4, 6, 7, and 14) and NACE Publication #6J162.[16] Data may also be found in the FHWA study on best current technology in Ray, Henton, and Rideout.[17] Some general requirements for specification content are discussed by Baker,[18] and valuable if dated material can be found in the standard specifications of the American Association of State Highway Transportation Officials (AASHTO)[19] and the *Construction Manual for Highway Bridges and Incidental Structures*.[20] Although they do not consider bridge-paint systems in particular, the three articles by Baldino[21,22,23] on guidelines for writing coating specifications have much valuable data on and insight into the form and language of the boiler-plate and general technical provisions of the specification.

Equally as important as the composition and language of the specification is, however, some understanding of the underlying technology, the materials used, and the job execution process itself. Some practical field experience is invaluable to the specification writer, as is direct contact with residents and paint inspectors and others familiar with the day-to-day operations of the painting contract. Again the various publications of SSPC, particularly the manuals and *The Journal of Protective Coatings and Linings* may be of great assistance here, as may be those of NACE (including the journal *Materials Performance*). Other specific articles,[12,24,25,26] while discussing paint systems for bridges (among other structures) in some detail, do not consider either the form or language of the specification. Still more basic are those texts that are more concerned with corrosion, surface preparation, application, and so on.[24,27,28,29] In this regard, Munger's paper is an excellent source of background data. Many application-equipment manufacturers also provide detailed booklets on technique. The booklet published by The DeVilbiss Company[30] on conventional spray application is an excellent example of this. Introductory texts that deal with basic coatings science and formulation[31–34] are more technical but provide invaluable background support. Data at this level, however, is more usually derived from one or more of the sources of the paint specifications themselves, which are discussed later in this chapter. The specification writer may also find valuable pointers from some of the texts of the inspection courses outlined in Chapter 7.

Finally, much additional data may be absorbed from existing specifications, although it is important that the would-be specification writer uses these materials as data sources rather than as fixed models and views a sufficiently wide selection of these documents to acquaint himself with all views and approaches. It is well worth remembering that many of these documents are imperfect, to say the least. The specifications of states such as Florida, California, and Louisiana, and authorities such as Massport, however, are well worth review.

Individual Paint Specifications

Today's specifications for individual paints,[35] from which the coatings manufacturer formulates his own interpretation, may be grouped loosely into one of three types: the compositional specification, the performance specification, and what is best described as a quasi-compositional specification.

Compositional Specification

The compositional specification is undoubtedly the most definitive as well as the most restrictive document. This specification both leads and polices the manufacturer. The compositional paint specification allows neither raw material suppliers nor manufacturers much leeway in formulation and, consequently, is hardly conducive to creative formulating. Its value is that it gives the specifying authority the greatest control over what goes onto the bridge, but it limits its quality to that specified and stifles growth. In some respects, the compositional specification assumes that the bridge authorities are better formulators than are the paint manufacturers themselves. This is, of course, rarely true, although when assisted with competent consultants and knowledgeable organizations such as the SSPC, bridge design engineers and specification writers may do a better job than the vast majority of manufacturers. Litigation abounds in cases where even the larger paint manufacturers have erred in their interpretations of compositional paint specifications.

There are, however, liabilities for the bridge authority also. Unless it can be proved that some deviation from the set specification occurred, in the event of paint failures involving the paints themselves responsibility lies with the state and not the manufacturer. Therefore, the compositional specification is best derived from compositions that have proven their worth over a period of time.

The compositional specification is generally divided into four substructure sections. These define the raw materials that may be employed, provide an itemized formulation breakdown in terms of both pigment and vehicle, and list a group of quantifiable parameters that may further channel the formulator toward the required product. These quantifiable parameters serve also as a list of requirements against which a testing agency may qualify the product. The fourth subsection adds performance controls.

The use of compositional specifications has declined. This decline has apparently been encouraged by the FHWA, which has advocated leaving formulation technology to the paint companies (in order to avoid duplication of effort) and has concentrated resources on evaluating coatings and developing test methods and criteria for performance characterization.[36] Unfortunately, the consistent prediction of field performance from accelerated laboratory tests remains elusive at this time and reliability of high-performance systems that are controlled by such tests does not match that of older systems having compositional control. Some authorities insist that a strict compositional standard, locking in the attributes of a well-proven formulation, is, in light of these failures, not to be dismissed lightly. As proprietary products become commodities and as specification writers learn more about the newer systems' formulations and their parameters of performance, the compositional specification could eventually return.

Performance Specification

The performance specification is a much less exhaustive document than the compositional specification and is primarily a policing specification. In its most pure form, the specification may not even specify the generic type of material to be used. The thrust of the performance specification is to qualify the product by a series of qualitative or quantitative tests that are keyed to the required function of the coating and not to its composition. Thus, a large coatings manufacturer that is reluctant to manufacture a special specification paint of fixed composition (and contend with cutthroat competition from smaller manufacturers) is more easily accommodated by the performance specification. This allows him to offer a line coating without special manufacturing and at a higher profit as long as the coating provides the necessary performance.

The trick is to come up with a group of accelerated tests with pass–fail criteria that can be correlated accurately with a minimum required level of performance in the field. Because there is much difficulty in correlating any specific accelerated test with field performance, this is not an easy task, but resolution is generally possible by using several different test criteria to qualify the product separately. An evaluation of test results against field performance for more acceptable high-performance systems that is derived from specifications that utilize the technique would show that enough correlation exists for meaningful specifications to be so designed. When system failures that are not predicted by test results occur, they occur most often for reasons not associated with a defect in the paint but from deficiencies in other areas (low film thickness, poor surface preparation, etc.). With time and experience with the system, better correlation of accelerated test results with field performance is acquired, and the adjusted specifications become more accurate and meaningful. Performance-type specifications are typified by the West Virginia Department of Highways specification for an inorganic zinc rich/vinyl system.[37]

By definition, the performance specification requires significantly more performance control by means of accelerated testing than does the compositional specification. Because many of the accelerated tests required in performance specifications take weeks or even months to complete, there are impracticalities in correlating testing requirements with job schedules. It is often necessary, therefore, to tag specific formulations with easily determined quantifiable constants at the time of coating approval and to use these to monitor batch-to-batch consistency. Furthermore, as noted by Ray, Henton, and Rideout,[17] attempts to assure material quality with laboratory performance tests are dogged by those same variables in test design, surface preparation, application of panels, and changes in test ambient conditions that haunt more formal studies that attempt to correlate lab performance with that in the field. Other, less satisfactory techniques of quality assurance involve certifications from the manufacturer.

Another major deficiency of the performance specification is that it assumes complete compositional conformity among individual batches of the product. This, unfortunately, is not always the case; intentionally or unintentionally, formulations may vary. Manufacturers may have alternative sources for raw materials and may, for reasons of cost, availability, delivery, or in-plant control, vary formulations; thus, the existence of two or more quite different formulations may be utilized to produce the same product. While most compositional formulations of the qualified products list (QPL) type require the manufacturer to notify the specifying authority of any changes in formulation, this is not always done because of the costs of resubmittal. The term "formulation change" is also quite relative. It is rare that some slight modification of formulation is not required for quality control, particularly adjustments for color, gloss, or viscosity, and "work-offs" of past batches are by no means unusual in the paint industry.

One aspect of the performance specifications that might be improved upon is the misdirected emphasis on testing individual products and not the complete paint system. Many failures, particularly with zinc rich systems, seem to result from the effects of the environment on the coating system as a whole and not on the individual elements. This is not only true for performance specifications per se but also for all paint specifications, except when single-coat systems are to be employed. Recoating criteria are most critical in many of the high-performance paint systems (zinc rich, coal-tar epoxy, wash-primer systems, etc.) and, by definition, specifications for total composite thickness and intercoat adhesion require a control of the total system and not just an individual paint.

While the realization that all elements of a given coating system should be purchased from the same manufacturer is now a cornerstone of good specification design, the manner in which midcoat, topcoats, and primers interact with one another viscoelastically is not as well considered, nor, perhaps, even understood.

Quasi-Compositional Specification

The quasi-compositional specification is often really a performance specification dressed up to look like a compositional specification, and it is typified by those documents prepared by paint manufacturers that wish to protect their formulations but qualify the product for use on public structures. They are generally not good compositional specifications but, with adequate performance criteria, are little different from performance documents. If the states wish to use a product and comply with bidding laws, a quasi-compositional specification designed by the product's manufacturer is often all they have, unless they wish to analyze the composition further themselves. Most usually, the paints they represent are excellent, for were they not, it is difficult to understand the state's motive in issuing such documents.

The quasi-compositional specification employs elements of both the true compositional form and the performance form, although the latter predominates. The specification is rarely so compositional that a given formulator without prior knowledge could easily create a coating solely from the requirements of the document. This type of specification is typified by many of the aluminized-epoxy mastic specifications, such as those of South Carolina.[38] This type of specification primarily polices, and leads the formulator only very broadly. The specification will channel the formulator not only by stipulating such things as general vehicle type (alkyl silicate, epoxy, alkyd, etc.) or pigment type (zinc dust, silicates, inhibitive pigments, etc.) but also by quantifying certain parameters ("weight of zinc in dry film shall be 81% minimum," etc.). It also may specify one or two less critical constants, such as viscosity, dry time, weight per gallon, and so on (which may be used with other determined constants to tag the product for batch-to-batch variations), but the bulk of the specification does much the same thing as does the pure performance specification, and often in as much detail. Quality-assurance difficulties are much the same in the case of the quasi-compositional specification as they are in the performance specification. Whether or not quantifiable tags are published with the specification or assigned in a QPL approval process, these are again used more successfully to monitor product compliance than are the long-term accelerated performance tests outlined in many specifications.

The quasi-compositional specification, after some years of satisfactory use, is an ideal candidate for in-house analysis and reformulation testing for added compositional depth by the states themselves. The apparent reluctance of most states to do this is, perhaps, related to a lack of formulating confidence, although a certain amount of apathy is also probably involved. Most small paint manufacturers recognize the quasi-compositional specification, if not by sight then by the lack of compositional interpretation when clarifications are requested of (but not provided by) the issuing bridge authority, and tend not to bid those jobs in which such materials are specified.

Additional discussion of the quality-control programs of all three types of specification is found in Chapter 7.

Sources of Modern Paint Specifications

Individual paint specifications today range from well-conceived, compositionally detailed documents based on systems that are laboratory tested and field tried before release by such organizations as Caltrans, the U.S. Army Corps of Engineers, and the SSPC to the "scissors-and-paste epics" of ill-schooled writers and flagrantly proprietary systems that do little more than lock in a single manufacturer. To some extent, of course, this reflects the change in the derivation of the paint specification from the simple lead-based system that evolved from decades of trial and error and from those products with which National Lead dominated the industry in the early sixties to the systems of today that are primarily derived from individual paint manufacturers.

Today's paint-system specifications for bridge structures appear to originate from one of several sources. Each source is considered below in terms of its strengths and its weaknesses.

Specifications Derived from Proprietary Products

Most of the newer paint specifications have been derived from the paint manufacturers themselves. Not surprisingly, they are rarely so detailed compositionally that other manufacturers could easily reproduce the products on which they are based. The parent paint manufacturers cannot be blamed for this position. It would seem a reasonable business practice to disguise a formulation with a list of vague constants that could be pulled from a simple evaluation of the wet paint by even the most poorly equipped testing lab while hiding key elements of the composition. Bidding laws are not conducive to the use of proprietary products, and such shams as "and/or equal" clauses are hardly effective because the definition of "same" or "equal" is debatable. The engineer awarding a contract is also unlikely to take a risk with a product type that all too often he knows little about himself. Inevitably then, the manufacturer who wants to get his paint used produces a document that can best be described as quasi-compositional, but which is boosted by a strong performance backup. The authority wishing to employ the proprietary product accepts this but reluctantly allows competition. There have been cases in which competition has been flagrantly discouraged by such techniques as letting specifications with bid-opening dates that would preclude the necessary performance testing unless the bidder happened to have the exact performance data already on file with testing performed by an independent lab. As further discouragement, required tests have not used standard test methods, such as those detailed by ASTM or Federal Standards Test Methods #141A, but instead have employed deliberate adaptations that are unlikely to have been used by organizations other than the manufacturer writing the specification. Requirements for test reports from independent labs further restrict open access. In other cases, requirements may include expensive procedures hardly necessary to the functioning product and which, on small or medium-sized jobs, would certainly preclude a would-be manufacturer from a successful bid if only because the required testing would be economically prohibitive. In other specifications of this type, "grandfather" clauses that accept an original manufacturer's product as a control without such test data are found.

The major advantage of this type of specification is that the state gets a known, proven product with an established performance profile, albeit at an inflated price (on initial jobs at least). The disadvantage is that the bidding laws are, conservatively speaking, bent to some degree or another, and the specification is "loaded" in favor of the original paint manufacturer.

Despite the analytical capabilities now possible in many laboratories, there is often little attempt by the bridge authority to expand a specification by input from their own analysis. Similarly, there is no attempt to strip the specification of unnecessary test requirements, to standardize tests to ASTM, or to substitute more cost-effective procedures. Almost always, the bridge authority plays it safe and allows the specification to go as written by the paint manufacturer. Decisions such as this appear to be motivated more from concern over the effects of change than a desire to protect the original manufacturer, and, again, they reflect the lack of paint-formulating expertise in most highway departments.

Specifications Derived from Raw-Material Suppliers

One of the more common methodologies for specification derivation (if only because it was the technique developed by National Lead to secure the widespread use of BLSC in the early 1960s) is that of the direct marketing of a completely developed formulatory technology by a raw-material supplier whose product is essential to the particular technology. This marketing is directed simultaneously at both the specifying authority and the paint manufacturer. The technique is continued today by the Mobay Chemical Company, with its urethane resins[39] and by the Halox Pigment Division of Hammond Lead Products, with its borosilicate and phosphosilicate pigments.[40] The technique again takes advantage of the lack of formulating skills on the part of the state highway departments and bridge authorities but produces a range of very detailed specifications based on well-developed and laboratory-tested formulations that the state employs first on test bridges and then more generally if the test bridges prove satisfactory. When released, the specifications are sufficiently detailed so that any would-be manufacturers can provide coatings and coating systems that comply, with little deviation from the original test product. The paint manufacturer, having no proprietary claims, simply provides the manufacturing service (although many such specifications do allow some opportunity for direct input in "finishing the paint off") and the market remains highly competitive. The monopoly of the raw-material supplier is avoided by licensing the raw-material manufacturing techniques with other raw-material manufacturers, and a measure of respectability is gained by shepherding specifications on the raw material through ASTM.

If paint systems so introduced into one state are successful, adoption by other states becomes easier and general acceptance follows. In general, the technique has worked well in the past and to everybody's favor except that of the large paint manufacturers, whose profit margins are too large to allow competition in what rapidly becomes a relatively cutthroat market.

The specification suffers from its thoroughness in that even the smallest manufacturer can supply the product. Very often such manufacturers have neither the capability nor the funds to provide the technical service required in the field. The specification is also more prone to the unscrupulous manufacturer who may attempt to adulterate the formulation to his own profit. Good analysis and control is therefore mandatory, but the fully compositional nature of the specification facilitates this and data on testing procedures are furnished, again, by the raw-material manufacturer.

The success of this approach is attested to by the success of BLSC paints, and the rapidity with which such specifications proliferated in the 1960s, notwithstanding the excellence of the red lead system that they displaced.

The "Borrowed" Specification

Perhaps the most common route to the adoption of a given paint specification by a particular state or bridge authority is, in fact, through the experience of another state or authority. Paint formulatory capabilities are conspicuously rare in the research and materials divisions of most state departments of transportation, and many authorities that have jurisdiction over a single structure have no in-house testing laboratory whatsoever. Few outside testing laboratories formally employed by such authorities have formulatory expertise either. Therefore, the most cost-effective manner (that bears the least risk) in which to introduce new paint specifications to the field is to research the experience of other states and adopt the more successful systems by issuing simple copies of these specifications. Such systems are already field tried and proven and, with luck, all of the bugs will have been worked out. Careful investigation of application experiences, failure analysis, and so on may be employed to refine even further those pertinent sections of the specification. The danger is that a coating system may work well in one part of the country on one bridge design or in one environment but may not work in another. Therefore, attention to the parent state and the type of structure previously employed is important. (Conversely, of course, a paint system that failed in one particular state may be quite suitable in a state with a less aggressive environment.)

This mode of specification development has been employed with most paint systems, irrespective of their original source. The final document as a specification, however, is, more often than not, no better or worse than the specification from which it is copied, and the technique can hardly be described as progressive specification development.

In-House Specifications

Probably the purest form of specification development is that practiced by those states and authorities, such as Caltrans, the Texas Department of Transportation, and the U.S. Army Corps of Engineers, who have qualified paint chemists on hand. Most such laboratories rely on raw-material manufacturers for some initial input (if only to introduce the raw material) and may even derive coatings from existing proprietary products. Here, the profit margin of the raw-material supplier and the paint manufacturer is at least diluted, and the final specifications that emerge are detailed compositional documents that may be used by all potential manufacturers.[41–44] Initially, of course, the cost of development is high. The authority has to employ paint chemists and maintain a stocked and equipped paint laboratory, but, in the final outcome, costs may be saved because of the more intense competition and lower material prices that ensue.

Specifications Derived from Other Industries

In many cases, bridge paint specifications may be derived from the use of identical materials in other industries. Again, these may move into bridge-painting practice as performance specifications, such as the aluminized-epoxy mastics,[38] as in-depth compositional materials, such as the Florida alkaline-silicate-zinc specification,[45] or, more completely, as the original three-coat epoxy paint system used on the Newport Bridge in Rhode Island. This last system had originally been used in the petroleum industry on off-shore drilling rigs.[46] As in specifications produced specifically for bridges, these documents may have been derived from a variety of sources, including raw-material suppliers, paint manufacturers, and so on.

In this type of specification development, it is important to ensure that the new coating system is as appropriate for bridge service as it was for the original use. Environmental effects, such as UV exposure and/or heavy salting, system tolerances for problems caused by access difficulties, and intricate surface configurations and mechanical effects, such as continuous vibrations, must be checked carefully.

Specifications Derived from Independent Sources

Good technical paint personnel are a closely held commodity, and most are found in the service of the individual paint and raw-material manufacturers. A comparative few are associated with specialty engineering firms or research and development organizations. Some have and will provide invaluable and in-depth assistance in formulating and/or designing paint systems for bridges, but most, as consultants, are expensive, and few bridge authorities have the foresight to employ such groups until their skills are required in litigation.

One practice among some states, bridge authorities, and engineering concerns is to probe for free advice among formulators (consultants or those employed by paint manufacturers, etc.). In most cases, however, if free advice is given, it is always biased to those the advisor best represents. No consultant works for nothing.

Some industrial organizations, such as the SSPC, the Federation of Societies for Coating Technology (FSCT), and the NACE, will also provide assistance in designing paint specifications, if only by referral or through their excellent range of publications. Fully detailed compositional paint specifications and even formulations may be found in Volume 2 of the SSPC *Manuals*, for example. Much valuable data is available from these organizations because each acts as a central clearing house for data from their broad memberships as well as from the many research and evaluation programs that are specifically related to the protection of steel. Rarely, however, will these organizations provide full specification-writing services.

Some bridge authorities with large bridges employ engineering firms for consultation in the structural maintenance of their structures. As part of their responsibility, they often design paint systems for the bridges. Unfortunately, it appears that few firms retain individuals with any expertise in paint technology, and these organizations, for the most part, have even less formulating depth than do the research and materials divisions of most state highway departments. Generally, their specifications are derived from the same sources that the individual states employ, although engineering concerns will more frequently seek to employ the services of other, more specialized consultants. The service they receive from such individuals and firms varies widely and ranges from carefully written, fully formulated compositional specifications of considerable merit, to the "and/or equal, scissors-and-paste" epics that might easily have come from the engineering company's own staff.

Qualified-Products Approach

One of the fairest approaches to system selection is the qualified-product-type specification. Here, any manufacturer wishing to have his product or paint system employed on bridges under a given authority must first submit that product or paint system for either lab or (preferably) field testing by that authority. The technique is possibly best exemplified by the process employed by the state of Florida for their inorganic zinc-based system.[47]

In Florida, the manufacturer is given wide latitude with regard to coating composition, and intermediate and finish coats of radically different generic types may be allowed. Systems are approved mainly on the basis of the field performance of the complete coating system on a bridge member over a period of three years. At least 250 square feet of surface is used for the trial, and surface quality, mil profile, and film thicknesses must conform to set requirements. In this case, both three- and two-coat (without tiecoat) systems are allowed and tested, but the aesthetics and performance must conform to the state's rigid standards ("there shall be no bubbling of topcoat on application, and no peeling, blistering, cracking or adhesive loss over the first three years").[47] Only 1% rust is allowed over the three-year period. If the system proves successful in the field and is approved for use on the state's bridges, the state retains a one-quart sample of each coating for subsequent laboratory examination, and constants are determined against which to measure future batches of the system.

GUARANTEES

One- or two-year guarantees are not rare in bridge paint specifications today, although the guarantees are often vague and poorly written and probably offer little real protection to the bridge authority.

Few guarantees are properly written or legally binding. Few define those types and degrees of deterioration that are acceptable and those that constitute a failure. In a typical contract, the onus of the guarantee is usually placed on the contractor or, perhaps, on the paint manufacturer, while the specification writer, the inspection authority, or even the bridge authority itself may have some culpability. With so many parties involved, it is not surprising that the net result of bridge paint failures is either a complex litigation, with each party suing the other, or the assumption of the loss by the state or bridge authority.

Yet the practice of guaranteeing paint jobs (in all industries) is increasing. The guarantee not only offers the owner advantages in improved quality assurance and establishes an exact cost per year but also allows methodologies other than low price in evaluating bids. The guarantee may also assist both reputable manufacturers and painting contractors by further qualification and, thus, may eliminate the less reputable competitor. For the paint manufacturer and contractor, guarantees are a powerful (if dangerous) public relations and advertising tool.

Guarantees are more common in Europe and Japan. In Germany, for example, large painting contracts have been underwritten by insurance companies as part of a "protocol" methodology.[48,49] In this method, a trial section of the job (representative of the whole) is painted by the contractor, according to the requirements of the manufacturer of the coating system (under the supervision of his representative), thereby establishing a standard of quality. The contractor must then execute the remaining bulk of the contract according to this standard. In the event of system failure, responsibility is judged by weighing the deterioration of the trial section against that of the bulk of the job. Jobs are underwritten by insurance companies and premiums were about 3% to 5% of the total contract price in 1980.

In the United States, guarantees are becoming more common in other industries and where environments are more predictable and applications and inspection easier (see Moore[50]). They have been offered between manufacturer and owner, manufacturer and applicator, between applicator and owner, and among manufacturer, applicator, and owner, with split financial responsibility between manufacturer and applicator. Guarantees may be written over a set number of years and may be extended by prorating over the final years.

Recent experiments in "turnkey" or single-point responsibility contract approaches (similar to those used in off-shore oil drilling markets or in the maintenance of chemical-processing plants) have been attempted. In this approach, one organization, usually the paint manufacturer, assumes total responsibility for the job as the general contractor (see Hower[51]). The manufacturer hires and trains both the painting contractor and the full-time inspectors in the application of the specific products to be used and provides specification-writing services, contract bonds, and other details. The paint manufacturer also assumes contract-management responsibility and provides the bridge authority with a blanket guarantee, with assured maintenance for an extended set period.

Such a system (based on inorganic zinc/epoxy midcoat/aliphatic urethane finish) has recently been implemented on the Newport Bridge in Rhode Island (M. Kulick, Rhode Island Bridge and Turnpike Authority, and T. Calzone, Carboline Co., personal communication, 1984). The only formal parameters of the job dictated by the bridge authority were that the existing paint be removed to the metal, and that the new system last at least ten years. Several manufacturers bid this job (others refused), and the authority and their consultant evaluated the bids for the theoretical soundness of the paint system, the warrantee, and the job cost. On award, the paint manufacturer became solely responsible for the quality of the paint system for the next ten years.

There are obvious theoretical advantages to such single-point responsibility contracts, although it still remains to be seen just how well such guarantees will hold up under test. While little used on bridges to date, the approach may herald a new trend. One advantage of this type of contract is that in the event of problems, the interparty disputes that often haunt coating failures are eliminated. In selecting the paint manufacturer as general contractor, there is also some advantage to be gained from his intimate knowledge of the product. In addition, if the inspectors are trained employees of the manufacturer, there are advantages in the fact that the inspector has himself applied the paint system in the past and is familiar with the parameters of its employment. Aware of his responsibilities, the manufacturer is also more diligent in his training of the painting contractor in the correct execution of the job, and training sessions are usually conducted before the job starts. Because materials are supplied directly by the manufacturer, the painting contractor realizes no advantage in keeping down the film thickness, although there is no real incentive against the application of too-high film builds. (Curiously, the specification for the Newport Bridge does not specify film thickness maximums.) Perhaps the most important advantage of this type of contract is that application bids are evaluated by the manufacturer, who is not bound to select the lowest bid (nor for that matter to advertise the bid generally). Instead he can select the painting contractor from a carefully compiled list of qualifed concerns with which he is comfortable in terms of work expertise, cost, and ability to satisfy the contract requirements within the allotted time frame. In general, the paint manufacturer is better equipped than most parties to make such evaluations and to choose a contractor prudently.

The advantages and disadvantages of single-point responsibility coating specifications are discussed by Pinney et al.[52] In this discussion, the advantages of the concept are advocated for large, complex jobs and difficult environments (which would seem to qualify most bridge projects), although the authors see less cost effectiveness over the traditional, less expensive contract methodologies for simpler work. The authors also point out that the contractor's allegiance in single-point responsibility contracts shifts from the owner to the manufacturer, and this shift may engender significant mark-ups for labor and extra charges for added efforts normally absorbed by the contractor to satisfy the owner and ensure good will for future contracts. Furthermore, the authors suggest that a third party, such as an engineering-based organization rather than a paint company, should conduct the program. This allows maximum utility of the widest selection of paint systems available, while moving the control of the program to a knowledgeable third party. The financial strength to support a guarantee may be more problematic on large jobs for an engineering company than for a large paint manufacturer, however.

The role of the independent consultant in the management of coating programs with or without the turnkey approach has been discussed in more detail by Trimber and Hull.[53] Pinney et al.[52] point out that, despite the obvious advantages of the turnkey approach, the methodology merely puts the manufacturer in the same position that the contractor has historically found himself, and, in order to get the job, the manufacturer's prices will have to be cut because the lowest total bid price scenario will inevitably remain. With the turnkey approach must come an awareness that not only are initial costs important but also the strength of the warrantee, the record of the paint manufacturer in these types of jobs, and his qualifying procedures for inspectors and applicators.

Notwithstanding these possible disadvantages, most agree that the single-point responsibility approach in the maintenance contract can rid the owner of the unwanted responsibilities that he is ill-equipped to deal with in-house. The owner gets his specifications prepared properly as part of the entire package and does not have to rely on in-house personnel (who are often poorly qualified) to manage the job or hire expensive specialists who are usually underused.

The major disadvantage to traditional bridge painting methodology would seem to be that such contracts will be confined to large bridge paint manufacturers who can muster sufficient financial capability to back the long-term guarantees. It is possible that in the future some kind of insurance on the guarantee, such as is found in Germany but that is not now generally available in the United States, may become available. But the recent reluctance of the insurance industry to underwrite engineering risks is not encouraging in this regard. It does seem, however, that there may be some future trend toward this type of contract, particularly on large bridges.

Chapter 6

Job Execution

Experience would indicate that, of the many aspects of the multifaceted discipline of bridge painting, the process of job execution is perhaps the most critical to satisfactory system performance. It is during this process that all the exactitudes of corrosion science and paint technology and all the ravings of the experts are reduced to reality and, to some degree or another, compromise is the invariable rule. The extent of such compromise is the measure of the job. On some jobs, compromises with the best practice may necessarily be forced by such specific conditions as leaky decks, access difficulties, and local contamination with industrial pollutants, but, for the most part, compromise with best practice usually results from insufficiencies on the part of the contractor and/or the inspector.

The varied degrees of success and longevity of similar paint systems used in similar environments attest to either the wide diversity of expertise among those painting contractors who perform the work or an equal diversity in the amount and quality of control exercised during the cleaning and painting processes by the inspectors who oversee the operations.

In practice, both factors affect job quality (see Table 6-1). As is discussed in Chapter 7, the quality of inspection is all too often questionable, and, as a result, the attitude and skill of the applicator becomes of paramount importance and his selection critical.

CONTRACTING METHODOLOGIES

Bridge painting contracts are normally of the fixed-price format, IFB (invitation for bid) contracts, with specifically advertised requirements. While several alternative contracting methodologies are possible, firm fixed-price contracts are usually the rule. In special circumstances, time-and-materials and labor-hour contracts may be used, but these are often confined to extras.

Other forms of contracts, such as cost-plus-fixed-fee awards, seem rare in bridge painting, although they are used in other industries. Single-point responsibility contracts have also been used and two-step formal advertising methodologies, involving technical proposals that are evaluated and accepted prior to invitation to bid, would also seem possible in bridge painting, particularly when complex containment methods are required.

Table 6-1
The Effects of Inspection Quality and Contractor Expertise on Overall Job Quality

Inspection Quality	Contractor Expertise	
	Poor	Good
None	Poor	Fair
Fair	Fair	Good
Good	Good	Excellent

Firm Fixed-Price Contracts

Unfortunately, most large and even many small bridges are owned and maintained by public bodies, and the applicator selection process is almost universally fixed by federal, state, and local bidding laws wherein the contractor offering the lowest tendered proposal gets the award.

Such bidding laws are not, in view of the present norms of job control, conducive to optimizing contractor selection. The problem has been eloquently summarized by Roth (R. B. Roth, Oliver B. Cannon & Son, Inc., personal communication, 1984):

> Practice has been, and I do not see it changing, to have the responsible authority or entity advertise for bids, issue a bid package to applicants who have the minimal bonding capacity required to tender, take the low price, award a contract, and, after that, it is no longer his problem. Meanwhile, back at the bridge, battle lines are forming, coupled with the daily prayer meeting, and the job stumbles to some sort of a finish. Eight to twelve years later (if one is fortunate), depending on the environment and more importantly the federal (sic) funding, the same scenario is repeated.

Although undisputedly open and fair, the "lowest tendered proposal" approach to contractor award inevitably favors the sacrifice of quality for price. Although it has been well established that a "lowest cost per gallon" approach to the purchase of a coating system is illusionary, and that a cost per mil square foot per year of protective service is, in fact, the bottom line, the implementation of analogous reasoning to the selection of a contractor is bedeviled by present bidding laws. In attempting to minimize the public cost, in fact, all too often such a bidding procedure results in an exorbitantly increased cost as structures are repeatedly painted because of protective-coating failures that result from poor job execution.

Other Types of Contracts

On large structures under independent bridge authorities, some experimentation with single-point responsibility contracts is being tried, although prices may be discouraging. These are discussed in Chapter 5. Time-and-materials processes may be used for extra-work orders on fixed-price contracts, the inspectors monitoring labor. Up-front agreements concerning the nature of extra activities (additional mobilization time, etc.) must be made if problems over interpretation are to be avoided later.

In-House Painting Crews

Comparatively little painting appears to be done by means other than contracting out to painting contractors. Few states now employ their own painting crews extensively, except for touch-up work or for small structures with easy access. In most states, these activities represent less than 10% of all bridge painting, although some states, such as West Virginia, do use their own crews more extensively, and at least one state, Maine, utilizes state-employed crews almost exclusively.

Some states, such as Louisiana, Connecticut, and Ohio, employ practically no state crews because of the high cost of wages and equipment necessary to maintain in-house crews, while other states, such as Massachusetts, cite high insurance premiums, pensions, and so on as important disuasives for the state-employed crew. Most state crews appear to be used for touch-up maintenance of small existing structures and serve to prolong the service life of existing paint systems between major repainting, which is contracted out.

It is not unusual for authorities operating large bridges to employ a maintenance painting crew for similar duties, although, again, major painting is almost always contracted out. In areas where painting is seasonal in nature, the problem of what to do with the state painting crews over the winter has also been cited as an obstacle to the retention of this mode of maintenance.

Curiously, most states do not feel that they get any better quality with their own painting crews than they do with the contract painters. Contract painters are usually better paid than state painters, and these differentiations do not encourage the recruitment of the best operatives by the state. Additionally, many states that employ their own crews do not appear to provide even the modicum of inspection and control that is afforded the contracted job, and this is considered most definitely depressive to work standards.

The value of the postpainting touch-up maintenance for paint systems, particularly the high-performance paint systems, which are more prone to localized premature failure arising from misapplication and other problems in job execution, was recognized by the federal government in its recent decisions to fund maintenance operations if certain conditions are met. One of these criteria is that the state must perform touch-up maintenance of the high-performance system during the early years of service, thus helping to support the system over its most vulnerable service period. Another criterion is that all systems shall be high-performance coating systems. There is, unfortunately, little real definition of the term "high performance," except that projected service life should be fourteen to eighteen years rather than three to eight years.

Arbitrary Enforcement of Specifications

Advocates of open bidding cite the inherent fairness of the process as one of its principal justifications. Many contractors, however, cite inequalities and inconsistencies in the subsequent enforcement of specification details as a major deficiency in the fairness of this low-bid award process. The inference is that, regardless of the controls called for in the specification, the unscrupulous competition can "low ball" the job and rely on the lack or inadequacy of inspection and/or the bridge authority's reluctance to support difficult inspection decisions as a means to return a profit through deviations from the specified procedures. Such deviations may range from minor compromises in film thickness or mil profile to flagrant shortcuts that may include radical compromises in surface preparation and even the "stealing" of entire coats.

The responsible bidder's quandary is this: does he bid the job as written and run the risk of losing the job to less scrupulous firms, or does he bid the job on the assumption that controls will not be enforced, only to win to his own disadvantage if the controls are, in fact, bona fide?

Because of the highly inflationary effects of containment, the financial risk is even greater in some of the newer contracts that call for containment measures for environmental protection. It is not unknown for containment provisions to be included in such specifications "in case containment is needed to satiate a potential public outcry." The reason is that to ignore such a potentiality by not including high-cost containment requirements in the contract documents would lay the omission at the specification writer's door if problems materialize. To include such clauses, and enforce them only if required, puts the onus on the bidders, at least one of whom will probably take a chance on nonenforcement and "low ball" the job. If all bidders come in high, the authority still has the contingency to throw out the bids and rebid the job.

Contractors also complain that they must often assess not only the bid document but also the specifying authority and the history of similar contracts previously issued by the same authority in order to make an intelligent bid based on which contract provisions will be enforced and which will not. Bid decisions forced by such practices belie the fairness of the open-bid procedure and leave the well-established contractor (with much to lose) at an even greater disadvantage to the unscrupulous contractor to whom business losses may be less consequential. Despite the inevitable difficulties that arise between contractors, inspectors, and bridge authorities, most of the better regarded contractors readily confess their preference for well-defined jobs over jobs with questionable requirements and applaud the propagation of good specification writing and inspection practices in order to eliminate the above inequalities.

THE PAINTING CONTRACTOR

As in any industry, there are good and bad bridge painting contractors, although the divergence of skill levels in the particular field seems unusually wide. They range from large and medium-sized organizations that are well equipped and staffed by schooled and experienced superintendents, foremen, and operatives, all with good reputations and long records of successful work with many and varied coating systems to small two- and three-man operations that are, in many instances, in and out of business on a relatively frequent basis. The worst of the latter organizations practice their art under a variety of corporate guises with the minimum of equipment and less expertise, but plenty of craft and "hearts full of larceny." Furthermore, this divergence appears to be widening, although it is possible that the apparent widening reflects as much change in the paint systems as it does actual deterioration in the expertise of the smaller contractors. The tolerant lead/oil systems could accommodate most application insufficiencies much better than the high-performance systems with which the contractor must now work. Unfortunately, there may be too few reputable houses and too many fly-by-night concerns as well as too little opportunity for those responsible for bid awards to differentiate clearly one from another legally and, thereby, make judicious selections.

Although distinctions of competence do not necessarily coincide with size, it remains true that the larger successful bridge painting concerns are generally reputable, skilled, and familiar with the newer coating systems, while many of the smaller contractors may be neither as experienced nor as reliable for quality workmanship. Large concerns with bonding abilities in the \$5 to \$10 million range, and who employ several hundred painters on multiple jobs nationwide at any one time are generally (although not always) employed on larger bridge structures. This is probably because the bonding restrictions of the smaller firms preclude their access to the larger jobs. On the smaller jobs, the small firms are more competitive and, therefore, displace the larger contractors.

If most failures are indeed related to application error (including surface preparation deficiencies), then the discussion above should indicate that the smaller structures are more vulnerable. Unfortunately, meaningful correlations of this type are difficult to establish. There are many more smaller structures than larger ones, although coating failures on larger bridges are more publicized. Compounding the difficulties is the fact that the majority of smaller structures are still painted and repainted with lead systems, which are both more familiar to the small bridge painting contractor and more tolerant of those application deficiencies that would produce certain failure with high-performance systems. (Many of the larger contractors report that over one-half of their jobs are now based on high-performance systems compared to 10% to 25% of their work prior to 1975. The smaller firms, on the other hand, still show a higher percentage of jobs using conventional lead or oleoresinous systems.) Finally, as noted by Ray, Henton, and Rideout,[1] there remains a conspiracy of silence concerning coating failures, and the extraction of data is difficult and sometimes misleading.

Undoubtedly, there is also some divergency of competence among fabricators employed in new construction. If this is so, however, it is neither as blatant nor as widely acknowledged as it is in field work. Nor is skill the only origin of observed divergencies. Even the best contractor or fabricator is primarily concerned with the profit motive and will take shortcuts if allowed to secure his profit at some expense to quality.

The issues of reputation and pride have a place in any discussion of paint contractors. Although few, if any, painting contractors or fabricators can boast an altruistic regard for quality, the better houses would as soon do a job properly as badly. It is as much in the applicator's interest for a job to go well as it is in the interest of others responsible for the job, for on the success of past jobs rests an applicator's reputation. This is beginning to be recognized as more important as prequalification clauses sneak into specifications. Increasingly, clauses requiring the contractor to supply a list of jobs of similar size and complexity on which similar paint systems were used are appearing in bridge paint specifications. Whether or not or to what extent such experiences are reviewed in the bid-award process is quite another thing, but such requirements do foster some self-service in the generation of a record of good performance. Also, many jobs now contain clauses that bind a contractor to restore deficiencies that may occur in his work for one or two years after job acceptance and this is also persuasive.

In some circles, the more cynical might argue that the difference between the good contractor and the bad one is that the former knows how far he may safely push his compromises with the specification and where to make compromises without discovery, while the latter knows no better than to "rape" the entire job without prejudice—to take his money and run. Still, painter pride, while a source of some derision in these same circles, is by no means rare. Experience would indicate that, while perhaps in a minority, there remains a certain percentage of journeyman painters who retain a very real pride (often with justification) in the quality of their work.

The Contractor and the Paint System

Data from the contractors confirm increased use of the inorganic zinc and vinyl systems. Most of the larger firms seem quite comfortable with this type of system, although such specific concerns as the tendencies of certain primers to dry spray in field applications, especially during high-wind conditions, or the difficulties in surface preparation are still voiced. Generally, smaller contractors do not prefer zinc, citing application problems and too great a chance of error. Most contractors like alkyd systems and cite the systems' tolerance for application error. Similar surveys also indicate a definite increase in the use of the aluminized-epoxy mastic, which some contractors describe as more forgiving than many of the other high-performance systems.

The type of paint system has a limited and indirect impact on the cost of the job, in the opinion of most contractors; cost impact relates more to mandated surface preparation and application costs than to material costs. Although oil and alkyd systems are perhaps only one-half as costly to apply as inorganic zinc, surface preparation costs are much more significant than application.[2] Differences are, however, magnified by other factors. Most contractors usually report that such factors as the design and location of the structure and, in field maintenance, the type and condition of the existing paint system, all of which may have significant impact on job production rates, have the most impact on the cost of the job.

The condition of the steel itself in field maintenance is also critical to job costs. Old corroded steel that is heavily pitted will require much more time to clean than will steel that is newer and still bears coating or mill scale. Certain coating systems, such as inorganic zinc, will not penetrate and wet these pits easily, and residual contamination within the pit will tend to hold moisture and may lead to premature breakdown of any paint film (see Figure 6-1). A red lead/linseed oil primer, on the other hand, readily wets all surfaces, blasted or not, and is most forgiving (see Figure 6-2). Inorganic zinc films are relatively intolerant not only of poor surface preparation but also of film thickness

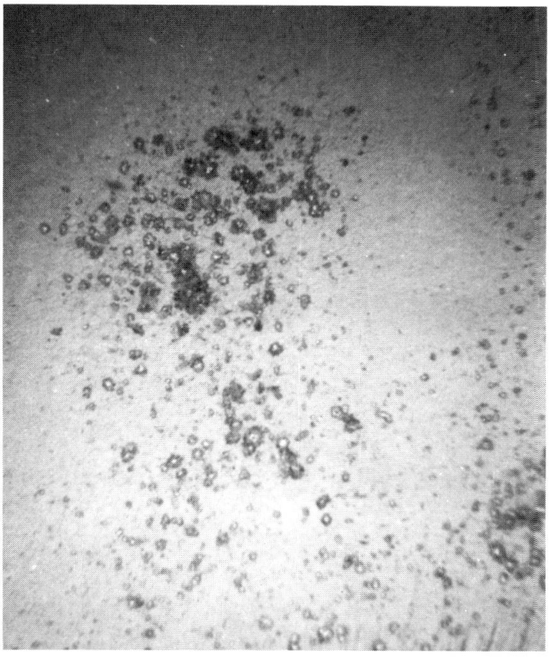

Fig. 6-1. Deliquescent Road Salts in the Pits of Inorganic Zinc-Painted Steel Holds Water and Results in Premature Zinc Breakdown

Fig. 6-2. Oil from Droplets of Red Lead/Linseed Oil Primer Are Sorbed into Surface of Old Chalking Film

variations, mudcracking in high film-thickness areas, and runs which require scraping, wire brushing, or even spot blasting (see Figure 6-3). The type of surface preparation may become very important to cost in some urban areas where the impact of environmental restrictions on blasting can be extremely costly.

Few large contractors consider the amount of inspection to be a significant factor in job costs, although smaller contractors from whom response was poor in our surveys, might not agree in this case.

The Contractor and Insufficiencies in Specification, Engineering, and Inspection

To generalize on the competence and motivation of painting contractors and fabricators as a whole is simplistic, ignorant, and often unjust. Yet, there remains much distrust among engineers, owners, inspectors, and even paint manufacturers of applicators in general, and this may have been spawned (certainly fostered) from individual experiences with less reputable organizations. To some extent this distrust may be cultivated by the larger contracting houses, which tend to exaggerate the work deficiencies of the smaller concerns.

Unfortunately, while all contractors must have one eye on the profits, there is a common misconception that the contractor is incompetent, a rogue, or a thief. In fact, on many jobs, the contractor may be better versed in what is going on than either the engineers, inspectors, or even the paint manufacturers and may serve to redeem insufficiencies in specification design and even inspection, if given the chance.

It is not at all unknown for a contractor with more experience in executing difficult paint jobs than the specification writer to recognize the inadvisability or inadequacy of a specification for the preparation or application of a coating system in a specific area. Given the lack of painting expertise of many specification writers and/or bridge engineers, this should not be surprising. The painting contractor, on the other hand, having worked in numerous locations and wrestled daily with rigging difficulties, surface preparation, and the application of coatings on all kinds of bridges with the gamut of design peculiarities, is indeed a specialist and might reasonably be expected to have much to offer in response to such inappropriate specifications, if given the chance. Similarly, in new construction, a fabricator may realize difficulties in some aspect of the specification (paint system or structural element) from past experience. Acknowledgment of the contractor's (or fabricator's) role in this regard and the opportunity for him to augment or adapt a specification seems to vary from job to job, from contractor to contractor, and from engineer to engineer. Whatever the response, in those cases where a contractor feels he is unable to endorse a specification to which he must comply, it is in his interest to bring the deficiencies to the attention of the engineer. While this should be done in writing for the public record, it is not always so recorded. Documentation of this type not only protects the contractor in the event of difficulties, but it seems more likely to exact some definitive response from the engineer. Larger contractors tend to

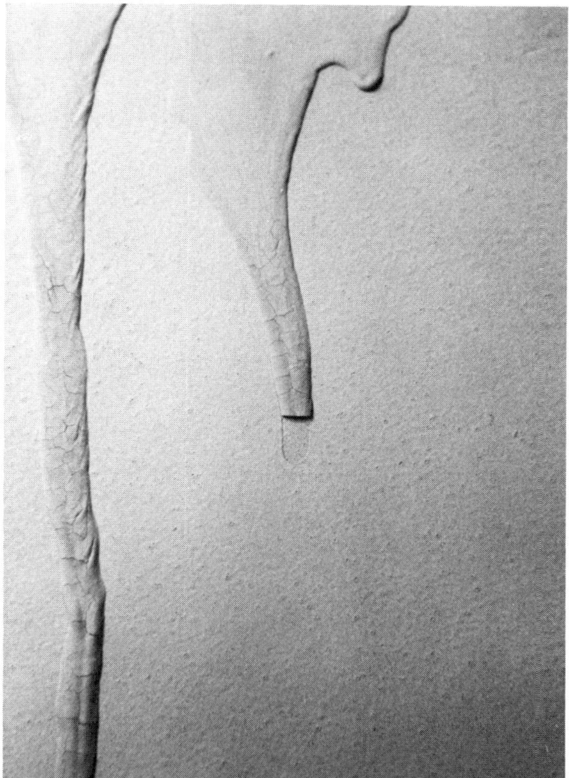

Fig. 6-3. Runs and Sags in a Newly Applied Inorganic Zinc Film Mudcrack and Show Poor Adhesion to Well-Blasted Steel

employ such procedures more often than smaller contractors, who remain silent, either because of lack of interest, the misconception that they will not be held responsible in the event of a problem, or because they themselves do not recognize the flaws.

The quality of inspection and lack of knowledge on the part of bridge painting inspectors are cited by many painting contractors as harmful to their cause. As Manta notes, "if I can find a paint inspector who knows what he is doing, it's dollars in my pocket. . . . "[3] The knowledgeable inspector may be an invaluable asset to the painting contractor in monitoring the work. Despite the inevitable problems that arise between contractor and inspector, a good inspector provides field guidance to operatives and serves as a sounding board during job problems. Not only will he recognize deficiencies at the time of their inception, thereby preventing expensive rework involving re-rigging but sometimes, if failures do not occur until after job completion, he may also save remobilization and travel costs. The presence of a good paint inspector will also normalize the competition and prevent the less salubrious contractors from getting away with a slipshod job.

On the other hand, the ill-trained inspector can not only cost the contractor money but can severely impede job progress. Lack of knowledge fosters reliance on the letter of the specification with all manner of silly and costly directives, including paint rejections, reapplications, and even system removal. Several interesting examples of this type have been noted by Manta,[3] by Kase and Hess,[4] and by Strauss.[5] As is discussed in Chapter 7, production hold-ups due to inadequacies in the quality control of materials and the interpretation of test data are also hurtful to the contractor's cause.

Contractor Licensing

While few formal attempts to technically qualify painting contractors have been implemented, in at least one state, California, licenses granted to contractors under the California State Licensing Board may be publicly revoked, suspended, or the contractor cited for a variety of causes. These last include the willful disregard of specifications and failure to complete the job in a workmanlike manner, failure to prosecute a job with diligence, abandonment of a project without legal excuse, and failure to comply with contract requirements, among some three dozen or so other causes. Revocations, suspensions, and citations are formally reported quarterly in a publication issued by the California State Licensing Board.

THE PREAPPLICATION PROCESS

Prequalification of Contractors

The qualification of bidders might to some extent address the inadequacies of the bidding laws, but, today, most if not all bidding requirements stress only such socioeconomic profiles of the contractor as ownership, bondability, credit rating, and capital and equipment. Only as a secondary measure do they take into account the expertise and past record of the companies making the tender. The bonding companies, which might provide some secondary control, unfortunately rely only on a contractor's financial resources as the primary criterion for bondability.

The technical qualification of potential bidders, while desirable, has received little attention in most paint specifications, and classification of contractors as good or bad has been described as "difficult, if not impossible."[1] Certainly the prequalification process would encourage both political pressures and legal disputes but, in view of the acknowledged drawbacks of the painting process, more evaluation of potential methodologies is probably justified.

The establishment of some sort of control would be welcomed by most reputable contractors and is probably easier to achieve than changing bidding procedures from low-bid awards to some kind of professional and technical evaluation of proposals. Brevoort and Thomarios,[6] for example, cite the lack of qualified applicator lists and procedures by many governmental and highway authorities as a major problem in today's practice.

> Properly established and maintained, such lists would make possible the bid and award of bridge painting contracts on a competitive basis, at fair prices, that permit quality work to be achieved and the intent of the specification to be met.

While the attention given to policies that give preference to disadvantaged, minority, and women-owned business enterprises is commendable, it seems unfortunate that some additional attention is not given to preference for those enterprises with a solid history of quality work, particularly in view of the recognized need in this area.

An obvious difficulty in the prequalification of paint-contracting firms on the basis of quality lies in the fact that only a limited number of operatives on any job are normally retained by the contractor on a full-time basis. Large contractors operating in all parts of the country transfer men across the country, but this is more expensive than hiring from union locals, and the option is not open to many firms with regional operations in those areas of the country where bridge painting is a seasonal practice. In these cases, the bridge painting contractors typically "let go" most of their crews over the winter months and hope to pick the best up again at the start of the new season. For many contractors, the majority of their painting crews are supplied by the local union. Only a few of those workers may be journeymen experienced in the use of the type of materials and procedures being employed on a specific job.

While a certain amount of training, either by direct instruction or by example, is provided to such operatives in the field by the contractor's superintendent, foreman, and crews, by paint manufacturer's field representatives, or even by the inspectors, and while much of the local union's less experienced help may be segregated to less critical tasks and support roles, the transience of manpower can only confound attempts at contractor qualification. It is, after all, the individual operatives and not the firm that are direct determinants of the success or failure of the job. Although the standards of quality acceptable to the firm are to a large extent controlled by the contractor through the job superintendent or foreman, actual performance may vary considerably from job to job depending on the strengths of the particular crews. In addition to the difficulty that this work-force transience leaves the responsible contractor trying to substantiate the viability of a continuous performance record, the variability may also be seized upon by less scrupulous concerns as an excuse for past failures.

Prequalification is, moreover, politically explosive and a headache not readily embraced by civil servants. Despite the inclusion of skills-qualification devices in specifications and bid documents, those responsible for the letting of painting contracts express serious doubts about the possibilities of low-bidder rejection on the basis of insufficient skills qualifications. Legal and public relations requirements will invariably preempt engineering considerations and may effectively emasculate such specifications prior to award or even bidding. It may be claimed, for example, that skills prequalification conflicts with such other requirements as equal opportunity clauses and affirmative action. While such requirements are not necessarily incompatible, if such a claim can be substantiated by the contractor, inevitably the skills requirement will be preempted.

The Bid Process and Award

After the job is advertised (in newspapers, such specialty services as the *Dodge Reports* and *Engineering News Record*, or even trade journals), interested bidders will purchase specifications and drawings, and, in the case of field work, are invited to examine the structure to be painted. On some jobs, such prebid inspections are mandatory and potential bidders not attending are disqualified from bidding. This may be a good idea, but it is not widely practiced. In most instances, attendance at a prebid conference that is arranged to outline the job to potential bidders, to highlight important aspects, and to interpret specifications and address the bidders' questions is not a prerequisite for bidding, although the prebid conference is valuable to the contractor and to the bridge authority alike. The prebid conference ensures that bidders are fully cognizant of all aspects of the specification and can tender a proposal that accurately addresses the job requirements at a price that is both competitive and sufficient. Nothing bedevils job success from the outset so much as the bid that is (intentionally or not) insufficient for the scope of work required. In addition, potential bidders may use the prebid conference to raise issues for resolution that may not have been fully appreciated at the specification-design stage. Response, often in the form of a specification addendum, inevitably improves the overall specification and contributes to the betterment of the job. The prebid conference is also used by contractors to discover and evaluate their competition, for the competition itself can markedly influence their own quotations.

Several weeks after the prebid conference, tendered bids are all opened publicly and the low bidder is announced, although the actual award may not be made until later. Bid openings may be attended not only by the bidders themselves but by paint manufacturers seeking to supply the materials for the job to the successful contractor. It is not unusual for paint manufacturers with an eye toward providing material-cost estimates to contractors to take out copies of specifications themselves, although, in soliciting a supplier, interested contractors or fabricators will usually supply manufacturers' copies of the technical specifications themselves.

Material Source Options

Paint may be produced directly by the bridge authority and provided to the contractor who is then solely responsible for the execution of the job or, more usually, the paint is supplied through the contractor, whose bid reflects the additional price of the material. There are advantages and drawbacks with either technique. If the painting contractor is responsible for the purchase of the material, he thereby becomes a customer of the paint manufacturer, in whose interest it is to satisfy the contractor if he is to get paid (a process that may be extremely protracted with some of the smaller contractors). After supplying one job, it is not unknown for such small contractors to hold a manufacturer's bill until they need more paint, whereupon they will pay for the first job to get paint for the next. This kind of practice will not work with all paint companies, of course, particularly the larger houses who will refuse the contractor's business. So long as open bidding on jobs remains the rule, the unscrupulous contractor will continue to take jobs by low bid, and there will always be a market for the smaller paint manufacturers who will serve such contractors and tolerate these tactics.

The more reputable paint manufacturers are less likely to adopt an adversarial position with a bridge authority or state department of transportation to protect a short-term business relationship with a contractor. If large amounts of paint are involved, however, or if there are established and lasting relationships between a contractor and a manufacturer, the position of even the large manufacturers may be less easy to predict.

Even with the larger contractors, there still remains a closer short-term business tie between paint manufacturer and contractor than between paint manufacturer and bridge authority, and, when disputes arise, this may encourage the paint manufacturer to serve his contractor before the authority. Together they may try to blind the authority with science, blaming the inspection or the specifications as inappropriate, impractical, or even, with new specifications, impossible, and seeking changes and deviations to the contractor's advantage. There is on record at least one instance in which a contractor grossly underbid a job and then made up his losses by actually getting (with the aid of lawyers and hired experts) the high-performance specification downgraded after job start-up. An experienced and independent inspection service may prove valuable to the authority in these types of situations, although the bridge authority must stand firm in the face of these practices (which does not always happen).

Other problems may arise in situations where the authority itself buys the paint. These generally revolve around job delays and shut-down problems occasioned by delivery and testing delays and application problems. Inevitably, a wary contractor will seek to capitalize on such difficulties by claiming extras for which the authority must seek reimbursement from the manufacturer.

In the event of short-term application problems or more severe long-term performance failure, the bridge authority may be caught between manufacturer and contractor, both of which will disavow responsibility. Again, the contribution of a qualified inspector independent from all parties can do much to interpret such problems for the authority and may even help to resolve the difficulties by either technical assistance or arbitration.

Job Preparation

The eventual award is published in the same journals as those in which the job was advertised. At the time of the award, the contractor will be formally notified of the earliest start-up date. By inference, from the specification, this will also establish the required date of delivery, after which the contractor may be assessed for liquidated damages. Overruns may be based on working days or calendar days, and, on large jobs of several million dollars, liquidated damages may amount to several hundred dollars per day.[1] Depending on job circumstances, these penalties may or may not be levied.

Samples of the paint materials intended for use on the job may then be requested for testing by the state or bridge authority prior to work commencement. Early delivery of such materials becomes essential if a paint manufacturer or a product is not qualified and performance specifications are employed. Performance testing may take several weeks and tags for the control of future batches must be set. Compositional specification materials will be sampled during manufacture at the paint plant or, preferably, on the job from actual stocks. Unfortunately, it is not unusual for jobs to begin without full approval of the product, on the understanding that removal of a product not meeting the specification will, if necessary, be carried out by the contractor, without charge to the bridge authority. The contractor may gamble on the reluctance of the inspector and engineer to advocate removal of an apparently good paint system for slight irregularities in material test data. After much interparty wrangling, the outcome is as liable to be nonremoval of the paint, with some understanding of additional obligation on the part of the contractor and/or manufacturer to redress consequent failures if they appear, as it is for the new system to be stripped. Additional surety bonds guaranteeing the disputed section for a set time have been used by contractors in some of these situations to avoid removal.

Mobilization and Preconstruction Conference

After notification of the award, the contractor will begin to mobilize and staff the job. A preconstruction conference will be held that affords all relevant parties an opportunity to evaluate each other and to discuss the project in detail. On small jobs, such meetings may be very brief informal affairs; on larger jobs, they can be most important for all parties and for the job in general. Present may be (with luck) the specification writer, the paint manufacturer, and the inspector, and certainly the engineer, the contractor and/or the fabricator, and the contractor's job superintendent.

Schedules, logistics, and equipment to be used are normally discussed as well as the interfacing of the project with other projects on the same structure or in the same shop. The various parties will voice their specific concerns and, while difficulties may or may not be resolved, they should be identified and where possible quantified. The contractor may require additional interpretation of details of the technical specifications or information concerning access, traffic patterns, lane closings, equipment placement, and the like, while the engineer may need details from the contractor concerning equipment, crew size, rigging procedures, containment, and so on. In new construction, specifics on delivery schedules, erection, and disruptions to road, rail, and water traffic may be discussed. In field jobs, a contractor may discuss concerns over specific practical aspects of the application with the engineer, specification writer, and/or manufacturer, who may in turn stress areas of concern over the technical provisions of the specification with the contractor and inspector. The contractor may also take this opportunity to go on record concerning inappropriate aspects of the specification and design difficulties (as perceived by him) that may affect the quality of the job.

Although certainly not all potential problems will be identified and dispatched in such preconstruction conferences, if properly conducted and attended by all parties, many of the difficulties and delays that can plague a job can be minimized in this forum.

Rigging

If necessary, the contractor will then bring experienced crews in to rig the job. The rigging of a structure is a vital prerequisite to any paint job and is a skilled practice in itself. Proper and adequate access to all bridge surfaces without overreaching or endangering operator safety is basic to the job process, and several excellent references are available that discuss in detail the many safety considerations in the design and use of scaffolding and rigging systems.[7–12]

Not surprisingly, the impact of Occupational Safety and Health Administration (OSHA) regulations on many aspects of the painting contractor's equipment and procedures have been considerable, and while operator safety has generally improved, productivity has inevitably suffered and overheads have increased. The impact of OSHA regulations on the painting contractor's operation are discussed in some detail by Fox.[13]

While such standard rigging features as swing scaffolds, single-point staging, spiders, and boatswain's chairs are used from job to job, bridge design features invariably require some adaptation. Many bridge painting contractors employ crews that specialize in rigging structures and some advertise their services to bridge authorities and engineering concerns that require rigging expertise for inspection and nonpainting maintenance (George Campbell, George Campbell Painting Co., personal communication, 1985). For many years these contractors have either designed their own equipment or modified that available on the market. The greater regulations noted by Fox[13] have encouraged more standardization, however, and, more recently, scaffolding companies have started to market a greater variety of equipment, including many of the adaptations first conceived by the contractors. All manner of apparatus is now available to all contractors.

Devices such as modular suspended staging decks that are supported from the flanges of longitudinal I beams, by bypassing cross beams, can travel the length of longitudinal beams without disassembly or leaving the working deck.[14] Cable mounted sections of a similar type have been used by Caltrans with considerable savings in production time.[15] Permanent cable connections securely anchored into concrete abutments or by means of beam anchors serve, after initial installation, for repeated maintenance operations that use cables that are slung and tightened using turnbuckle assemblies.[14] Such features as these will often compensate for the lack of rigging facilities provided by the bridge-design engineer at the time of bridge erection.

Scissor lifts (see Figure 6-4), hydraulic boom lifts (see Figure 6-5), and "snooper"-type devices have also extended easy access. Snooper devices, such as those manufactured by Paxton Mitchell, Omaha, Nebraska,[16] can put a man forty feet in under a bridge deck from the roadway itself (see Figure 6-6). Scissor lifts, controlled by an operator from the working platform, can raise a man and 400 pounds of equipment forty feet in the air. Hydraulic booms, which are similarly controlled, can reach over twice that height, although with reduced loads.

Smaller structures with lower clearance (overpasses, etc.) are often painted from the ground by pipe staging rigs (see Figure 6-7). These versatile assemblies are easily modified to any height and shape and may be fitted with wood or metal decking,

Fig. 6-4. Scissor Lift

Fig. 6-5. Hydraulic Boom Lift

complete with adjustable outriggers if required. Castors (which lock during the painting operation) add portability. In some cases, bedladders and pipe-staging assemblies may be mounted on vehicles (see Figure 6-8). Pipe staging may also be used on bridge legs, bent supports, and columns and can serve as tarpaulin frames for containment methodology. High supports, legs, and suspension towers are more often than not painted with boatswain's chairs or spiders (see Figure 6-9) or power-hoist staging. Ladders (used for access) are rarely used on bridges as painting platforms.

A more detailed review of rigging technology is beyond the scope of this volume. Additional valuable data may be gleaned from the literature of those scaffolding companies specializing in systems designed for bridge work.

New Construction

In new construction, job start-up procedures are essentially similar to those in field construction, although there may be more subcontractors involved, including painting contractors for priming and field application, and haulers, riggers, and so on. There may also be several inspection agencies involved in a particular project. Scheduling becomes more important and the fabricator must coordinate the receipt and fabrication of the steel (which invariably will precede priming) with painting and shipments to the field, erection schedules, and field painting. Such aspects as the storage, handling, and protection of finished pieces in the shop, in transit, and in the field can be critical operations to both job quality and profits.[17]

Consideration of painting requirements, including coating types, amounts, and painting schedules, are extracted from the overall job specifications, and details and drawings are sent to the paint shop that will interface with purchasing to secure bids from suitable paint manufacturers. On large jobs, the fabricator may contract with the manufacturer at a fixed price over several years, and a schedule of deliveries will be set up. Many larger fabricators now employ a paint specialist on their staff or as a consultant to advise in matters pertaining to the specification and use of coatings as well as surface preparation and related matters.[18] Whenever difficulties in execution or the appropriateness of specifications exist, the better fabricators, like the better contractors, utilize these specialists to work out procedures whereby such difficulties may be rectified and to negotiate extras or financial compensation adjustments rather than addressing the difficulties by recommending practices not conducive to good job quality.

Start-up Procedures

At the inception of any job, there is inevitably a period during which the job team (engineer and specification writer, contractor, fabricator and crew, inspector and manufacturer) becomes acquainted with both the paint system and the specification, the bridge or piece design, and with each other. With the more common paint systems in shop work or on relatively simple structures, the process may be rapid and the learning

Fig. 6-6. "Snooper" Articulated Inspection Apparatus. (Courtesy Paxton Mitchell Co., Omaha, Nebraska)

Fig. 6-7. Pipe Staging Rigging for Painting Undersides of Overpasses

Fig. 6-8. Vehicle-Mounted Bedladders

Fig. 6-9. Spider-Type Rigging

curve short. On more complex structures, particularly in the field, with new paint systems, or with an inexperienced crew, the education may be more protracted and involve several false starts. Rigging and scaffolding methods are tried and revised, and containment methodologies are debugged, and traffic patterns are verified.

It is during this period that the presence of the paint manufacturer's experienced technical representative (not a salesman) is invaluable. He must demonstrate and train the applicator in the use of his system and educate the inspector to the nuances of good and bad application. The quality of this type of service may be good or bad depending on the individual manufacturer and the contractor receiving assistance. In general, it seems that technical service of this type from the larger manufacturers is of a higher caliber than similar service (if available at all) from smaller manufacturers. In other service respects, such as delivery and response, the smaller manufacturer often seems better. Recent constrictions on profits of many of the larger manufacturers may result in some future cutbacks in the extent of this type of extensive field support.

Interpretations of surface quality and mil profile are made during this stage, and the applicators gain the feel of the coating, determine the equipment necessary (pumps, hoses, tip sizes, etc.) and the type and levels of thinner required for satisfactory application. They develop a familiarity with application procedures required to produce reasonably constant ranges of surface quality and film thickness on all features of the structural design (webs, edges, riveted and bolted sections, etc.). Interpretations of the specification are assessed for both compliance and practicality and, where necessary, adjustments in techniques or even in the letter of the specification may be made (with the input of the paint manufacturer, specification writer, engineer, and painting contractor) to resolve difficulties most effectively and secure the best job. At this stage of the job, some degree of flexibility is important and closed minds on the part of the engineer and specification writer can be counterproductive. Application of all elements (coats) of the complete system are made (with particular regard to practical recoating schedules) and results are assessed for quality, film thickness, and adhesion.

In the first weeks of the job, the rhythm of the job will be established, and, during this time, the input of all those responsible for the job is essential. All elements of the design should be evaluated, but especially critical areas, such as heavily bolted connections, recessed and back-to-back angles, boxes, seams, and the like should be evaluated for methodology. Striping techniques by brush, by conventional spray, or by other application methods should be determined, and schedules should be practiced so that the technical requirements of the job may be accomplished with a minimum of rigging and the least delay. Specific problems, such as the possible use of mist coats and zot coats are evaluated at this time.

As the job progresses, a rhythm of practice normally develops, production rates increase, and the overall quality improves. Rarely, however, will a job proceed without periodic difficulties that must be resolved in order to proceed further. These may be quite varied—some minor, some serious—but typically involve curing difficulties and recoating schedules with inorganic zincs, the resolution of cleaning and painting procedures in areas of virtually impossible access, local severe contamination of drying films with road soils, or shop-blasted steel contaminated with shop dirt before zinc priming.

Containment methodologies in which large areas of steel are contained with tarps or solid enclosures and blasting and priming proceeds over several days may result in contamination of newly primed sections with blast dust, which must be washed down or vacuumed before finishing. Normal blowing-down procedures in these cases are less effective and often more time consuming than power washing the dirt away. Washing also helps ensure the cure of alkyl-silicate-based zinc primers.

Other problems include (1) areas of excessive pitting and metal loss, (2) weather-related recoating delays such as the subsequent salt contamination of newly applied zinc rich primers from leaking decks or from deicing practices in cases of snow and ice storms, (3) the application of incorrect materials and their removal, and (4) the peeling of newly applied paint systems from old interfaces. The contamination of newly primed surfaces (particularly porous zinc films) with bird droppings, soils, and debris can also cause severe problems as can oils, antifreeze, battery acids, and soot from breakdowns, fires, and traffic accidents. All can occur as part of the daily operations of bridge painting, and all must be resolved in order for the job to progress. Questions invariably arise as to who pays for such rectification and the exact procedures necessary. The most effective forum for such discussions and resolutions is in the periodic job meeting.

Job Meetings

On large jobs in particular, the regular job meeting can be invaluable. Such meetings allow the engineer, contractor, inspector, paint manufacturer, and consultants to keep track of job status, schedules, and procedures. These parties may interpret specifications, report on the appropriateness of existing procedures, discuss problems, and resolve most difficulties through this forum. Rarely, if ever, will a specification accurately anticipate all conditions that will be met in practice, and often such conditions cannot be anticipated at the preconstruction conference. Often, in order to maintain quality and production, some interpretation and modification of the specification will be necessary. Rather than dealing with these matters informally on the job, they are best discussed in open meetings wherein all parties may represent their own position and the consensus can be noted. Here, the discussion may be recorded, procedures clearly delineated, and, where necessary, formal requests for extra-work orders made.

In the event of serious difficulties after the fact, including job claims, coating system failures, and litigation, the minutes of such meetings may be an invaluable aid in resolution, discovery, and/or settlement. The minutes may also be of great value as part of the technical job record in the postpainting monitoring process.

THE MODERN APPLICATION PROCESS

Like paint systems, the painting process (including surface preparation) has over the last several decades evolved from what was very much an art to what is, if not a science, a relatively complex technology, requiring both practice and understanding. For the most part, the spray gun has replaced the brush and roller and the centrifugal blaster and blast nozzle have replaced the scaling hammer and wire brush. In itself, however, such a summary is simplistic, for both brushes and scaling hammers are still used as auxiliary equipment today and, on some jobs, hand- and power-tool cleaning with brush and roller application remain. Today, their employment is more selective and judicious, however, and responds not only to the economically induced preferences of the contractor but to the real engineering dictates of a quality job, factors that are not necessarily opposed to each other.

The modern applicator, therefore, has not so much abandoned the old practices as enlarged upon them. He must not only be familiar with an ever-increasing diversity of coating types but also with the many specific types of equipment that their employment demands and the fine points of the application techniques that these systems require for optimum results.

One does not apply an oil paint by brush with the same technique that one would use for a vinyl touch-up, nor would one apply wash primer with an airless gun in the same manner as one would apply an epoxy finish. The application of high-build vinyls or coal tar epoxies at acceptably uniform film thicknesses in touch-up or in restricted or intricate areas is all but impossible without care and experience. Primers, tiecoats, and high-build mastics all make their individual demands on any applicator, as do individual systems within the same type of generic product makeup. One zinc rich primer with a fast solvent system, for example, will require utmost gun control if dry spray is to be avoided; a second may require a continual balance of speed, gun motion, and distance from the work if an overly wet film is not to sag. Adjustment of tip sizes, pump pressures, and selection of types and quantities of thinners are essential in today's application if satisfactory film-thickness control is to be maintained. As the realization that the parameters of acceptable film thickness require maximum as well as minimum values, so the demands on the applicator's knowledge and skill increase further. Design and steel configurations will also influence the method of application; large, flat, continuous areas require the long, continuous sweeps possible from airless equipment, while edges, bolts, and intricate sections can only be coated successfully with short, restrained gun motion and careful trigger control, both of which are more easily accomplished with air spray than airless techniques (although, even in these cases, airless equipment is more usually used by skilled operators).

Climatic conditions and work locations, especially in field painting, will have a significant effect on application technique, even to the point that they may preclude application. Effects of this type are not necessarily limited to temperature or dew-point restrictions, but include problems of wind speed and direction, which can exacerbate dry spray and may carry wet overspray for a mile or more downwind from a structure, contaminating property and spawning insurance claims.

Along with the selection and use of new equipment has come the necessity for familiarity with support equipment, not only the obvious compressors, heaters, agitators, and traps, but also metering devices and controls, cleaning equipment, scrubbers, and the like. In the shop application of specialty features (fasteners, gratings, railing sections, lighting fixtures, etc.), electrostatic applications, powder coatings, and even newer methodologies may be employed. Thermal-spray-applied zinc and aluminum systems, now enjoying increased use on bridges, utilize their own equipment—oxy-fuel and electric-arc (wire-fed) guns, plasma (powder-fed) guns, and power sources, which call for even more exacting application experience than does conventional equipment.

Abrasive blasting has widely replaced hand- and power-tool cleaning not only in Zone 2B and Zone 3 areas where high-performance systems are employed but more universally, even in spot maintenance where lead systems are still employed. Rideout, Ray, and Henton, in their 1982 survey,[19] report that almost all states have now adopted blast preparation (usually SSPC-SP-6 or -7) and only in the rural desert Southwest are any major holdouts found. This is also confirmed by Leyland.[20] Engineering considerations related to abrasive blasting are reviewed extensively in Chapter 3. Discussions of equipment and techniques are found in Chapters 2.0 through 2.9 of Volume 1 of the *SSPC Manual* and in many other sources.[21–25] Further discussion here would be superfluous.

While a complete treatment of the use, advantages, and disadvantages of the various application methodologies used in bridge painting is itself beyond the scope of this volume, a review of the engineering considerations of the basic application methods is offered below. Both spreading and atomizing techniques are utilized in several forms on bridges. They are discussed below and compared in Table 6-2.

Spreading Techniques

Spreading techniques employed in bridge painting are brush, roller, and, occasionally, mitt application. All are somewhat limited by the holding capacity of the tool and by the fact that there is no gain in volume solids in moving paint from source to steel. This severely limits film-build opportunities compared to atomization techniques and increases a tendency for films that are applied by spreading technique to sag compared to the same systems that are spray applied at the same wet-film thicknesses. These considerations make it necessary to apply coatings at the highest possible volume solids, and the elimination or minimization of thinning on the job is vital.

Occasionally in cold weather, viscosity elevation of the paint may require some compensatory thinning in order to facilitate application and control film thickness within the required parameters of the specification (while preventing such film defects as sagging). In hot weather, the application of fast-drying systems by spreading techniques may be especially difficult because of excessive solvent evaporation and drag. Alternate solvent systems for application in summer as opposed to early spring and late fall may be necessary for some systems.

Roller Application

From an engineering standpoint, the application of structural-steel coatings by roller is possibly the least justifiable technique. It does not provide the controls that are possible with other spreading techniques, and the high-build capabilities of the atomizing techniques cannot be achieved with a roller. Neither does roller application give the improved surface-working opportunities that are inherent in brush application. In most cases, roller application gives film thicknesses lower than does brush application, although the use of heavier naps will help.

On the other hand, film uniformity from a roller application is reasonably good

Table 6-2
Relative Cost and Engineering Characteristics of Common Application Methodologies

Application Technique	Cost per Square Foot (Production Rate)	Equipment Cost	Control	Film Quality and Uniformity	Overspray and Waste	High-Build Capability	Suitability for Use with Fast-Drying Lacquer	Suitability for Use with Slow-Drying High Build
Brush	High (Poor)	Low	Excellent	Fair	Low	Fair	Poor	Poor–fair
Roller	Medium (Fair)	Low	Fair	Fair	Moderate	Poor	Poor–fair	Poor
Air spray	Medium (Fair)	Medium	Very good	Very good	Very high	Good	Fair–good*	Good
Hot air spray	Medium (Fair)	High	Good	Excellent	Very high	Very good	Fair*	Very good
Airless spray	Low (Excellent)	High	Good	Good	High	Very good	Excellent	Very good
Hot airless	Low (Excellent)	High	Fair	Very good	High	Excellent	Good*	Excellent
Air-assisted airless	Low (Excellent)	High	Very good	Very good	High	Excellent	Excellent	Excellent

*May cobweb.

with experienced applicators, and, with the exception of highly thixatropic coatings, leveling is usually good. Coal tar epoxies, which are not easily applied over large areas by other spreading techniques, have been applied very successfully by roller.[26] Most coatings, in fact, can be applied by roller, although most can benefit from alternative application methodologies. Some coatings tend to bubble (from air entrapment) during application, and this can lead to insufficiencies in film continuity that can create serious difficulties with the thinner films.

The major advantages of roller application lie in the excellent production rates possible with low-cost equipment, particularly on large, uninterrupted areas if very wide rollers are employed. This results in relatively low-cost figures per square foot for application. The use of "man-helpers" or roller extensions as a practice to avoid rigging is frowned on by the industry because this encourages overreaching, poor appearance, and still further reduced film thickness. Nonetheless, the technique is still employed.

Roller application is particularly suited to the coating of chain link fencing and some open gratings on bridges. These are painted with wide-nap rollers from both sides more efficiently than is possible with other techniques. In general, however, roller application is not well suited to complex shapes; edges, corners, and intricate areas are normally cut in with brushes. Specialty rollers that are utilized in segmented sections mounted on a sprung spindle have been developed for pipe coating but do not appear to have been widely employed on bridges.

Equipment requirements for roller applications are low, and, despite several modifications involving variations in automatic material supply, the tray or roller-bucket feed are still most widely used. Roller naps will vary from short (less than 1/4 inch) to very long (greater than 1 inch), the longer naps being used over irregular surfaces, the shorter naps over smooth surfaces. Equipment and application techniques are well described elsewhere.[8,27–30]

Brush Application

As one of the slowest and, consequently, one of the most costly application procedures, brush application finds its clearest advantage in the amount of control that the applicator has over the work. It is particularly suited for access into small boxed-in areas and for the painting of intricate design features. It ensures reasonable coverage within the tolerances of the amount of material that can be carried by the brush and the volume solids of the material. For the striping of seams, edges, rivets, bolts, and recessed angles prior to a more general spray application, brush application is particularly valuable. This not only ensures coverage of these difficult areas but reduces the possibility of excessive film build from multiple-angled passes over and around these features. Brush application of many zinc rich coatings can be difficult, not only because of the inappropriateness of the technique to these fast-drying coatings but because of the difficulties in maintaining sufficiently well-agitated pots for good homogeneous application.

Brushing is least suited for application to large areas or for the application of fast-drying systems and low-build lacquers. The one possible exception is the application over large areas of poorly prepared or heavily pitted steel, where the ability of the brush to reach into the pits and effectively wet out the steel is most desirable. The working of the brush into the surface of the steel also assists wetting and adhesion by displacing and blending any contamination into the paint instead of merely laying the paint over it as would be the case with spray application. Even with many vinyl specifications, brush application is still required for the primer coats for optimum wetting of newly blasted steel or for the rectification of excessive pinholing that is often encountered in spray application.

Uniformity of application by brush is not generally good, and, with faster drying lacquers, excessively pigmented systems, or systems with poor flow, brush marks can lead to thin spots, streaks, and holidays. Systems designed for brush application are often formulated specifically for that purpose with slow solvents and carefully adjusted rheologies. Brush techniques and equipment are reviewed in several sources.[27–30]

Mitt Application

The paint mitt,[29] a reversed lamb-skin glove with an impermeable inner lining that is dipped into the paint and then daubed across areas of the steel, is used for painting certain odd-shaped sections and for application of high-build systems below water. Not unlike brushing, the procedure provides excellent surface wetting and is used for the application of high-solid systems of extremely high viscosity that are not well applied by other techniques. Control is excellent, and with these types of materials high-build capabilities are possible by mitt application. The technique is rarely used other than as an auxiliary technique except for underwater applications. Application rates are usually faster than by brush, but rates will depend upon the material used and the steel configuration to be coated.

Atomization Techniques

Air-spray (conventional and hot spray), airless and hot airless, and, more recently, air-assisted airless-spray application techniques are all employed in the painting of structural steel for bridges. Electrostatic applications and powder-coat applications by means of electrostatic powder-coating techniques or fluidized-bed procedures may be employed on small items in the shop; these are not discussed here. Interested readers may find details on these application methods elsewhere.[30,31,32] Similarly, thermal applications of zinc and aluminum coatings, while enjoying increased use on bridges, are not treated in detail in this volume. The interested reader can find more detailed considerations of these less commonplace application methods elsewhere.[33,34,35] The more common spraying techniques are, however, discussed from an engineering standpoint in some detail below.

Conventional Air-Spray Application

In conventional air-spray application, a stream of paint is fed to the gun under air pressure and then atomized by a second stream of air (see Figure 6-10). Various types of gun and pot configurations exist; bleeder/nonbleeder, internal mix/external mix, and pressure-fed, siphon-fed, and gravity-fed guns. Details of the equipment may be found elsewhere.[31,36–39] In maintenance applications, pressure-fed, nonbleeder, external-mix units are most widely used, with several guns operated from a single pressure pot.

Although less common than airless application, conventional spray is still employed on bridge structures where greater control than that offered by airless application is required. The method delivers much lower volume rates of paint than does airless application and requires coatings of lower viscosity, which usually translates to reduced solids. This allows better control of film thickness and provides greater versatility in the coating of complex geometries. This can be important where high-film builds must be avoided, such as in the coating of heavily riveted areas, rails, and tight angles. In addition, the air gun itself is well designed for control (see Figure 6-11). Not only are there a wide range of air caps, fluid caps, and needles, but fluid-needle and spreader-

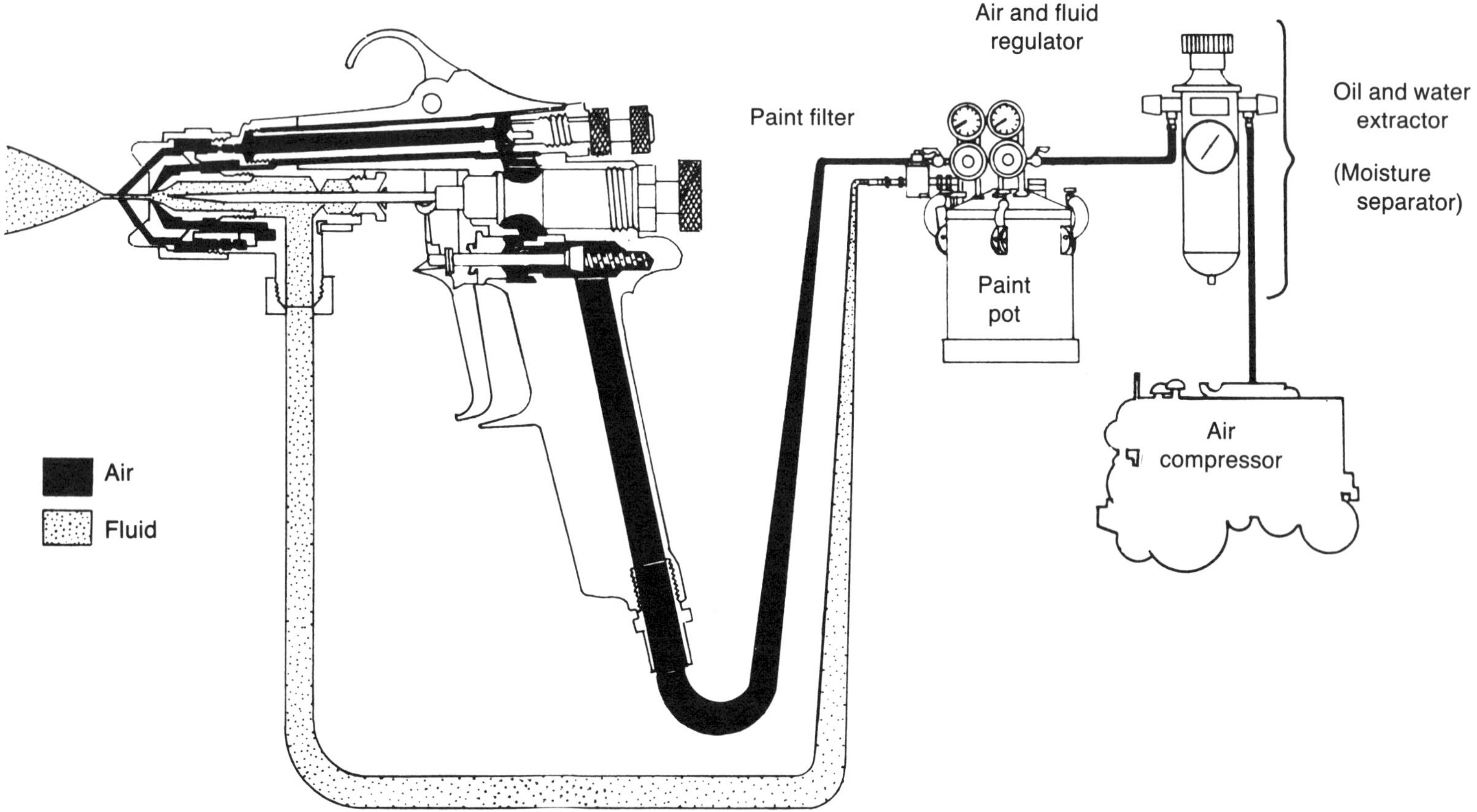

Fig. 6-10. Air-Spray Principle. Conventional air-spray setup. Notice that the fluid enters only the front part of the spray gun. (Not drawn to scale.) (Courtesy Graco, Inc., Minneapolis, Minnesota)

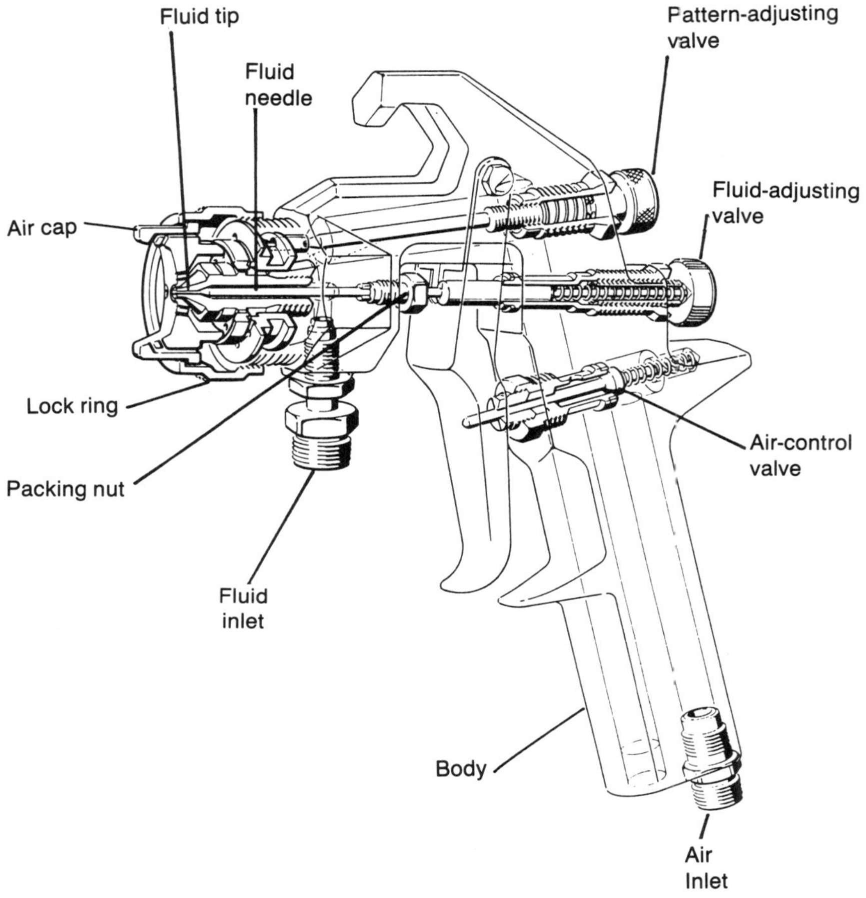

Fig. 6-11. Air-Spray Gun. (Courtesy Graco, Inc., Minneapolis, Minnesota)

adjustment valves provide control of the actual amounts of paint passing through the fluid tip and the air to the wings or horns of the air cap control the spray pattern. Adjustments of both gun (air) and pot (fluid) pressure will also, of course, allow still further control.

While growth in the utilization of air spray is not encouraged by the trend toward application of high-solids/high-viscosity materials, the increased need for the control of these materials in specific applications has been adequately satisfied by air-spray support of more general airless application.

For the contractor, air spray is costly compared to airless application. It results in reduced production rates and greater material losses (20% to 40%). The process is dirty, with much more overspray than in airless application, and this can result in higher insurance claims for property damage, greater fire hazards, and more environmental and health hazards. Risks of overspray contamination of completed work are significant, and additional masking may be required. In poorly skilled hands, dry spray can be problematic, particularly with systems rich in low-boiling solvents. The loss of solvent by evaporation between gun and work can lower material temperatures by as much as −12 to −11°C (10–12°F) below ambient. This may cause moisture contamination of films when temperatures are only slightly above the dew point. Higher solvent contents both increase the shrinkage of applied films on drying and make solvent attack on an existing film more likely. Air spray may also be precluded by high-wind conditions because of dry spray, particularly in exposed areas or in high-temperature conditions. Air spray techniques are described in the Louisiana Department of Highways's training unit, "Structural Painting,"[9] and elsewhere.[29,30,36,38–41]

Hot Air-Spray Application

Hot air-spray application utilizes temperature (as well as solvent) to reduce paint to application viscosities. Necessary air pressures are reduced, which eliminates some of the overspray difficulties of cold conventional spray, although, unless solvent systems are carefully designed, dry spray problems may be compounded. Higher viscosity systems may be employed in hot air-spray application and consequently the technique is more suited than cold air spray to the modern high-solids coatings. Despite higher equipment costs, production rates are faster than conventional air spray, with fewer passes required for a given film build. High production values result in somewhat lower control. Higher application temperatures also eliminate the dangers of the depression of material temperatures below the dew point during application, as well as aiding in the wetting of, and adhesion to, steel substrates. Increased temperature can also assist flow and leveling and reduce such film defects as pinholing and cratering. Lower solvent content also reduces shrinkage and the risks of attack on existing older films by solvents. On the other hand, wind effects in exposed areas may cause more severe problems from dry spray with many coatings.

Despite the better production rates, equipment is more expensive and jacketed lines, being more bulky for operators, may compromise increased production rates somewhat. More important, many advantages realized by hot air spray are more easily and less expensively achieved by airless application. This method is discussed in detail elsewhere.[8,30,31]

Airless Spray Application

Airless spray is by far the most common of all application techniques employed in shop and field applications for bridge steel, providing production rates four and more times that of conventional spray. Atomized by forcing a stream of paint under hydraulic pressures of up to 4000 pounds per square inch and more through very small orifices (see Figure 6-12), there is no air involved in conventional airless spray and consequently the process is, despite the high volumes of material delivered, both cleaner and more economical than air spray. As air pressure is only used to drive the airless pump, which magnifies pressures by many orders, smaller compressors may be used in airless application, although this advantage (often touted) is not of much consequence to the field applicator who already requires large compressors for blasting operations. While small electric, direct-drive units are available, the larger pumps powered by air compressors are normally employed in bridge painting, particularly where high-viscosity systems are applied. A single material hose is less cumbersome for the operator, who can apply large volumes of paint over large work areas rapidly and with little material waste. Dry spray and overspray are both reduced compared to conventional air-spray techniques, and there is less bounce-back from corners and the like, which reduces masking requirements.

Most important, higher pressures at the gun provide easier atomization of the higher viscosity materials, and this is highly conducive to the higher solids systems favored in modern practice. Pump ratios (ratios by which the airless pump multiplies the

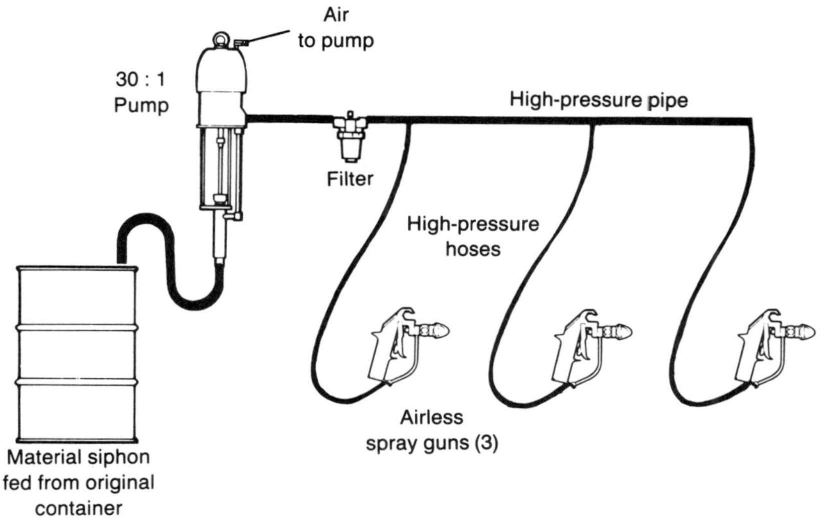

Fig. 6-12. Airless Spray Principle. Cold airless spray systems paint at ambient temperatures. The system consists of a pump, filter, hose, airless spray gun, and tip. It is recommended for use with coating materials that atomize readily at room temperature, such as architectural and protective coatings. The pump transfers the material from a container and propels it under high fluid pressure through the filter and hose to the gun and spray tip, where it is atomized and directed to the workpiece. Pressures of cold airless systems are from 1500 psi. to 4000 psi. (Courtesy Graco, Inc., Minneapolis, Minnesota)

driving pressures from the air compressor) of up to 45 : 1 are now possible. As airless pump sizes increase, the application of increasingly viscous systems of progressively higher solids becomes easier, and equipment trends in this direction will probably continue. Unlike air atomization, there is less solvent loss between gun and work. This allows a more economical use of available solvent in flow and leveling, and with careful control of the thixatrope/solvent system, coatings of high aesthetic quality in very high film builds are possible from airless techniques. Reduced solvent loss by evaporation between gun and work also reduces material temperature depression and virtually eliminates blushing and the moisture contamination problems that can accompany conventional spray techniques.

The major drawbacks of airless spray application lie in the loss of control that is inevitable in the delivery of high volumes of paint. Control can be particularly sensitive where paint systems that have critical maximum-build requirements (zincs, moisture-curing urethanes, etc.) must be applied to such areas as rivets or angles, requiring multiple passes from different directions. These areas are probably best striped using other techniques before applications of the principal coats by airless spray. Airless guns themselves (see Figure 6-13), moreover, have fewer control adjustments than do conventional guns and, while a full range of tips are available producing versatility in both spray patterns and material delivery, material flooding and tailing limit the range of tips that may be used with any specific material. There is also a tendency for operatives to overuse tips, which can rapidly become worn with some coatings (zinc primers). Fine tips also encourage feed problems, such as clogging (despite filters and screens), and,

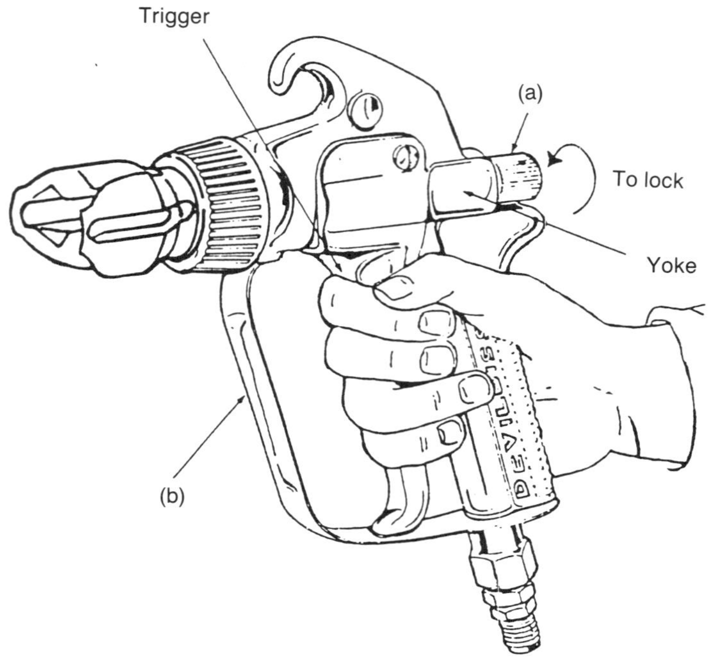

Fig. 6-13. Airless Spray Gun: *a,* safety lock protects against accidental spraying; *b,* guard protects operator and trigger. (Courtesy The DeVilbiss Company, Toledo, Ohio)

although this has been reduced by the use of reversible tips, blockages can still severely impede job continuity. Control problems can also be reduced to some extent by solvent dilution. Other handy innovations in airless technology include adjustable tips that eliminate the necessity for tip adjustment by continual tip changes.

Despite the elimination of compressed air from the spray fan, air entrainment in airless-applied films can still be problematic, particularly in such high-build lacquers as chlorinated rubbers and vinyls. Pinholing is not an uncommon headache with these applications, and the presence of air pockets or vacuoles in the dried film is a subject that has been discussed elsewhere.[42] Airless equipment and techniques are also discussed in greater detail elsewhere.[9,29,30,31,43,44,45]

Hot Airless Application

Far less common in practice than cold airless application are hot airless techniques,[8,30] which further reduce the energy required for atomization of a given coating. This results in advantages similar to those noted in the progression from cold to hot air spray: increased film build, faster production (as much as 20% to 25% over airless application), and faster drying. Thinning is also less necessary because viscosity control is achieved with temperature, and this allows application of yet higher solids systems. Alternatively, lower pressures (below 1000 psi.) are required for atomization, and this may be employed to apply lower solids systems with more control. As in hot spray, wetting and adhesion are better (except on very cold steel), and aesthetics are usually improved because of improved flow, which also reduces air entrainment and pinholing. Unlike conventional hot spray, little solvent is lost between gun and work, although more is lost compared to cold airless application.

Lines for hot airless equipment are more cumbersome than for cold, and control is generally poorer because of the delivery of higher volumes of solid coating. Application in intricate areas or with coatings that are not suited to application at high film thickness are not advocated. The process is not as well correlated with increased solids thermosetting systems as it is with thermoplastics because great care is required with the use of two-component systems of high solids in order to avoid equipment damage through temperature-induced premature solidification of the product. To resolve these difficulties, multicomponent guns with metered feeds have been introduced, although their use in bridge painting is not at this time common. Although dry spray is less of a problem with airless than with air spray techniques, hot airless techniques will tend to be more problematic than cold airless unless careful attention is paid to the solvent system of the material used.

Air-Assisted Airless Application

In air-assisted airless application, special air-fed "airless" guns that direct a compressed air flow into or around the thin film region of the airless spray fan, just outside the airless orifice, are used (see Figure 6-14). This improves the efficiency of the airless atomization and eliminates the tails that plague airless application at low fluid pressures. The net effect is to reduce the fluid pressures necessary for atomization from 3000–4000 psi. to 700–1000 psi.

The process allows much greater versatility in control, and it is claimed that production rates may be adjusted (by control of fluid pressure and tip size) to meet those normally achieved with either air or airless application.[46] It is also claimed that the process provides better transfer efficiency than airless (less overspray and fog) and produces film quality that is almost as good as that produced by air spray. Claims are also made that air-assisted airless application can utilize slightly larger orifice tips than airless application for the same paint, and this allows higher production rates and reduced clogging.

More important, while production rates are comparable to or better than airless,

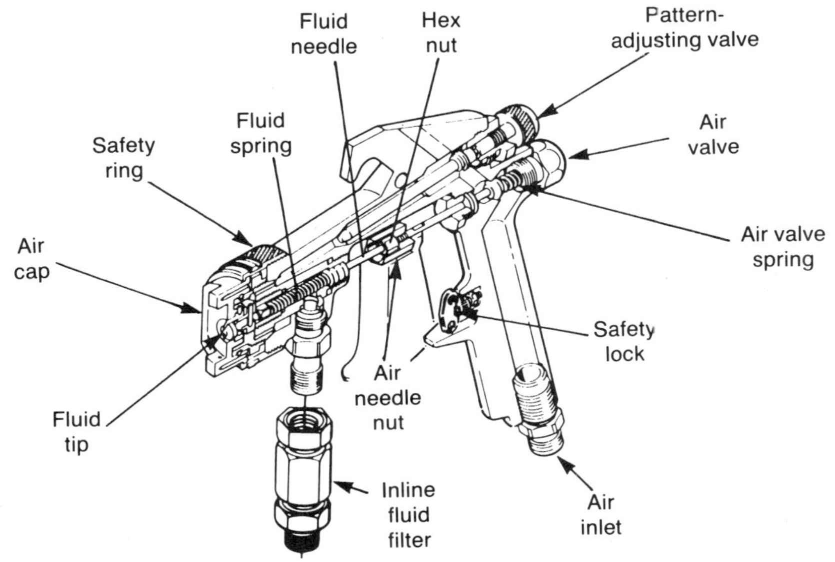

Fig. 6-14. Air-Assisted Airless Spray Gun. (Courtesy Graco, Inc., Minneapolis, Minnesota)

the same equipment may be adopted to produce flow-control rates comparable to conventional air-spray applications. Furthermore, as with air-spray techniques, pattern adjustments are possible with air-assisted airless application, without changes in airless tips. Like conventional spray application, the applicators are again encumbered with multiple lines. Equipment is surprisingly small and light compared to regular airless units, and is reviewed in detail elsewhere.[47]

PRACTICALITIES OF THE JOB PROCESS

Washing and Cleaning

While the theory and techniques of surface preparation are discussed elsewhere in this volume, certain practical aspects should also be included. On spot-cleaning jobs in the field, for example, where old films will inevitably become the substrate for the new coating system, preblast cleaning is an important precursor to actual surface preparation (either blast, hand, or power cleaning). The process is vital on jobs in industrial areas or in the snow-belt states, where salt contamination of paint films can cause serious problems if not removed. Gross[48] has demonstrated well the amount and variety of chemical deposits that can accumulate on bridge structures and has shown these to vary with bridge geometries, geographical location, and the type of crossing.

The washing procedure is particularly important in early spring when rains have not had a chance to naturally clean the bridge, although in many areas, natural cleaning is prevented by design features. As part of the yearly maintenance procedure of many structures, the removal of road sand, grit, and curbside debris is often implemented by washing on a regular basis. This type of washing is rarely satisfactory enough for painting, however. Only after this spring cleaning is complete should the structure be washed prior to painting, but it is important that scheduling of yearly maintenance be made with the painting contractor.

Generally, the contractor should start washing the structure at the high points and work down so that lower decks and legs are not recontaminated by salt and debris from the upper levels. Steam cleaning, high-pressure detergent washes, or water and solvent mixtures are possible cleaning techniques. It is important to realize that water alone will not remove nonsoluble organic residues (such as oils, tar, grease, and other road soils), while solvent will not solubilize salt deposits (see Table 3-3, p. 53). Neither solvent nor water will effectively remove chalk deposits and some mechanical assistance is often required—hence the use of pressure sprays and the like. Certain repaint systems are more forgiving of heavy chalk contamination on old film substrates than others; oil/alkyd systems (System Groups I and III) will require less attention than, for example, latex systems (System Group V) in order to render the surface suitable for repainting. Increasing awareness of the problems of salt nests in severely rusted areas have highlighted the benefits of cleaning these areas, particularly for nonblast applications.

Blast containment of large areas with tarps may also introduce a need for a postblast washing of primed steel before finishing. Where large containments of structures require several days' work within the same enclosure, blast debris from one day's work can severely contaminate surfaces primed earlier. So heavy is the debris on such films that it becomes as cost effective and certainly as technically wise to wash the primed steel after containment is removed as it is to blow it down.

Surface Preparation

The practical difficulties in blasting and cleaning often derive from problems of access or steel condition. On many structures, certainly the older ones, invariably there are areas of poor access. Certain design features of expansion bays, small boxed members, column interiors, and steel near abutments may preclude access, despite a general bridge design that is not prejudicial to the most exacting coating system. In these areas, it is better to utilize a less demanding paint system (e.g., bituminous material or possibly a wax coating) that is laid on heavily over whatever surface preparation is possible than it is to attempt to carry the coating system into the closed-in areas. The aluminized-epoxy mastic may also have some utility here.

On older structures in the snow-belt states and in industrial areas, particularly in expansion bays or on lower flanges and the like, the steel can be very heavily pitted, and such extensive deterioration may not be visible until after the scale has been removed and blasting commences. Here, mil profile requirements become quite meaningless, for the pits are often tens of mils in depth. Inevitably, the bottoms will retain salts that in wet areas may produce blistering and premature breakdown of the coating after finishing. Cleaning by blasting is difficult and slow, and results even then may be questionable. Although the contractor may balk at the production rates of this type of work, there can be no compromise in blast quality and the steel must be cleaned. Wet blasting with water/abrasive mixtures may give the best results. The situation is not unlike that found on salt-deteriorated weathering steel, and, at this stage, the jury is still out on how best to deal with it. High-wetting zinc systems (if applied by brush) may help, as may the use of high builds of barrier systems, such as epoxies. Where aluminized pigmentation is allowable, the aluminized barrier mastics may serve well on this type of steel.

Such steel may also bedevil the application of primers, for its very porosity will produce the same bubbles and pinholes that are seen in topcoats over porous zinc films. Furthermore, such holes may telegraph from one coat to another. Uncleaned debris from the pit bottoms may hold onto water after rain, and this may result in premature breakdown of even zinc films. As noted above, brush application may assist in coping with such areas. Red lead and oil primers, ideally suited to penetration into such difficult surfaces, are all too susceptible to early breakdown by those same agencies that produce the pitting in the first place (salt water) and the consequent attack of cathodic alkali.

Priming

The usual problems of priming involve the maintenance of production and job schedules and balancing these with the weather. As surface preparation is virtually always the rate-determining step of any operation, the contractor will blast for as long as he can in any single working day, leaving the last two hours or so for reblast, blowdown, and priming. In good weather there are few difficulties. In high humidity or when rain threatens, the scheduling of this blast/priming operation can become quite difficult. Steel turns more rapidly and the contractor may lose large sections of blasting before he can prime-in. While the reblast of such areas is never as difficult as the first blast operation, production rates are inevitably reduced. Similarly, newly primed films may be overblasted or compromised by contaminating abrasive if they are not dry. Problems may be compounded as the use of water-based systems increases. These are prone to flash rusting, rain removal, and/or protracted drying times in high humidity. Where maintenance schedules are such that the contractor must work long hours in order to satisfy a client's requirements, the interval between blasting and priming may necessarily push beyond the normal eight-hour maximum. Here, a knowledgeable inspector is vital to determining the extent of touch-up blasting necessary for the maintenance of the blasted steel exposed in the early hours of the blast day.

Other weather-related factors that may confound the contractor's operation pertain to temperatures, particularly in areas of the country where the painting season is foreshortened in spring and fall. Applications of latex paints, oil and alkyd systems, and, particularly, epoxies are badly compromised by cold weather, and the contractor must plan his application schedules carefully if cure is to be maintained. Not only must he adhere to temperature limitations at the time of application but he must adjust his application schedule to the expected delays of surface temperatures achieving minimums in early mornings, and he must allow enough time for cure before evening temperatures again drop below minimum. When two or three coats of epoxy must be applied and cured within a set recoating schedule, the judgment of weather conditions over two and three days can become particularly exasperating.

In the practical application of all primers, holidays, skips, thin areas, and areas developed by overblast are unavoidable. With the older paint systems, these required little more than simple field touch-up. With the new systems, things are not as simple, and touch-up procedures for inorganic primers must be established before the fact, although there is some disagreement on touch-up methodologies. Some manufacturers of inorganics require sweep blasting of the original coat before a second coat may be applied;[49] other manufacturers require little or no preparation but instead a heavily thinned touch-up coat.[50] Still other methods involve touch-up with an organic primer that is compatible with the inorganic zinc and with the intermediate and finish coats to be used. This may require the formal specification of a special organic zinc rich touch-up primer. Application of an organic touch-up primer is best done after local scarification of the zinc-primed steel by at least power cleaning. Areas of ferric corrosion on zinc-primed surfaces are probably best reblasted and reprimed with the inorganic. Care should be taken not to build up the inorganic in the areas peripheral to the reblast. Heavy single-coat builds of inorganic zinc will also mudcrack and should be removed by reblast. Isolated runs, however, can be scraped, wire brushed, and touched up with an organic.

Spray techniques using zinc films that have critical film-thickness requirements need careful development by the crew and must be adjusted for both material and the job in order to coat both intricate areas (bolts, rivets, etc.) and large unbroken sections (webs) with enough paint to afford protection but not so much that mudcracking and/or total cohesive failure of the topcoated system is possible.

Contamination of primers (particularly alkyl-silicate-based zinc rich primers) before recoating can also be a problem. In maintenance applications, the dichotomy between the requirement for adequate cure of zinc rich primers and the avoidance of surface contamination can be vexing. Most experts will agree that some delay in the recoating of such systems will reduce the possibility of incompletely hydrolyzed films being trapped beneath the finish coat and, therefore, mitigate the possibilities of lateral cohesive failures in the partially cured film after topcoating. In certain organic zinc rich systems, slow solvents (e.g., cellosolve acetate) may require days before films are substantially solvent free (some solvent may be retained in the film for months). Unfortunately, delays of even a week can result in surface contamination under certain conditions. This may not be readily apparent on dark, zinc rich films. Heavy zinc-corrosion product resulting from heavy rains can usually be removed by washing or water blasting of the zinc. Certain organics with thermoplastic vehicles may be broomed free of loose material and recoated readily if good solvent welds are possible between the subsequent coat and the applied primer. More difficult scenarios are created when inorganic zincs or thermosetting organic zinc films are so contaminated. Difficulties are also created by the contamination of partially finished structures with deicing salts late in the season, when zinc rich films can be impregnated with chlorides. In these cases, even water blasting will not totally remove the chlorides and other contaminants from the porous films, and intercoat adhesion problems, osmotic blistering, and the early corrosion breakdown of the system are possible.

The use of foreign steel in the fabrication of new bridges is also increasing. When this steel is shop-applied at the manufacturing plant and dispatched by ship from such places as Korea and Japan (often stored on deck with minimal protection), more severe contamination of the primed surfaces can complicate the field painting of the steel after receipt at the job site. Also, there is often more field touch-up required after long sea voyages, especially if the steel is not erected immediately. Even in cases of domestic fabrication, where zinc-primed steel (including galvanizing) is freighted to the job by sea or over salt-ladened highways during the winter, the field painting contractor must first contend with the removal of excessive zinc salting before field application may proceed. It remains questionable how well these techniques are normally executed (by water blasting, etc.), how well total removal of all such soluble contamination from the steel is performed, and/or what the effects of this contamination will be on the long-term performance of these structures. The state of Louisiana now recommends that a wash-primer tiecoat be applied over their epoxy primer and before the application of their vinyl finish if the steel is more than four months old.[51] These types of problems confirm the wisdom of Michigan's full-system shop-application approach.

Tiecoats and Midcoats

Application of tiecoats and midcoats to primer films should only be made after primer cure is complete, although it should be made before surfaces become too contaminated with road dust and the like. Cure is not generally a problem with thermosetting systems, unless temperatures have been misjudged. In that case, epoxies give more problems than urethane primers. Lack of induction time or application at too low a temperature can produce sweating, blushing, and amine carbonate formation in many epoxy systems, which, if recoated, will inevitably cause intercoat failure. The removal of such exudates with isopropanol or some other recommended solvent may be necessary before topcoating. The recoating of coal tar epoxy systems may be particularly difficult at both temperature extremes. Not only will low-temperature applications produce exudate but high-temperature applications may catalyze the cure so much that intercoat adhesion may be difficult to realize because of the hardness of the surface. Twenty-four hours may be the maximum possible recoat interval between coats under certain conditions. Beyond this, sweep blasting or solvent softening may be required prior to recoating. In the application of new paint systems directly over old, hard, epoxy coatings, sweep blasting of the old coating may also be necessary.

With thermoplastic primers, solvent release may be slow when large amounts of high boilers are employed either in the formulation or in field reduction. Zincs may be problematic because of their weak cohesive nature, their porosity, and their propensity for contamination. Coin-hardness tests and MEK (methyl ethyl ketone) rub tests are most important with inorganic zincs before recoating, particularly in single-pack systems.[52]

Where mist coating techniques are employed, detailed procedures should be established (by the manufacturer's representative, if necessary) before widespread application begins. Care should be taken to establish correct gun technique, appropriate schedules for the mist pass and for the full pass (to establish the appearance and opacity of the mist coat), and the degree of cure or setup between the two coats. Similar procedures are necessary in zot-coat applications, although here the amount and type of reduction solvent must be controlled carefully. It is not unreasonable to find that changes in reduction solvents may be necessary (with all coats) because temperature conditions may vary from day to day, and the contractor, engineer, and inspector should be aware of what conditions dictate which reducer to be employed.

Tiecoats of the WP-1 washprimer type require even greater control, specific attention being necessary in agitation of the base component (settling may be heavy) in mixing (too slow or rapid a combination of components may result in premature gelation) and in use. The application of low film thicknesses from an undiluted mix may be difficult with airless application, particularly in tight areas or around intricate configurations, and substantial postmixing thinning using one of the lower alcohols may be required. Around bolts and rivet heads, brush application may be preferred for additional control of film thickness. Thickness may be controlled by monitoring the opacity of the applied wash primer. Very thick films on galvanized surfaces have been observed to result in subsequent delamination of the wash primer and total system. It should be stressed that the wash primer will not gel, but its effectiveness will lessen progressively after mixing. It may be very difficult to convince an inexperienced contractor of the need to dispose of a still-liquid wash primer left over after a prior day's application. Recoat schedules must also be carefully controlled after washprimer application. The washprimer film becomes very hard, and the recoating of films after twenty-four hours may produce intercoat problems between the finish coat and the wash primer.

Finish Coats

The finish coat must not only provide protection but must also address aesthetic concerns. Uniformity of gloss and color are particularly important. Poor gun techniques can result in pinholes, overspray, and patchy, blotchy, and uneven films, even in relatively flat finishes, and can compromise the effectiveness of what otherwise might be an acceptable job. Such difficulties as these are often more of a problem with fast-drying systems than with slower ones, although nonuniformity in film thicknesses that result in sagging, color float, and windblown contamination can compromise slower systems.

High-build intermediate and finish coats, even with a mist or zot coat, may still give such application problems as bubbling, frothing, and pinholing over porous zinc primers. Many coatings tend to telegraph craters in midcoats to finish coats. Excessive frothing and pinholing will not only reduce the effectiveness of protection but will compromise the gloss and color uniformity of the system. Curiously, these difficulties may vary with the same product from job to job or application to application, and may even vary in intensity from day to day, depending on wind conditions, temperature, and so on. In these situations, input from the manufacturer in the redesign of the reduction solvent system may be essential.

Urethanes (particularly the slower drying polyester type) are also very sensitive to humidity variations. Their applications on structures where sudden mists are prevalent can result in loss of gloss and the occurrence of surface bubbles.

Wrinkling of oil-based finish coats may produce an apparent patchiness when viewed from a distance, as may the results of application in late fall and early spring.

Although these observations rapidly become less objectionable as the films age and chalking sets in, they may certainly compromise the aesthetics of the newly finished job.

Epoxies must be very carefully controlled when applied under borderline temperature conditions. Much below 50°F, reduced cure will tend to leave films sticky and easily contaminated until cure can again resume at normal temperatures. While amine carbonate formation caused by the reaction of uncured epoxy film with moisture and carbon dioxide from the air will not cause recoat problems in the case of the finish coat (as it does in primers and midcoats), the resultant yellowing and discoloration will badly compromise the appearance of the job. Although carbonate formation is not generally a problem in bridge work, such discolorations can compromise the overall finish coat aesthetics in shop-painted steel, when artificial heat with carbon dioxide-producing devices is used to maintain temperature, or in the field, when artificial heat is used within enclosures to prolong the painting season.

Readers interested in a more complete discussion of the practicalities of the application process may consult several sources;[26,29,30,53,54,55] application defects are discussed in greater detail elsewhere.[30,56–59]

NOTE ON LITERATURE SEARCH

In view of the critical role of the contractor in the bridge painting discipline, the lack of a published literature that explores the subject from the contractor's viewpoint is surprising. Equipment manufacturers specializing in surface preparation and application have produced texts that adequately cover those aspects of the job process, and other texts deal with such subjects as rigging, safety, and so on. There is a plethora of texts on paint materials and coatings systems, but little is written by painting contractors that is directly concerned with their day-to-day operations and the difficulties with which they must contend. Apart from two or three chapters on surface preparation and painting costs, the painting contractor's contribution to the *SSPC Manual* is conspicuous by its absence. Some pertinent articles may be found in the journals *American Painting Contractor* and *Painting and Wall Covering Contractor*. Many of the publications of the National Association of Painting and Decorating Contractors of America seem not specifically directed at heavy-duty maintenance or bridge painting, and in some cases, these publications seem overly simplistic. The instructional manuals *Wheels of Learning: Painting One* and *Wheels of Learning: Painting Two* by the Associated Builders and Contractors, Inc. are more comprehensive, and demonstrate at least some of the expertise that is apparent in the field.

To construe this silence as evidence of disinterest or inarticulateness on the part of the painting contractor is a gross error. As one contractor confided, "Unlike consultants and paint manufacturers who receive direct return from exposure by publication, the painting contractor rarely benefits from this" (George Campbell, George Campbell Painting, personal communication, 1986). If given more formal exposure through publication, the practical and common-sense approach that many painting contractors privately display in their understanding of problems that are the subject of many overly theoretical treatises would do much to bring balance to our understanding of the job execution process.

Chapter 7

Quality Assurance and Quality Control

Those aspects of inspection that are directly associated with job execution—shop and field inspection and the quality assurance/quality control of coatings—are considered in this chapter. Prespecification (systems option) inspection is considered in Chapter 5 and postpainting inspection is discussed in Chapter 8.

JOB INSPECTION

Consistent quality inspection of the painting process is without doubt the most vital key in the entire process whereby the specification, as conceived by the engineer, specification writer, and/or the paint manufacturer, is transformed into a well-executed job. On this point there is broad and general agreement among all those having any kind of expertise in any aspect of the bridge painting discipline. More important, there appears an equal conviction that the most fundamental deficiency in the entire bridge painting process is the absence or inadequacy of such control. Responses to questions on the subject are in astonishing agreement that the single most effective contribution toward improved effectiveness in bridge painting would be some dramatic improvement in the quality and extent of inspection. Representatives from all facets of the industry, including inspectors themselves, seemed well aware of the inspection problems and the extent of inspection deficiency.

Perhaps more surprising than the unanimity of acknowledgment is the apparent lack of response to the problem by most state and bridge authorities. There is much discussion of the difficulty, and many states (as well as the federal government) have embarked on various education programs and training courses designed to acquaint the bridge inspector with his job (the monitoring of surface preparation, paint application, etc.) and the tools of his trade, as well as to provide some background in corrosion science and the way coatings function. Despite this, the coatings inspector remains the low man on the team. There seems to be much talk of the need for inspection but little commitment to fund the job and create a professional force.

Profile of the Typical Inspector

While there are some good paint inspectors, in many states the bridge paint inspector is poorly trained, undedicated, and ill-motivated. Too many are uninterested in their job, and most spend as much of their time as possible off the job or in their field vehicles reading newspapers. They are hardly encouraged by the attitude of their peers. Those individuals that do show promise are more often than not promoted away from inspection, a process that continually weakens any in-house force before it can develop.

The best inspectors are often found to have been painters,[1] painting foremen, or sandblasters at some time in their careers, and these individuals are probably most valuable because they know the tricks, devices, and foibles of those they must control.

As early as 1953, the California Division of Highways realized the value of such individuals and actively sought to develop inspection crews from bridge painters. As Woods states in a recent article, "Not only could such individuals relieve resident engineers [but they] know what to look for and where to find it."[2] Two categories of inspectors were set up by California: Grade II inspectors administer contracts and assume many of the duties normally performed by a resident engineer, while supervising Grade I inspectors, who do only the bridge inspection.

For the most part, however, ex-painters are found less widely among state government inspectors than they are among the ranks of the few professional paint inspection companies, where the financial rewards are greater. State highway department inspectors usually have less painting background and may range from steel or welding inspectors to general laborers and even political appointees (reaping the meager benefits of some obscure political obsequence). In state government, paint inspection is often a part-time occupation (particularly in northern states) because of the seasonal nature of the painting practice. This means that the inspector often has other duties that call for other unrelated skills, some of which are perhaps found more satisfying. Even during the painting season, it is not unknown for paint inspection duties to make up only part of the state inspector's total job assignment.

In shop painting, inspection often must be performed by those responsible for the inspection of other aspects of the overall fabrication (welds, etc.). While the importance of the expertise of these individuals in their primary function is often well appreciated, skill in painting inspection is neither acknowledged nor encouraged and is all too often virtually nonexistent. Given this scenario, it is hardly surprising that there is little professionalism among the ranks of paint inspectors.

Inspection Training: Courses and Literature

In one of the best treatises on inspection, Dunkley[3] suggests:

> The inspector should have an appreciation of the fundamental aspects of corrosion and factors affecting its control, together with a basic knowledge of the nature of the materials employed as well as the influence of the environment and conditions at the time of application. His training should be progressive, and linked with career structure, which in turn implies the principle of continuous long term employment.

More on-the-job training of apprentice inspectors by skilled individuals is essential, and greater care in the selection of apprentices is warranted. To attempt to motivate the typical paint inspector solely by means of classroom sessions on corrosion science and paint technology and without a balance of practical training is probably futile.

While several courses exist for the training of paint inspectors, few deal specifically with bridge painting. Limited courses have been offered by individual states (California,

Florida, Louisiana, Massachusetts, etc.) and, in some cases, inspection methodologies are actually built into state transportation departments' standard specifications.[4,5] At least one text is available.[6] A revised FHWA course[7] is presently being implemented. Inspection courses are also offered by NACE,[8] the Institute of Applied Technology,[9] British Gas, and the Institution of Corrosion Science and Technology,[10] although these are not specifically directed toward bridge work.

The NACE inspection course[8] may be the most comprehensive general course and is divided into three separate courses, each about one week long. Most of the important instruction as far as bridge painting is concerned is found in the first session (1985 cost $1500 per participant). The other two courses (1985 cost $750 per course per participant) are more concerned with other aspects of inspection, such as the newer application methodologies, powder coatings, linings, and substrates other than steel. Some of the subject matter in the second course, such as centrifugal blasting, human relations considerations, safety, and certain testing methods, is, however, very relevant to bridge painting.

The Institute of Applied Technology course[9] is good but is quite heavily directed toward nuclear-power generation. Nonetheless, there is much relevance to bridge inspection in the course, and its text makes a valuable reference.

The original FHWA *Bridge Paint Inspection Training Manual* has been rewritten.[11] While ostensibly directed toward bridge-paint inspectors, the original course considered the processes of surface preparation and application more generally than it did the particulars of these disciplines as they are developed and modified by the specific difficulties of bridge-painting practice. The original text has also been criticized as being overly theoretical for the typical bridge-paint inspector, with too great an emphasis on ancillary aspects of paint technology rather than on the more practical aspects of inspection. The decision to have the material rewritten by authors who have direct field experience has now corrected these deficiencies, and the revised course is being delivered to state transportation department inspectors by these new authors in some twenty to thirty sessions.

The ASTM *Standard Guide for Paint Inspectors* is another very valuable text.[12] Other reference texts with valuable training data are also available.[13,14,15]

There is a tendency in some inspection courses to emphasize the more easily taught concepts (basic corrosion science, coatings technology, etc.) and to demonstrate the range of instrumentation available rather than to discuss those aspects of inspection that relate to the recognition and detection of untoward conditions and practices on actual structures (the kind of practical treatments noted by Bayliss[10]). Perhaps the most valuable aspect of such courses is the familiarization of the apprentice inspector with (1) surface preparation techniques and qualities, (2) the use and misuse of application equipment, and (3) much of the terminology of both corrosion science and paint technology. Discussions of coating deficiencies (particularly those occurring during or just after application) and their recognition, terminology, causes, and rectification are also particularly important.

Some specific treatises on paint film deficiencies are available.[7,16–19] Pictorial representations of various types of coating failures may be found in relevant specifications of the ASTM *Federal Standard Test Methods #141A* and in other specific sources.[20,21] Possibly the most comprehensive coverage is to be found in the original and revised editions of Hess's *Paint Film Defects,*[16,17] which segregates failures that occur during application from those failures that occur in service. Excellent glossaries are available,[22] many of which may be found in several of the sources above, although the most authoritative reference in this respect is probably the *Paint and Coatings Dictionary.*[23] Spray equipment manufacturers also publish reference materials on finish problems related to spray application.[24,25,26] Munger also discusses application problems.[27]

Professional inspection companies, however, view formal classroom-type training only as supplementary to job experience and background in painting bridges. Several have reported a reluctance to invest in this type of training because of doubts as to its usefulness and concerns over a degree of transiency among their personnel that makes the financial commitment a risky investment. Apparently, there is a lot of hiring and firing in the process of establishing a satisfactory inspection staff.

Some of the independent inspection companies provide experienced inspectors who will conduct on-the-job training courses for staff apprentices of bridge authorities. This type of service is expensive but might be effective in conjunction with classroom training. Additional valuable tips by experts in inspection are often included in the special section "Maintenance Tips" of *The Journal of Protective Coatings and Linings.*

Even workshop training sessions with instruments may make little sense, for in many states the bridge-paint inspector remains as poorly equipped to do his job as he is poorly skilled to do it. Some states and authorities have now included clauses in their job specifications that mandate that the contractor shall supply the inspector with so many wet- and dry-film thickness gauges (type often specified) with which to maintain thickness control. The gauges are nonreturnable, and, in this manner, many state inspectors are becoming equipped with Tooke gauges and magnetic film-thickness gauges, if nothing else. That the states must resort to such practices as this to equip their inspectors in itself says much for the value that is placed on inspection in many state transportation departments. Psychrometers, surface thermometers, mil profile comparators, blast standards, holiday detectors, and so forth for the most part remain lacking, and cameras and recording devices—even indicator papers—are rarely used at all except by private inspection companies.

Compared to the literature that details the actual practice of bridge painting inspection, excellent sources of data on the types of instrumentation used for inspection are many.[15,28–31] Excellent in-depth scientific treatments of specific aspects of inspection (film-thickness determination,[32] mil profile,[33] etc.) are also available. Some essential inspection tools are as follows:

1. Psychrometer, surface thermometer, anemometer
2. Mil profile comparator, photographic standard of required surface preparation, ion-indicator papers
3. Wet-film thickness gauge
4. Magnetic dry-film thickness gauge, Tooke gauge
5. Holiday detector, flashlight, mirror, moisture meter
6. Utility knife, Testape™*, painter's hammer, chisel, marker
7. Paint remover, solvent, cotton, 4-ounce jars, distilled water, clean pint cans
8. Camera (35mm and/or Polaroid-type), video camera, binoculars
9. Safety gear (belts, hats, boots, etc.)
10. Walkie-talkie or citizens' band radio (large multilevel jobs)
11. Cassette recorder, notebooks, envelopes (for samples)

*Testape is a trademark of Clive H. Hare, Inc.

Private Inspection Companies

In some cases, state highway departments or bridge authorities may utilize outside inspection for painting bridges, although this seems more common in shop application than it does in the field. Even in those companies offering such inspection services, the "paint inspector" may be primarily skilled in other disciplines, checking paint and surface preparation as part of his overall responsibilities.

A handful of companies, widely separated geographically, specialize in the inspection of coatings and the coating process. These are the truly professional organizations, and it is here that the real expertise is found. Unfortunately, field rates for this type of inspection are high (generally $25 to $35 per hour plus expenses) and beyond the budgets of most state public works departments, although perhaps not beyond those of toll-bridge authorities. While the rate for paint inspectors is perhaps no higher than for steel inspectors, the need for both types of inspection in effect doubles the rate.

Even fewer paint inspection companies specialize in the inspection of the bridge painting process. Compared to job assignments in such other industries as nuclear-power generation, off-shore drilling, and water tank construction, the opportunities for work in bridge painting inspection have been less than widespread. Although it is claimed with some justification that painting inspection is much the same whatever type of structure is involved, this holds true only for the basic process. The experienced bridge paint inspector is more aware of the relationship of bridge design to deterioration and better appreciates such areas of greater and lesser criticality as flanges of floor beams located near end connections and bearing points, expansion bays, horizontal surfaces of lower chords, rocker pins and rollers of bearings, and the anchor-chain gallery hardware of suspension spans. As noted above, there are few detailed and specific reference sources to assist the general painting inspector unfamiliar with bridge structures, although such texts as the *Bridge Inspection Training Manual 70*[34] may serve to acquaint him with bridge design features and their relative vulnerability and so compliment his basic knowledge of paint inspection.

Value of Inspection

The value of paint inspection is well explained by Kaminski and Allen,[35] who hold that inspection costs (less than 2% of the total job) represent a premium on insurance against the cost of rectifying a premature paint failure. Much the same position is taken by Roebuck and Brevoort,[36] who quote inspection fees at a higher percentage of job costs, generally 3% to 8%. The analogy is not unreasonable. The primary role of the inspector, however, is not only to ensure that a premature failure does not occur but that the projected service life of the coating system is realized and, insofar as possible, exceeded. Inspection ensures that a bridge authority gets as close as possible to a dollar's value for every dollar spent in job execution. Measured with the cost of the paint job on a cost-per-year-of-service basis, the cost of inspection becomes insignificant compared to its return.

Kaminski and Allen also demonstrate the cost advantages that are realized with professional inspection compared to nonprofessional (in-house) control, or to no inspection at all. They show that over the life of the job inspection may save, in additional service, as much as the original cost of the job.

So many premature bridge-paint failures can be directly traced to questionable practices during surface preparation and application that the dollar return for good inspection should be obvious. Moreover, as paint systems have increased in complexity, the requirement for good inspection has grown to the point that the presence of the in-shop or field inspector is no longer simply advisable but critical.

Duties of the Inspector

The inspector's role is multifold and is discussed elsewhere[3,10,15,31,37] as well as in the inspection training texts.[7,8,9] Inspection is necessary at every stage in the painting process—preblast cleaning (if necessary), blasting (or other surface preparation), priming, midcoat, and finish coating. The contractor must satisfy the inspector at each stage of the process before proceeding to the next.

The exact duties of the painting inspector vary with the paint system being applied. Peculiarities of the job may also affect both duties and their scheduling. The various inspection requirements of the typical job necessarily follow the specification and may be classified according to the individual job processes that make up the specification (i.e., preblast cleaning, surface preparation, priming, midcoating, finish painting, etc.). Many of the inspection requirements in today's best practice follow. Unfortunately, inspection on most jobs is so deficient that these requirements might reasonably be considered to reflect the aspirations of inspection practice rather than its reality today.

Inspection Requirements

General

SPECIFICATION ENFORCEMENT AND JOB CONTROL MAINTENANCE

1. The inspector protects the bridge authority's interests as the field representative of the engineer. He ensures that all requirements of the specification are continuously maintained. The inspector controls the job with firmness but fairness and accurately reports all activities in a timely fashion to the engineer.

2. The inspector reads, understands, and fully appreciates all articles of the technical specification as well as such other pertinent documents as blast standards, pictorial representations, and the coating manufacturer's product data sheets. Where possible, before the job starts the inspector will fully familiarize himself with the equipment being used as well as with the individual coatings, their properties, and application characteristics.

3. The inspector is fully cognizant of the intent of the specification and areas critical to job execution (e.g., blast requirements, acceptable film-thickness ranges, cure requirements, recoating schedules, etc.), and realizes the consequences of specification deviations. He examines the structures to be painted prior to job start-up and anticipates critical areas that may be related to design features, location, and/or exposure.

4. The inspector provides counsel to the engineer on any design feature whereon the specification may be compromised or inappropriate because of access, steel condition, or other factors.

5. The inspector knows and understands all inspection procedures required by the specification, is familiar with the use of the instrumentation involved, and is a climber, comfortable with working on steel irrespective of height and type of rigging. In the field, particularly on high structures, inspection is even more critical than in the shop, for it is in these hard-to-access locations that many contractors see an advantage over the inspector and an opportunity for deliberate shortcuts. For this reason, one of the

primary requirements of the painting inspector is familiarity and comfort in climbing on high steel structures and an ability to move with no less agility than the paint crew.

6. The inspector is fully acquainted with the safe practices of rigging, equipment handling, and materials safety and will render counsel to the engineer concerning the same where necessary, without delay.

JOB-PROCESS MONITORING AND REGULATION

The inspector's principal duties involve the control of cleaning, surface preparation, and application of all coatings. These requirements are outlined below.

INTERFACING WITH ENGINEER AND CONTRACTOR'S REPRESENTATIVE

1. The inspector is capable of clearly communicating with the contractor, superintendent, foreman, and bridge engineer concerning all aspects of the job process.

2. The inspector maintains a daily log of all operations detailing crew size and makeup, hours worked and activities, climatic conditions, and measurements taken (film thicknesses, application schedules, adhesion data, etc.). He records the daily productivity rates, the job process from bent to bent, the records, dates, and makeup (including batch numbers), material delivery, storage conditions, and amounts of paint and abrasive used daily. All activities are illustrated with a photographic record, especially areas where problems, disputes, or anomalous conditions occur.

RESPONDING TO EXTRAORDINARY CONDITIONS

The inspector is sufficiently familiar with operations to recognize unusual and possibly untoward phenomena at the outset, reporting the same to the engineer before the problem is compounded by recoating and so on.

OTHER FUNCTIONS

Depending on the particular job, the inspection requirements may also involve containment and traffic control. The inspector may also be required to certify the job for payment.

Preblast Washing

1. Verification of washing schedule with reference to general cleaning and yearly bridge maintenance

2. Establishment and supervision of geometrical order of washing (upper truss, upper deck, lower deck, legs, etc.)

3. Verification of acceptability of washing procedure and washing solution makeup

4. Verification of rinsing operation

5. Verification of suitability of washed surfaces to receive new coating

Surface Preparation

BEFORE OPERATION

1. Verification of surface-quality standards with engineer and contractor

2. Verification and control sampling of contractor's abrasive (type and size) as suitable to achieve required anchor pattern and profile depth; estimation of mix ratios of grit/shot in shop work

3. Preparation of practical standards (photographs or plastic sealed panels) that represent the required quality of cleanliness as called for by the specification and interpreted by the engineer

During operation

1. Verification and recording of climatic conditions as suitable for the cleaning process (temperature—steel and ambient, relative humidity and dew point, wind speed and direction)

2. Examination of abrasive composition and mix (shop work), and comparison to control

3. Examination of traps, separators, and air to ensure no contamination by oils or water

4. Examination of blast quality and profile at intervals convenient to the contractor (lunch breaks, etc.) and marking noncleaned areas and areas of insufficiency for reblasting; on hand- and power-cleaned jobs, examination of cleaned areas and peripheries for loose rust, film removal, and feathering of edges

5. Examination of day's blast before blowdown

6. Establishment of areas requiring reblast

7. Verification of blowdown and steel condition prior to priming

8. Use of potassium ferricyanide indicator papers or other means (white latex paint) to verify removal of soluble iron salts

9. Recording data and time of start and finish of blasting operations and blowdown

10. Photographing typical areas, bad spots after rejection and after reblast, special conditions, disputed areas, and untoward practices

11. Recording number of blast nozzles being employed and daily production rates; in-shop work, recording square footage prepared by centrifugal blasting

12. Marking and photographing for structural examination and repair any areas of excessive metal loss, perforation, rivet loss, tears, and so on uncovered by the blasting process

13. Documentation of all extraordinary conditions, excessive or recurring contamination, or deeply pitted steel, clearly recording their location and extent

14. Monitoring and recording time, materials, manpower, and equipment used (especially when addressing rectification of any extraordinary conditions that may fall outside the scope of the specification)

In jobs where the blasting process is contained with enclosures, it may also be necessary for the inspector to monitor enclosure efficiency, cleanup, and spent-grit storage. Actual sampling of fixed and movable air-quality monitors is more likely to be performed by other specialists.

Priming and Pretreatment Coats

Before operation

1. Verification and recording of existing climatic conditions as suitable for priming as per specification

2. Review of weather forecasts and determination of reasonable probability that primer may cure safely without weather-induced compromise

3. Recording batch numbers and checking for batch approval

4. Examination of thinner (solvent) for suitability and sampling if necessary

5. Inspection of boxing and mixing procedure and sampling if necessary

6. Verification of addition of thinners and thinning ratios; recalculation of theoretical wet-film thickness required to achieve specific dry-film build after thinning

7. Checking contractor's tips for wear and checking lines and equipment

DURING OPERATION

1. Monitoring climatic conditions

2. Verifying that in-pot agitators are in place and working during use

3. Verification of the application process, adequacy of rigging, correctness of technique, number of passes, and freedom from dry spray, bubbling, sagging, and so on

4. Monitoring wet-film thickness and counseling contractor on expected dry-film thickness range with reference to the specification

5. Sampling if necessary

6. Recording date and times of start and finish of priming operation

7. Ensuring that correct induction times are observed and that no paint is applied beyond an allowable pot life (may be particularly important where pot-life termination does not result in high viscosity increase and gelation, i.e., wash primer

AFTER OPERATION

1. Checking film cure, quality, and thickness

2. Marking up holidays, skips, and thin spots for touch-up or application of make-up coat

3. Marking up runs, sags, and wrinkled and/or mudcracked areas, and other areas of deficiency for rectification

4. Marking up areas of excessive film thickness for engineer's judgment on acceptability

5. Marking up areas of excessive film contamination, including zinc salting, for cleaning prior to application of midcoat or finish coat

6. Ensuring that proper procedures (including surface preparation and coating type) are known and followed by operator in application of make-up or touch-up coat

7. Monitoring and recording time, materials, manpower, and equipment used (especially when addressing rectification of any extraordinary condition that may fall outside scope of specification)

Intermediate Coats, Midcoats, or Tiecoats

BEFORE OPERATION

1. Verification that all required touch-up of primer has been successfully completed

2. Verification that condition of primed surface both as a whole and in touch-up areas is suitable for recoating (adequately cured, adhesive, free of contamination, etc.)

3. Verification and recording of existing climatic conditions as suitable for coating application

4. Review of weather forecasts and determination of reasonable probability that new coat will cure safely without weather-induced compromise

5. Recording batch numbers and checking for batch approval

6. Examination of thinning solvent for suitability and sampling if necessary

7. Inspection of boxing and mixing procedure and sampling if necessary

8. Verification of addition of thinner and thinning ratios; recalculation of theoretical wet-film thickness required to achieve dry-film build after thinning

9. Checking tips for wear and checking lines and equipment

10. Verification that contractor's recoat schedules are in compliance with specification (*note:* ensuring that touch-up areas as well as general sections have been given adequate time to cure)

DURING OPERATION

1. Monitoring climatic conditions

2. Verifying that in-pot agitators (if applicable) are in place and working during application

3. Verification of the application process (adequacy of rigging, correctness of techniques, number of passes, and freedom from dry spray, bubbling, sagging, etc.)

4. Ensuring that any zot-coat or mist-coating operation is performed according to manufacturer's requirements regarding amount and type of dilution, film thickness, and interval between the zot or mist coat and full coat application

5. Checking wet-film thickness and counseling contractor on expected dry-film thickness range with reference to the specification

6. Sampling coatings if necessary

7. Recording data and time of start and finish of midcoat operations

8. Ensuring that correct induction times are observed and that no paint is applied beyond the allowable pot life (may be particularly important where pot-life termination does not result in high viscosity increase and gelation, i.e., wash primer)

AFTER OPERATION

1. Checking film for cure, quality, and film thickness (Tooke gauge is used for repeated checks of magnetic film thickness gauge on multicoat systems)

2. Marking up holidays, skips, and thin spots for touch-up or application of a make-up coat

3. Marking up runs, sags, and wrinkled, bubbled, pinholed, or other areas of deficiency for rectification

4. Marking up areas of excessive film thickness for engineer's judgment on acceptability

5. Marking up areas of excessive contamination for cleaning prior to finish coating

6. Ensuring that proper procedures (including surface preparation and coating type) are known and followed by operator in application of make-up or touch-up coat

7. Monitoring and recording time, materials, manpower, and equipment used (especially when addressing rectification of any extraordinary condition that may fall outside scope of specification)

Finish Coats

BEFORE OPERATION

The same before-operation inspection requirements apply to the finish-coat process as apply to intermediate, mid-, and tiecoats above.

DURING OPERATION

1. Monitoring climatic conditions

2. Verifying that in-pot agitators (if applicable) are in place and working during application

3. Verification of the application process (adequacy of rigging, correctness of technique, number of passes, and freedom from dry spray, bubbling, frothing, sagging, flooding, and flotation of color, etc.)

4. Checking wet-film thickness and counseling contractor on expected dry-film thickness range with reference to the specification

5. Sampling coatings if necessary

6. Recording date and time of start and finish of topcoat application

7. Ensuring that correct induction times are observed and that no paint is applied beyond allowable pot life

After operation

1. Checking film for cure, quality, and film thickness (Tooke gauge is used for repeated checks of magnetic film thickness gauge on multicoat systems)
2. Checking total system for adhesive and cohesive integrity
3. Verification of uniformity of color and gloss; marking up deficient areas for recoating
4. Marking up holidays, skips, and thin spots for touch-up or application of a make-up coat
5. Marking up runs, sags, wrinkles, and frothed, bubbled, pinholed, or other areas of deficiency for rectification
6. Marking up areas of excessive film build for engineer's judgment on acceptability
7. Ensuring that proper procedures (including surface preparation) are known and followed by operator in the application of touch-up and make-up coats
8. Verification that total film thickness of coating system is in compliance with the specification
9. Monitoring and recording time, materials, manpower, and equipment used (especially when addressing rectification of any extraordinary condition that may fall outside scope of specification)

Continuous versus Noncontinuous Inspection

The willingness of states to commit funds for painting inspection of larger structures is more noticeable than their commitment on small structures. This is probably related to both the visibility of the structure and the increased number of crews that are employed on the larger bridges and the better worker–inspector ratio that this provides (resulting in a more justifiable inspection cost per job). On smaller structures, a single inspector may serve several different crews on a series of overpasses that may be several miles apart. There is some disagreement among inspection companies as to whether continuous on-site inspection is necessary, or whether a single inspector may control several different crews adequately. The majority of professional inspection companies prefer continuous inspection and will cite examples of failure that support the necessity for constant supervision. On the other hand, with suitable caveats and releases, these inspection companies have little problem with the intermittent inspection concept as long as procedures prohibit progress from one process of the job to the next without inspection and approval of the first process.

It may be argued that an inspector can assess the quality and anchor pattern of a blasted-steel section after the fact as well as he could by standing over the operation. With good record keeping and close attention, each job (even recoat times) might be controlled by noncontinuous inspection. In such a case, however, difficulties with oil and water contamination of the surface during or after blasting (from blowdowns, etc.) may go unnoticed. Paint-handling techniques, including mixing solvent additions, constant agitation of zinc, insufficient induction time, or applications beyond the limits of the effective pot life may all escape detection as may such inconsistent spraying procedures as excessive overreaching. Without the on-site monitoring of wet-film thicknesses at the time of application, in some cases (e.g., soft thermoplastic films) it may be difficult to assess film thickness properly before the second coat is required, although the use of shims can help in this respect. With instruments such as the Tooke gauge, it is now possible to discover many irregularities after the fact, although it is necessary that specifications be designed so as to facilitate such instrumentation (tinted zinc primers, clear color differentials between coats, etc.). In some cases, the spot use of paint removers may be necessary.

Difficulties from noncontinuous inspection practices are also possible. These may involve claims from the contractor of production losses (e.g., repeat blasting necessitated by delays in inspection of blasted areas that result in the "turning" of steel before priming). Claims may be made that it is the responsibility of the bridge authority or the inspector (as the authority's agent) to bring specification deviations to the attention of the contractor without delay and certainly before such deficiencies are compounded by the application of additional coats, the removal of which would cause unnecessary hardship to the contractor. Despite the cost, the most efficient way of controlling the paint job is to have the inspector on the job continuously, although he may control several full crews with ease if they are in close enough proximity. Consequently, inspection costs can be trimmed by limiting the distance between such areas. This should be a part of every specification, although, in practice, it is not.

Assistance to Contractor and Engineer

The professional inspector may also serve as a source of invaluable assistance to both contractor and bridge engineer in matters pertaining to job execution. His background and training will often serve the job well when difficulties arise involving specification interpretation, the use of materials and equipment, and the cost estimation of extras, particularly where either the contractor or the engineer is relatively inexperienced (by no means an uncommon occurrence).

The experienced professional inspector is well equipped to interpret the intent of the specification and not simply its letter. Such interpretations, coupled with sound judgment, can save both contractor and bridge authority unnecessary delays, and this, compared to the inflexible adherence to the letter of the specification that stamps the inexperienced inspector, inevitably results in reduced frustrations, added savings, and improved overall job confidence. Similarly, in the recognition and documentation of the unusual and untoward, there is no substitute for experience in decisions involving an assessment of a procedure's importance to quality against the effects of shut-down, reblast, and recoat on job production schedules. Under such circumstances, without the benefits gained from experience, the untrained inspector is too vulnerable to pressures from the bridge authority, or persuasions or "re-education" by the contractor.

In considerations of safety, containment, and rigging, the bridge-paint inspector may also have much to offer. Most professional companies reserve the right not to work on jobs that are improperly rigged.

All of these ancillary benefits come without added costs, and the contribution of such experience to the overall outcome of the job has an indisputable value.

Record Keeping

Slowly, the use of simple inspection instrumentation (thickness gauges, mil-profile comparators, psychrometers, etc.) is becoming more widespread. But the use of recording instrumentation is not, except perhaps on those jobs controlled by the professional inspection company. A daily log of job progress and activities is kept by most inspectors, but the completeness of such logs varies widely. The daily monitoring of weather conditions is not unusual (although there is often confusion concerning exactly where ambient and the more important surface temperatures should be taken). There is also too little concern with dew point, which is as important, if not more so, than relative humidity.

The inspector's log may, however, be a most important document in the event of a claim or a premature failure and should contain many other data from which job progress and difficulties experienced may be recreated and studied as part of the post-painting monitoring of the paint system's service. Consequently, it is necessary that the times and dates of application of the various coats to specific areas of the bridge be recorded and keyed to batch numbers of the materials used. Recoating delays, surface contamination, and difficulties in surface preparation and application (including equipment difficulties) should all be recorded as well as pump and tip size and condition, air pressures, and condition of traps. The type and amount of solvents employed in thinning should also be recorded. Notes concerning the skill of the operative employed may also be useful. Changes in operators of spray guns and blast nozzles can have a great affect on job quality. Such records of the everyday job process may be done almost routinely and require little experience on the part of the inspector.

Wiser still, however, is the inspector who uses such recording devices as cameras (Polaroid shots are very convenient) to record job progress. Shots should be dated and, if necessary, witnessed daily. Tape-recorded notes from the field may be transcribed easily later, and this makes good practice except where excessive traffic noise confounds the recording.

More recently the advent of video cameras (with their excellent close-up attachments) have added significantly to control documentation, although many contractors will object to such recording devices and clam up in their presence. Unfortunately, the expense of this type of equipment precludes its use by all but the professional inspection companies, which are readily embracing it. At this time, this type of equipment is still more common in failure analysis and in postpainting monitoring of jobs than it is in normal progress inspections despite its indisputable value.

MATERIALS ASSURANCE AND CONTROL

Internal Laboratory Control

Control of the quality of the paint itself is significantly more refined than the control of job execution. For the most part, the laboratories of state departments of transportation are well equipped with all the simple instrumentation required for the control of the basic compositional constants (total solids, pigment content, nonvolatile vehicle content, and weight per gallon) and regularly maintain a close check on these as well as on such physical characteristics as viscosity, dry time, fineness of grind, color, and gloss. Because many state laboratories are responsible not only for paint but for soils and water analysis, asphalt, concrete, and so forth, the analysis and testing facilities are more complete than probably would be the case were separate paint laboratories employed. Relatively advanced instrumental analytical facilities (infrared spectroscopy, gas chromatography, atomic absorption, etc.) are available in many cases. As a result, with any reasonable degree of expertise and with the necessary software (spectra libraries, etc.), it is possible for most state transportation departments not only to control paint composition but to analyze composition and, where formulating skills are available, to reconstruct the specific coating.

Unfortunately, laboratory staffs of most highway departments seem infinitely more skilled in the specific disciplines of basic chemistry (organic, analytical, and inorganic chemistry) than they do in such applied technologies as paint chemistry. In very few states (California, Texas, etc.), do there appear to be qualified paint formulators.

External Laboratory Control

Some states utilize independent laboratories for testing purposes. While some of these laboratories may boast paint chemists and formulating staff, in the majority of cases, the extent of their formulating prowess seems to be as questionable as their understanding of coatings and coating systems. Some may rely on formulating consultants (retired paint chemists who are listed as staff members or staff consultants), but many external laboratories specialize in quite narrow disciplines, more often than not disciplines other than paint technology. The profit motive, however, encourages the advertising of a broader spectrum of skills whether or not they have been mastered. Consequently, where such specializations are found in paint chemistry, the levels of expertise in all aspects of testing, analysis, and formulation can be very high (well beyond that of most highway testing laboratories and even some paint manufacturers). Where the specialty is not coatings, however, the same difficulties that haunt the poorly qualified state laboratories are prevalent here too.

Very often, the practice among such external laboratories is for one laboratory with a different specialization to subcontract the paint testing to another laboratory that may or may not have an improved degree of expertise and may or may not provide satisfactory services. Such subcontracting, however, inevitably increases the cost of testing.

Laboratory Expertise

Formulation skills are, of course, not necessary in order to control coatings. Control involves the testing of the coatings and/or coating systems against a set of predetermined parameters that theoretically at least, reflect those formulating extremes that will provide coatings giving the required performance. It may be argued, then, that it is analytical testing skills alone that are necessary to control coatings. In practice, however, this assumption can lead to much lost time and energy and unnecessarily elevated costs. A sound knowledge of paint chemistry allows the chemist (or technician) to interpret his findings (deviations) in terms of their effect on the formulation of the system, its application, and, consequently, its performance.

The experienced paint chemist will use the specification as a guide, while the inexperienced analyst will go by the book, irrespective of the job problems that this may cause. The experienced technician will easily recognize sampling errors, contamination, or other irregularities well before the testing is complete. Differences in color, odor, drytime, or gloss (although not sufficient to compromise the letter of the specification) may signal the alert technician to more basic deviations.

Too often, job schedules are interrupted (at cost to the contractor, if no one else) while paint manufacturers and testing laboratories argue over minor viscosity deviations or fractions of a percentage point in compositional constants. Even more blatant examples relate to sampling and procedural errors in testing. The example of an external testing laboratory using the wrong procedure to determine pigment content and then advocating that the nonvolatile vehicle content of an obviously cohesively adherent coating was "zero" is typical. In this case, it took weeks to convince the resident engineer with an equally inadequate knowledge of coatings technology of the impossibility of such a result. As in job control, it is more often than not in the case of the specification deviation that the skilled control technician proves his worth.

On the other hand, a new specification may not anticipate all aspects of a paint system that might result in serious consequences. Here, constant monitoring of the system by an experienced paint chemist pays off. Alert to the dangers of undocumented phenomena, he may be able to interpret their effect before the job is severely compromised and act to prevent a field problem. By example again, the increased destabilization (dehydrochlorination) of an improperly stabilized chlorinated-rubber coating, using Rule 66-type solvents, resulted in severe underfilm attack on a metal substrate by the paint system itself. The paint was totally unsuitable for use, although the original performance specification had been passed and the compositional attributes of the coating remained within the bounds of the specification. By rejecting what was ostensibly a satisfactory material, a serious problem was minimized. The experienced technician will be prepared to investigate apparent irregularities by testing that which is not actually called for in the specification but which may reveal a hidden flaw in the coating under test.

It is possible that the lack of formulatory skills in state department of transportation laboratories may, in part, have fostered the change from the compositional paint specification to the performance specification. States and authorities with formulative experience (California, Texas, U.S. Army Corps of Engineers, etc.) still adhere to the compositional format and provide perfectly adequate specifications. The argument may be made that the performance specification is a better methodology, providing as it does more latitude to the paint manufacturer, but it may also be argued that the performance format is encouraged by the low levels of formulatory expertise in state transportation departments. There are certainly few disadvantages to adding a strong compositional control to any performance specification.

It seems curious that few state laboratories have been able to analyze more fully and thereby provide compositional breakdown specifications of such commodity items as high-build vinyls and inorganic zincs on which they have now amassed twenty years of successful field history.

Materials Testing Methodologies

The several different types of specifications used by bridge authorities across the country require different quality control approaches. The most straightforward format is the compositional specification in which analytical procedures that evaluate specific quantitative compositional and physical parameters may be performed on each batch as it gets to the job. Tests are not overly time dependent and decisions as to suitability or nonsuitability can be rendered relatively quickly.

Compositional Testing

The testing of compositional specifications relies almost exclusively on the comparison of determined values of certain key compositional constants of a sampled product with specific values derived from a given formulation that produces acceptable performance. Most compositional testing is done on a weight basis, although all the vital parameters of the paint formulation (PVC, CPVC, volume solids, etc.) are volumetric and more difficult to determine experimentally. Armed with the standard compositional parameters—percent total solids by weight, percent pigment by weight, and weight per gallon—with atomic absorption for elemental inorganics, infrared spectrophotometric scans for organic and inorganic vehicles, and gas chromatographic analysis for solvent makeup, the essential formulation can, if necessary, be virtually recalculated by the experienced chemist who has knowledge of raw materials and their constants (weight per gallon, oil absorptions, etc.).

The tests necessary for the compositional breakdown of a coating formulation are as follows:

Qualitative breakdown

1. *Vehicle typing.* Infrared spectroscopy of smear from centrifuged paint. Plasticizers may be similarly determined after primary vehicle precipitation.

2. *Solvent typing.* Gas chromatography of collected distillate. Check if necessary with infrared spectroscopy of individual fractions.

3. *Pigment typing.* Atomic absorption of centrifuged pigment. Also, microscopy, x-ray diffraction, and spot tests may assist.

4. *General.* Color, smell, drying mechanism, dry-time, pot life, solvent sensitivity of film, pyrolysis of dried film, and component mixing ratios. Material Safety Data Sheet (MSDS) and label information may also assist.

QUANTITATIVE BREAKDOWN

1. Use weight per gallon to calculate weight of 100-gallon formula.
2. Use total solids, pigment content (and consequent nonvolatile vehicle content) to provide skeletal breakdown of formulation essentials.
3. Use qualitative data and raw-material data on specific gravities, oil absorptions, evaporation rates, solvent content of resin solutions, equivalent weights, and viscosity/solids profiles to determine detailed makeup of vehicle, solvent, and pigment.

These methodologies, however, will not be sufficient to provide a commercially acceptable product, which requires the addition of thixatropes, catalysts, stabilizers, wetting agents, and the like, and laboratory development.

While such minor ingredients as driers, stabilizers, and thixatropes may remain elusive (even plasticizers may be readily separated by vehicle precipitation techniques), it is quite possible to recalculate the all-important volumetric parameters from such rebuilt formulations. In less experienced hands, blind control of the compositional specification by means of the weight parameters is also quite straightforward and only marginally less effective.

In spite of this, the compositional analysis of a paint is no guarantee of field performance unless the viability of the compositional specification has been firmly established beforehand, and, here, this type of control breaks down. Where a specification results from years of trial and error in the field, the viability is proven, and compositional control is simple and effective. With newer, untried systems, in-depth compositional control will not necessarily equate with satisfactory field performance.

Performance Testing

Performance testing takes much longer to carry out than does compositional testing (1500 salt-fog hours is approximately nine weeks), and, to avoid job-scheduling delays, such tests are usually performed prior to job commencement. Some state transportation departments will prequalify coatings,[38] testing the performance properties on a specific sample before allowing the use of the coating. Some tagging of an original sample (known to provide the required performance profile) with compositional tests (percent pigment, percent total solids, percent nonvolatile vehicle, weight per gallon, infrared spectra, etc.) and physical tests (dry-time, cure time, pot life, viscosity, etc.) may provide a reasonably rapid quantitative-control mechanism, without having to resort to elaborate performance testing on every batch.

Batches of incoming material are thus compositionally evaluated against such tagged tests to guard against product changes from batch to batch during the job. Long-term performance tests may be repeated, although, if the tag tests look good, the product is usually used before performance testing is complete.

Other authorities will prequalify certain large national manufacturers before the job and may not perform batch-to-batch testing during the job or even on subsequent jobs. Typical programs are described elsewhere.[39,40] Apparently the justification here is that, unlike small manufacturers, the large national houses with proven products never change their formulations, maintain continuous and scrupulous quality control and quality assurance programs, and have greater technical depth. While quality assurance and product uniformity is generally good in the large paint companies, this philosophy does many small manufacturers a marked disservice because many of these companies also have excellent quality assurance and quality control programs. The assumption that size means quality and consistency is not universally well founded. Formulation changes in a qualified product that are necessitated by technical, toxicological, raw-material-availability, cost, and even delivery considerations are by no means restricted to the small local companies. More important, when changes do occur, they may not be disclosed to the specifying authority by the paint company. In some cases, unbeknown to specifying authorities, two or more quite different formulations may be deliberately used for a single product. Although the large national paint manufacturers do not have an exclusive on technical competence, the technical capabilities of the large manufacturer will generally be more proficient than the smaller company.

The testing of performance specifications is discussed in some detail by Tinklenberg.[41]

Performance Testing Procedures

One of the more serious drawbacks of the performance specification is the lack of correlation of laboratory and field performance. While such tests as the ASTM B-117 salt-fog test are widely used, there is much debate over their relevance to practice.[42,43] A heated, atomized salt-water exposure, for example, is hardly typical of conditions to which the normal bridge structure is exposed, and there appears to be absolutely no correlation between hours of salt-fog performance and years in the field. More important, certain paint systems are significantly more vulnerable to this test than others, and their vulnerability changes as the paint ages before testing.[44] In the case of bridges in inland southern states and rural states outside of the snow belt, it is difficult to see how the salt-fog test has any relationship at all to paint deterioration in the field.

Some improvement in salt-fog-testing methodology might come from the addition of periods of drying off, which, together with normal exposure, would stimulate more accurately some of the cycling stresses that eventually result in adhesion loss from the internal strain produced by expansion and contraction of the film as it absorbs and desorbs moisture and/or heats up and cools down. Noncontinuous salt-fog testing of this type has been adopted to some extent by British Rail (BR77). In this test, the normal salt-fog cabinet is replaced by a rotating drum similar to a standard weatherometer, in which panels are sprayed with a Harrison mixture[45] of 0.40% ammonium sulphate and 0.05% sodium chloride. These concentrations of salt do not nullify the inhibitive action of anodic inhibitive pigments as do high salt concentrations, and this, together with the moisture/dry-off cycling, is said to produce considerably better correlation with a normal exterior environment than does either salt-fog or the sulphur dioxide/humidity testing procedures, such as the Kesternich test.[46,47]

The Prohesion test methodology, outlined by Timmins,[48] which is more properly a testing philosophy that utilizes a series of selected tests rather than a single-test methodology, employs the BR77 procedure in conjunction with a low-temperature impact test (BS3900 E.3). These measure relative physiochemical agency changes after an initial screening test over steel surfaces that have a high distribution of rogue peaks with profiles of up to 8 microns. It is possible that this type of test may see increased use. While it is claimed that this test is more realistic, Funke[49] points out that caution is required in assessing a coating's resistance to low salt concentrations when normal use involves at least sporadic contact with high levels of deicing salts or contact with sea water, such as may occur on many bridges.

The practice of scribing panels for salt fog introduces a whole series of deterioration mechanisms that are not present in a paint system free of such holidays and abrasions, and, while admittedly accelerating failure, such failure may not necessarily be totally reproducible in the real world.

Water-immersion tests are, again, hardly appropriate for bridge structures except under temporary, relatively localized conditions. Accelerated weathering tests, such as carbon arc or zenon arc, and fluorescent UV concentration tests may make more sense, although these tests seem less able to stimulate corrosion as paint film deterioraton and, again, are weighted in favor of some paint systems and against others. This type of test has more relationship to sun-belt performance than performance in the Northeast, where road salt seems to be the more major factor in system deterioration.

Limitations of Performance Testing

There are those who view any test program short of actual test performance on bridges as potentially meaningless. Even evaluation of metallic coupons in the field does not duplicate the types of stresses often found on large bridge structures, and test-rack exposures only accurately duplicate the resistance to UV with, at best, some mild salt exposure. Similarly, while accelerated natural weathering tests such as the equatonal mount with mirrors for acceleration plus aqua (water) (EMMAQUA)[50] intensify the natural effects of dew (wet/dry) cycling and sunlight effectively, these duplicate neither the exposure to salt common to all snow-belt bridges nor, more significantly, the vibrational stress common to all bridge structures. Virtually absent from the present performance specification are tests that attempt to duplicate the short- and long-term mechanical-cycling stresses that bear on bridge-paint systems. Some newer tests, such as the KTA-Enviro test methodology and cycling tests being evaluated by the state of Michigan and the New York/New Jersey Port Authority, are promising steps in response to this deficiency. This type of induced strain is typified not only by the response of the coating/substrate system to thermally induced dimensional changes but also by the continual stressing produced by traffic vibrations as a bridge works in service. More important, there seems to be a real need for a test methodology that evaluates the paint system's response to these cycling stresses in the presence of moisture, oxygen, ultraviolet light, and (particularly for coastal substrates and structures in the snow belt) the effect of chlorides. In industrial areas, the contamination by sulphurous and nitrogeneous oxides might also be considered. Because typical environmental conditions change from state to state, some methodology by which elements of this type of testing could be added or subtracted at will would be valuable.

Undoubtedly, these more complicated testing procedures would increase the cost of the quality-control process, but the return-on-investment from a more accurate prediction of the service life of the coating system under real conditions would make the added cost worth it.

In much of the performance testing today, money appears to be wasted and one must question the motivation for certain tests. Such test procedures as carbon-arc and zenon-arc weatherometer testing are inordinately expensive compared to weatherometer tests, such as the fluorescent UV condensation test, and both do much the same job. Claims are made for one in favor of the other, but because neither exactly duplicates field conditions it is difficult to make a definitive case for either one on purely technical grounds.

Some performance tests often seen in today's performance specifications appear superfluous and simply may be leftovers from specifications originally designed by manufacturers for applications other than bridges. The high-temperature quench test seen in many inorganic zinc rich specifications may be a holdover from specifications for the nuclear industry, for example. In the same specification, there is a 1000-hour weatherometer exposure test to prequalify the inorganic zinc. This exorbitantly priced test seems hardly necessary in those applications in which the zinc is to be recoated. Arguments may be made that the test establishes the inorganic nature of the vehicle, but faster, less costly, and more accurate verification of this is possible by examining the infrared spectrum of the vehicle. It is also possible that the original test was added by the manufacturer who first designed the specification to inflate the cost of testing and so dissuade prequalification by other manufacturers (who generally foot the testing bill), especially on smaller jobs.

Full-System Performance Testing

Finally, it makes more sense to conduct performance tests on the total system rather than on individual coats (unless the individual coat represents the total system). The importance of purchasing all elements of a total paint system from a single manufacturer is well established, and, although specifications are littered with such vague, catch-all phrases as "the finish coat shall be compatible with the primer," much of performance testing is still done (at least partially) on individual coats. Laboratory testing of the total coating system is exemplified by procedures used by the Louisiana[51] and Michigan[41] departments of transportation as well as in some other newer specifications. While multicoat systems respond less rapidly to the effects of testing, they are subject to severe internal stresses that are absent or reduced in single-coat systems. In multicoat systems, each coat will respond somewhat differently to any given applied stress, and these differences themselves may compound the strain in the system as a whole. Many potential failure modes, such as lateral cohesive failure of zinc, bubbling tendencies, intercoat adhesion loss, and certain blistering failures, are predictable only from tests on the total system.

Because of changes in the availability of zinc for galvanic protection at cuts, scrapes, and holidays (effects of topcoat on the catchment-area ratio of the zinc anode to the steel cathode), the scribing and testing of untopcoated zinc film tells us nothing about the susceptibility of the system to galvanically protect such damage in topcoated systems with much less available anode.

An important corollary to this type of thinking is that prequalification should be done on the total paint system instead of on a single primer or topcoat. Any change in either primer or topcoat or in their relative film thicknesses would require renewed prequalification of the total system.

Sampling

Irrespective of the type of quality control, effective control of materials is totally dependent on the care and accuracy of sampling. In the field, it is normally the job inspector's duty to sample paint and, while there seems to be more awareness on the part of the inspector of the necessary methodology, sampling accuracy is still apt to be inconsistent. A sample retrieved from an incompletely homogenized pail is no more meaningful than a sample of the wrong paint, and it can cause unnecessary costs and job delays. To eliminate these difficulties many bridge authorities sample materials in the manufacturer's plant immediately after batch completion, or remove unopened five-gallon pails from the job site for mechanical homogenization in the testing laboratory. Not infrequently, a state transportation department will request the actual batch ticket from a supplier and will send an inspector to the manufacturing facility to observe the manufacturing process and to sample both raw materials and the finished products from the plant. An excellent example of this procedure is that set up by the Texas Highway Department.

Theoretically, the technique should give the transportation department complete control of the product, but much will depend on the savvy of the inspector himself. It is not unknown for unscrupulous manufacturers to deliberately confound an inexperienced inspector with a wide range of tricks. These may include changes in manufacturing schedules, deliberate distractions, encoded raw materials, bags and resin drums containing off-grade material, preset pumps, and so on. It is not uncommon to see a paint manufacturer represent a 500-gallon, high-speed dispersion kettle as one with a 1500-gallon or 3000-gallon capacity. They may also work from dummy tickets and even prepare "true samples" of a product in the lab that are then skillfully switched with batch samples practically right under the inspector's nose, with practiced skill.

Although Ray, Henton, and Rideout,[52] citing Law and Anania,[53] consider one sample per batch insufficient to judge conformance, it is rare that more than a single sampling per batch (or per delivery at least) is employed in practice. Law and Anania suggest that a 500-gallon batch of paint packaged in 100 5-gallon pails requires 8 separate samples for a statistically based acceptance sampling. Most individuals in the paint industry who were interviewed on this point considered so many samples too expensive and unnecessary.

Without a doubt, the most effective point for sampling is either in the fabricator's shop or in the field, directly from the painter's bucket itself. This precludes cheating, including "difficulties" that might occur between the manufacturing plant and the contractor's yard or between the opening of the can on the job and the application of the product to the bridge steel. Shop or field sampling is often done by the shop or field inspector, who must be well versed in sampling procedures (if only because any deviation from standard sampling practices will be used by the contractor and/or manufacturer as an excuse for any specification deviation). In the case of two-pack systems (epoxies, zinc, etc.), pot sampling is not possible unless testing facilities are close at hand (extremely rare). In this case, the best and most often followed procedure is to sample by removing entire pails of each component (unopened and randomly selected) from the job.

Due to the large volumes (from the laboratory point of view) involved in mixing two-component samples in the field, it is often necessary for the state laboratories to mix small aliquots. In certain cases (such as the mixing of systems with odd mixing ratios or systems filled by weight, such as zinc powder components, etc.), this procedure may be complicated. Manufacturers discourage the splitting up of batches because of inaccuracy in the field and do not normally provide a weight-to-weight mixing ratio. Weight-to-weight ratios are available to the testing laboratory from most manufacturers on specific request, however, and are recommended for laboratory mixing of small batches of samples taken from larger containers.

On the other hand, it is often impossible to evaluate the field application properties of a particular batch without testing it with the equipment used in the field—usually airless spray. Because airless spray application would require at least two to three gallons to fill the lines, the mixing of larger batches makes considerable sense. Unfortunately, few testing laboratories have airless spray equipment, and application difficulties are usually problems that are not anticipated in the laboratory quality-control methodology.

SUPPORT OF INSPECTION DECISIONS

Frequently in bridge painting, paint testing (particularly performance testing) is not complete before job start-up. The job is often allowed to proceed at the discretion of the contractor, with the understanding that if difficulties arise concerning coating compliance with the specification, the paint will be removed. In practice, when such difficulties do occur, the paint is not always removed, and the problems are sometimes resolved by "negotiations." This usually means that the bridge authority accepts the deviation as a fait accompli, although, more rarely, special performance bonds protecting the authority's interest over an extended period of time have been provided.

Backup support on the part of resident engineers and bridge authorities of inspection decisions that respond to deviations in both job and materials control is mandatory. In practice, however, backup is not always assured. Many quality-control decisions (particularly field decisions relating to temperature and humidity limitations) inevitably affect job schedules and may be as unpopular with the bridge authority as they are with the contractor. Again, removal of newly applied finishes is psychological anathema to all who are associated with job execution, and directives to this effect by the inspector naturally are unpopular. Responsible inspection is often, therefore, a thankless task, and, on too many occasions, the inspectors themselves are overruled by those they represent. Rules are broken in the form of costs, meeting schedules, keeping the job going, avoiding litigation, and so on. The failures that follow have no parent. Not without some justification, the practice has recently been dubbed by some inspectors as "the shuttle syndrome."

It does no good to complain about unprofessional conduct among inspectors if they are compromised by the resident engineers, owners, and bridge authorities. Not only does the practice defeat the intent of the specification and lead to less-than-perfect execution in the immediate situation, but the precedent places the inspector at a serious disadvantage in his attempts to control the job thereafter. While it is, of course, the responsibility of the inspector to defer to decisions made by the resident, any situation in which inspection decisions are overruled should be carefully recorded in the job record. Outside inspection companies, while rarely held responsible for job quality, will, to protect their own reputation, document such instances as a matter of course. The state-employed inspector, with less motivation, may not.

With unprofessional inspection particularly, difficulties such as these can also create an adversarial relationship between the inspector and contractor, resulting in a tendency for more arbitrary inspection decisions and, perhaps, leading to the replacement of the inspection staff, again to the detriment of the job.

Chapter 8

In-Service Monitoring of System Performance

The in-service evaluation of the applied coating system is a necessary procedure not only in order to determine the suitability and efficiency of the system, assess the mode and rate of its deterioration, and establish service life and long-term costs but also to signal recoating schedules and procedures.

The constant feedback of data on the performance of the newer coating systems from the field is invaluable assistance to the system-design engineer in his refinement of the paint system and the parameters of its employment (acceptable surface preparation, film thickness ranges, etc.). By such means, both the claims of the manufacturer and/or the promises of laboratory and test panel data may be checked and modified. Weaknesses may be highlighted and responses developed through formulation changes (vehicle molecular weight changes, PVC adjustments, pigmentation alterations, etc.) or specification adjustments (additional film thickness, revised recoating schedules, etc.). The coating system is thereby continually fine tuned to produce maximum dependable performance where possible, or alternatively, to be eliminated from applications where data reveals inherent insufficiencies.

In addition, the year-to-year recording of the gradual deterioration of the coating system on a given structure allows semiquantification of deterioration and, through extrapolation of resultant deterioration curves, permits reasonable predictions to be made of the intervals wherein the various recoating (maintenance) options retain viability and cost effectiveness. Not all areas of a given structure deteriorate in the same manner or at the same rate. Careful examination of the rates and patterns of deterioration and the correlation of these patterns with bridge-design peculiarities, exposure, traffic flow, drainage, and dry-off characteristics can reveal those areas of a bridge that are more vulnerable to attack and where protection should therefore be increased in future specifications. It can also reveal those areas that are less susceptible to deterioration, where savings may be afforded by reduced coats or lower film thicknesses.

Such data as these have far wider applications than solely on the particular structure from which they were generated because patterns of deterioration are often duplicated on similar structures with the same design configurations in a given geographic area. With enough data from enough structures, an accurate correlation of cause and effect and patterns of vulnerability may with some difficulty be predicted on totally different structures with different exposures.

EXISTING PRACTICES

Unfortunately, at this time, regularized control procedures for monitoring the time-related deterioration of a new paint system on any given bridge after job acceptance are for the most part ill-designed, poorly executed, and, in some cases, nonexistent. In some states, even the awareness of the necessity of the practice is absent. In other states, the yearly monitoring of paint systems is done as part of the general maintenance checks, and, even then, by different divisions from those responsible for the specification design of the coating system or by individuals who have no knowledge of coatings and coating systems and who are not well versed in the terminology of paint-system deterioration. It is not unusual for descriptions and representations of types of bridge-paint failures to be both inexact and totally misleading. Reports that a coating system is peeling, for example, provides little useful data at all. Detailed reporting of the site (within the paint composite) at which peeling is occurring, its prevalence across the structure, and the nature of the exposed interfaces can reveal much about the failure mechanism and its causes.

More often there is little or no exchange of data between those who are responsible for the collection and recording of the year-to-year observations of paint-system condition and those who are responsible for the conception and design of the specification in the first place. In a majority of cases, where records of coating-system service are available, they are, like inspection records, buried in dead files and seem rarely reviewed by the specification-design engineers. In many instances of bridge-paint failures, in fact, because of the poor quality, unreliability, or unavailability of inspection and postpainting data it is often as easy for the failure analyst to reconstruct the failure from his own observations and interpretations of data available at the failure site (substrate quality, mil profile, film thickness, etc.) than it is to rely on such inspection and postpainting records. Only where severe and obvious failure occurs soon after job completion does the specification-design engineer have occasion to monitor the performance of his system; in these cases, the abandonment of the specification is as likely as its refinement.

In less extreme cases, it is not unknown for experimental paint systems to be left and even forgotten. Certainly, available data are often confused and inexact, especially after several years of exposure. Better records are kept only when this type of monitoring is done by self-motivated chemists and engineers in addition to their assigned duties or by the paint manufacturer eager to see the adoption of the system by the state or authority on a more formal, widespread basis. Quality monitoring of experimental paint systems by many states and bridge authorities is often related to the dedication of a particular individual rather than to the success of any formalized, official postpainting control program. Unfortunately, personnel replacements, promotions, and changes in the requirements and marketing philosophies of manufacturers have a marked negative effect on such individualized monitoring.

In Great Britain, [1,2,3] the control and categorization of the performance of coating systems seems more systematized, with failures categorized by the British Department of Transportation in terms of the severity of the problem. Their categories I and II represent failures of a local nature or where the finish coating only has succumbed to the natural effects of weathering. Categories III and IV represent those instances in which finish coat (III) or the general system (IV) have deteriorated before the expected term of successful performance. In the later categories, a prespecification inspection of the condition and a determination of the causes of the deterioration by qualified paint technologists is mandated before a new specification is considered.

Panel-testing programs and more formalized span-testing programs receive much better attention (continual, careful monitoring of systems) than do actual full-scale applications on bridges. Keane, in his NCHRP Report #74A,[4] discusses many such limited evaluations. Other test programs are many and are reported widely in the literature.[5–11] These test programs are more formalized, with more companies and individuals having interest in their outcome, and, perhaps more important, are evaluated by the same departments and individuals that had input into their design. Such programs as these provide system-design engineers with much valuable and precise data concerning UV resistance (gloss and color retention), corrosion resistance, film embrittlement, and intercoat security from which rust and/or deterioration curves may be compiled and extrapolated. Unfortunately, as noted earlier, limited spans and, especially, panel-testing programs are not as representative of actual service as are the trial applications of systems on entire structures. The relevance of both patterns and rates of panel deterioration (and their extrapolation) may not hold for the system's performance on a complete structure under real-life conditions. Panel tests are rarely applied in the field by bona fide contractors, and even short-span tests are often weighted in favor of the system by the presence of manufacturers, better-than-normal inspection at the time of application, and, often, better-than-average conditions. In some respects, a system may be particularly vulnerable to certain bridge-work service that is not necessarily present on a particular span exposure.

Even less representative are panel tests of systems designed for spot maintenance or for use over rusted surfaces. Rarely are panels available that are truly typical of these types of substrates, and available substrate uniformity is inevitably problematic. Finally, it is rare that panel tests can duplicate the stress conditions (vibrations, etc.) that are always present on working bridge structures. Full-structure trials done on a bid basis with conventional inspection (or lack of it), with all the details and intricacies of difficult areas, and without the constant presence of manufacturer's representatives are more in keeping with the real bridge-painting world.

FEDERALLY ENCOURAGED PROGRAMS

New efforts to formalize postpainting monitoring are now being encouraged by the FHWA in the form of federally funded programs. Such programs include cases in which new paint systems being tried by state public highway departments are monitored in the field both during and after job execution. In these studies, the evaluator makes a prepainting evaluation of the structure, noting its location, design, and environment, including details of the exposure (proximity to marine and industrial pollutants, extent of the use of deicing salts, nature of the crossing, etc.) as well as details of the type, history, and condition of existing paint systems. Patterns of existing deterioration are noted and correlated with obvious causes and the type of deterioration (corrosion, cracking, blistering, etc.) is recorded. The evaluator then examines the specification and, if necessary, the coatings to be employed and is present on the job to monitor the performance of applicators, inspectors, engineers, and the paint manufacturer. As the job begins, observations on the readiness with which the theoretical specification is translated to practice are made and the operator's and inspector's learning curves are noted. Records are made of equipment employed, unit prices, surface quality, mil profiles, film thicknesses, dilution solvents, and time intervals before recoating. Difficulties related to job execution are noted and related to materials, bridge-design peculiarities, or personnel. Several unscheduled inspections are made during the course of the job and, upon completion, the evaluator makes the first of several inspections that will document the service produced by the coatings over the next several years. In this manner, performance can be accurately correlated not only to the coating system itself but also to the quality of both job execution and inspection, as well as to exposure and bridge-design considerations. More important, it is hoped to define more accurately the mechanisms and causes of any subsequent paint failure, in order to facilitate rectification of specification design, application, or formulation, or, in the extreme case, with reasonable authority, to abandon the coating system.

DIFFICULTIES WITH PROPRIETARY SYSTEMS AND PERFORMANCE SPECIFICATIONS

The value of such postpainting monitoring is, of course, reduced somewhat with performance specifications based on proprietary products. Unless substantial formulating data is made available to the bridge authority by the paint manufacturer, it is impossible to correlate the observed deterioration with formulation, and, in these cases, the postpainting monitoring benefits the paint manufacturer more than anyone else. The QPL-based system has supposedly had all the bugs worked out in proving tests, but proving tests do not always anticipate undisclosed minor changes in formulations or color, which may themselves affect service.

A typical example of this sort of difficulty is found in the chalking failure seen with certain high-build vinyl coatings used in many states. Authorities in Maine, for example, while observing substantial differences in the early chalk resistance of several bridges painted with different batches of the same color of the same product, had no way of correlating this data with formulation, unless they initiated substantial and costly failure-analysis procedures (D. Leyland, Maine Department of Transportation, personal communication, 1984).

The paint manufacturer, on the other hand, can learn much from the failure records and, if cost considerations allow, can revise the formulation for use on subsequent structures. Depending on the type of problems and specifications, such formulation revisions may or may not be reflected in a specification change, but proprietary considerations do not favor the dissemination of such data. The state or bridge authority, therefore, for the time and money spent in recording the failure, may or may not learn the real cause of the failure and may remain totally ignorant of the formulating remedy. Other manufacturers, with similar interpretations of the same specification, are left to repeat the same mistakes.

QUANTIFICATION OF DETERIORATION

Even in experienced hands, the accurate quantification of paint-system deterioration and the establishment of deterioration patterns is not easy. The problem is typified by considering the most obvious aspect of deterioration—rusting. While several organizations have attempted to organize scales of deterioration (ASTM-D-610, SSPC-PA-4, federal standard test methods #141A, method 6451), application of such scales to the assessment of real situations is somewhat difficult. Rarely are rust patterns uniform across an entire structure and, while it may be very valuable to report patterns section by section (i.e., interior webs of support beams, exterior girder plates, flanges, etc.), patterns are rarely uniform across even a single beam. Appleman[12] proposes a scheme for total-area rating that considers the structure in terms of individual elements and then subdivides such elements into individual sections that may themselves be more definitively rated. This approach fits well with the "specific-utility" concept of system design discussed by Hare.[13]

Evaluations are, at best, subjective, and differences between observers may add to the difficulties. Photographic monitoring, which provides a visual record of the same section of the structure over a period of years and from which a more accurate rust curve may be prepared, is valuable. Although it resolves many of the rating difficulties and provides a more satisfactory record of the rates of system deterioration, which may be studied at any time, photographic recording is expensive, cumbersome, and is rarely used.

Compounding these difficulties is the fact that rusting is not the only form of deterioration. Oxidizing systems, which are subject to autoxidation and chain scission with age, embrittle, check and crack, and lose adhesion (particularly over old mill scale). While rust curves might indicate that spot repair would be possible, such paint films would not make suitable substrates for new paint systems, particularly where spot repair involves blasting. Damage to the areas peripheral to the spot blast will weaken the old system, resulting in loss of adhesion after recoating, especially where the recoat films are strongly cohesive. Similarly, severe topcoat chalking, delamination, or erosion, without accompanying corrosion, may be reasonably considered criteria for repainting, especially where aesthetic requirements are important. As a result, it is the evaluation of the condition as a whole and not just the individual components of deterioration (e.g., rusting) that is essential in the assessment of system performance and in the determination of required procedures for rectification and renewal. Quantitative representations of such overall patterns are much more difficult, although rough qualitative ratings have been attempted by the British Standards Institute.[14]

SYSTEM SERVICE LONGEVITY

Perhaps the most important long-term benefit of the postpainting monitoring process is an accurate establishment of system longevity. The lack of a more definitive appreciation of coating system service life at this time may, in part, be related to the disorganized state of the postpainting monitoring process and the consequent dearth of measured mean deterioration curves. Sound scientific principles exist to establish the optimum time for repainting in terms of rate and extent of deterioration and the consequent cost of repainting at any specific epoch.[12,15] In practice, repaint maintenance decisions are more arbitrarily based. What one man sees as a maintenance requirement, another might not. Because the availability of maintenance dollars has never kept pace with the maintenance requirements for steel-bridge painting, the overall tendency has been to protract the interval between paintings well beyond the service life of the system. This practice tends to inflate expectations of service life based on past experience, as well as making rectification more difficult and expensive.

In fairness, of course, there are many other factors that confound good quantification of system service, not the least of which are the wide divergencies in the quality of specifications, job execution, and inspection noted in earlier chapters. Bridge design, the quality of the substrate (steel, rust grade, old paint condition, etc.), especially in maintenance painting, as well as the appropriateness of the degree and type of surface preparation are also important factors. As the concept of the substrate as part of the coating system has grown and coating systems have been tied to specific surface-preparation criteria, the effects of widely divergent surface-preparation techniques on expected service life have lessened, although, within a range of acceptable surface quality (e.g., SSPC-SP-6), enough room still remains to have a marked effect on system longevity.

It is the environment in which the coating system must do its job, however, that has the most profound effect on service life and the mode of deterioration. The SSPC environmental-zone approach is helpful in the semiquantification of the type of environment. However, within these general classifications, there are inevitably degrees of exposure not only from one part of the country to another but from bridge to bridge and even from section to section of a particular bridge, depending upon location, type of crossing, bridge design, condition, and traffic volume.

Most snow-belt structures undoubtedly should be classified as 2B, and that classification should worsen in the expansion bay areas or where deck leaks occur. Sheltered underdeck areas of bridges in good condition over nonwater crossings might easily be classified as a rather mild 1B environment. Over inland waterways, a 2A rating might be more appropriate for the same underdeck steel and, over a busy well-salted highway, splash back from the highway below will intensify the immediate environment beneath the bridge to class 2B (especially on the bottom flanges). Similarly, on bridges in the Northeast with leaking decks, an environmental zone similar to or even worse than a conventional class 2B rating will result beneath the deck in locations exposed to the leaking salt water. Salt trapped within the deck will leak across parts of the underdeck support structure, not only in winter but whenever it rains or snows.

These things being so, the question of expected system service life (although reasonable) can often be somewhat simplistic. Without data on the same or similar structures in the same or similar environments it can be quite difficult to answer. Usually, the best that is possible, even with well-established paint systems, is an educated guess based on the presumption that the structure is in good condition and that the specification is appropriate and followed without appreciable deviation.

It is, then, with some misgiving that the service-life estimates in Tables 8-1 through 8-3 are offered. They are based on numerous interviews with highway departments, paint manufacturers, contractors, engineers, and other specifying authorities across the country, together with data from a few available published sources.[10,15,16] Unfortunately, many such sources reflect the use of these same coatings in industries other than bridge painting, and considerable divergency is very apparent. Consequently, the service-life values given in Tables 8-1 through 8-3 are much less conservative than Roebuck and Brevoort's, which are based on in-plant service. For the most part, the present figures represent the time to first maintenance of newly applied systems on new bridge steel, maintenance being performed at 10% to 15% system deterioration, which may itself be conservative in many states. Somewhat reduced figures (proportionately) may be expected from the same systems reapplied as in field maintenance. Many of the values, especially for the newer systems, have had to be projected (this is indicated in the tables with a letter P). With some systems (e.g., system group XIIIA), the maximum 1B environment values of forty years may be too conservative. Inevitably, these figures will be universally offensive and will invoke criticism, especially among those whose products are typified. Any attempt to refine them is to be both applauded and solicited. But only by consistent and accurate data collection from an ordered postpainting monitoring process can such figures be properly modified and their credibility refined.

Table 8-1
Service Comparison of Inhibitive Coating Systems for Steel Bridges

System Group	System Description	Surface Preparation (SSPC-SP)	Film Thickness (Dry Mils)	Resistance to Water	Resistance to UV	Resistance to Alkalies	Resistance to Acidic Pollutants	Resistance to Abrasion and Impact	Service in Environment 1B—Dry Exterior (Years)	Service in Environment 2A—Fresh-Water Wet (Years)	Service in Environment 2B—Salt-Water Wet (Years)
I	*Primer:* red lead/linseed oil	2	2.0	Good	Good	Poor	Good	Fair	30	15	10
	Intermediate: red lead-iron oxide/oil-alkyd		1.5								
	Finish: aluminum/alkyd or phenolic		1.0								
	Finish: (alternative) BLSC/oil-alkyd		1.0								
II	*Primer:* zinc chromate/alkyd	3 or 7	2.0	Good	Good	Fair	Fair	Good	25	13	8
	Intermediate: zinc chromate/alkyd		1.5								
	Finish: aluminum/alkyd or phenolic		1.0								
	Finish: (alternative) BLSC/oil-alkyd		1.0								
III	*Primer:* BLSC/linseed oil-alkyd	3 or 7	2.0	Good	Good	Fair	Fair	Good	30	15	10
	Intermediate: BLSC/linseed oil-alkyd		1.5								
	Finish: BLSC/linseed oil-alkyd		1.5								

(continued)

Table 8-1
(Continued)

System Group	System Description	Surface Preparation (SSPC-SP)	Film Thickness (Dry Mils)	Resistance to Water	Resistance to UV	Resistance to Alkalies	Resistance to Acidic Pollutants	Resistance to Abrasion and Impact	Service in Environment 1B—Dry Exterior (Years)	Service in Environment 2A—Fresh-Water Wet (Years)	Service in Environment 2B—Salt-Water Wet (Years)
IV	*Primer:* nontoxic inhibitive/oil-alkyd	6	2.5	Good	Good	Fair	Poor	Good	20P	10P	6P
	Intermediate: nontoxic inhibitive/oil-alkyd		2.5								
	Finish: nontoxic inhibitive/oil-alkyd		2.0								
	Finish: (alternative) aluminum/alkyd or phenolic		2.0						22P	12P	8P
V	*Primer:* inhibitive latex	6	2.0	Fair	Very good	Fair	Poor	Fair	20P	10P	6P
	Intermediate: inhibitive latex		2.0								
	Finish: noninhibitive latex (2 coats)		3.5								
	Finish: (alternative) noninhibitive latex aluminum/latex		1.5								

P=projected service life.

Table 8-2
Service Comparison of Zinc-Based Coating Systems for Steel Bridges

System Group	System Description	Surface Preparation (SSPC-SP)	Film Thickness (Dry Mils)	Resistance to Water	Resistance to UV	Resistance to Alkalies	Resistance to Acidic Pollutants	Resistance to Abrasion and Impact	Service in Environment 1B—Dry Exterior (Years)	Service in Environment 2A—Fresh-Water Wet (Years)	Service in Environment 2B—Salt-Water Wet (Years)
VIA	*Primer:* alkyl silicate zinc *Finish:* high-build vinyl	10	3.0 5.0	Excellent	Very good	Excellent	Excellent	Very good	40P	25P	15
VIB	*Primer:* alkyl silicate zinc *Intermediate:* WP-1 vinyl wash primer *Finish:* high-build vinyl	10	3.0 0.3 5.0	Excellent	Very good	Excellent	Excellent	Very good	40P	25P	15
VIC	*Primer:* alkyl silicate zinc *Intermediate:* epoxy/ polyamide (high build) *Finish:* aliphatic urethane	10	3.0 4.0 2.0	Excellent	Excellent	Excellent	Excellent	Excellent	40P	25P	17P
VID	*Primer:* alkyl silicate zinc *Intermediate:* epoxy/ polyamide (high build)	10	3.0 5.0	Excellent	Fair–Good	Excellent	Very good	Excellent	40P	25P	16P

(continued)

Table 8-2
(Continued)

System Group	System Description	Surface Preparation (SSPC-SP)	Film Thickness (Dry Mils)	Resistance to Water	Resistance to UV	Resistance to Alkalies	Resistance to Acidic Pollutants	Resistance to Abrasion and Impact	Service in Environment 1B—Dry Exterior (Years)	Service in Environment 2A—Fresh-Water Wet (Years)	Service in Environment 2B—Salt-Water Wet (Years)
VIE	*Primer:* alkyl silicate zinc	10	3.0	Excellent	Excellent	Excellent	Excellent	Excellent	40P	25P	15P
	Intermediate: high-build aliphatic urethane		4.0								
VIIA	*Primer:* alkaline silicate zinc	10	3.0	Excellent	Very good	Excellent	Excellent	Very good	40P	25P	15
	Intermediate: WP-1 vinyl wash primer		0.3								
	Finish: vinyl		2.0								
VIIB	*Primer:* alkaline silicate zinc	10	3.0	Excellent	Very good	Excellent	Excellent	Very good	40P	25	15
	Intermediate: vinyl		1.0								
	Finish: vinyl		1.5								
VIIIA	*Primer:* phenoxy zinc	6	3.0	Excellent	Very good	Excellent	Excellent	Very good	35P	13	10
	Intermediate: WP-1 vinyl wash primer		0.3								
	Finish: vinyl		5.0								
VIIIB	*Primer:* phenoxy zinc	6	3.0	Excellent	Excellent	Excellent	Excellent	Excellent	35P	13P	10P
	Intermediate: epoxy/polyamide (high build)		4.0								
	Finish: aliphatic urethane		1.5								
IX	*Primer:* chlorinated rubber zinc	6	3.0	Excellent	Very good	Excellent	Excellent	Very good	35P	15P	10P

	Intermediate: chlorinated rubber (high build)		3.5								
	Finish: chlorinated rubber		2.0								
X	*Primer:* vinyl zinc rich	10	2.0	Excellent	Very good	Excellent	Excellent	Very good	35P	20P	12P
	Intermediate: vinyl		3.0								
	Finish: vinyl		3.0								
XIA	*Primer:* epoxy/polyamide zinc rich	10	3.0	Excellent	Fair–Good	Excellent	Very good	Excellent	35P	20P	12P
	Intermediate: epoxy/polyamide		4.0								
	Finish: epoxy/polyamide		2.0								
XIB	*Primer:* epoxy/polyamide zinc rich	10	3.0	Excellent	Excellent	Fair	Fair	Very good	35P	20P	12
	Intermediate: epoxy polyamide zinc rich		1.5								
	Finish: vinyl aluminum		3.0								
XIC	*Primer:* epoxy/polyamide zinc rich	10	3.5	Excellent	Very good	Excellent	Very good	Very good	35P	20P	12
	Intermediate: epoxy/polyamide red lead (marine environments only)		1.5								
	Finish: vinyl toluene/acrylic		1.5								
XIIA	*Primer:* uralkyd zinc rich (2 coats)	10	3.5	Excellent	Very good	Excellent	Very good	Very good	30P	15P	10
	Finish: vinyl toluene/acrylic		1.5								

(continued)

Table 8-2
(Continued)

System Group	System Description	Surface Preparation (SSPC-SP)	Film Thickness (Dry Mils)	Resistance to Water	Resistance to UV	Resistance to Alkalies	Resistance to Acidic Pollutants	Resistance to Abrasion and Impact	Service in Environment 1B—Dry Exterior (Years)	Service in Environment 2A—Fresh-Water Wet (Years)	Service in Environment 2B—Salt-Water Wet (Years)
XIIB	*Primer:* moisture-curing urethane zinc	6	2.0	Excellent	Very good	Fair	Fair	Excellent	35P	25P	12P
	Intermediate: epoxy/ polyamide		2.0								
	Finish: aliphatic urethane		2.0								
XIII	*Primer:* galvanizing	8		Excellent	Excellent	Good (pH 9.5)	Fair–Poor	Excellent	35	25	10
XIIIA	*Primer:* galvanizing	8		Excellent	Very good	Excellent	Excellent	Very good	40P	30P	18P
	Intermediate: WP-1 vinyl wash primer		0.3								
	Finish: vinyl		5.0								
XIIIB	*Primer:* zinc metallizing	5	8.0	Excellent	Very good	Excellent	Excellent	Very good	40P	25P	15P
	Finish: vinyl (carboxyl-ated)		2.0								
XIIIC	*Primer:* zinc metallizing	5	8.0	Excellent	Very good	Excellent	Excellent	Very good	40P	25P	15P
	Intermediate: WP-1 vinyl wash primer		0.3								
	Finish: vinyl (hydroxyl-ated)		2.0								

P=projected service life.

Table 8-3
Service Comparison of Barrier Coating Systems for Steel Bridges

System Group	System Description	Surface Preparation (SSPC-SP)	Film Thickness (Dry Mils)	Resistance to Water	Resistance to UV	Resistance to Alkalies	Resistance to Acidic Pollutants	Resistance to Abrasion and Impact	Service in Environment 1B—Dry Exterior (Years)	Service in Environment 2A—Fresh-Water Wet (Years)	Service in Environment 2B—Salt-Water Wet (Years)
XIVA	*Primer:* vinyl (carboxyl-ated)	10	1.5	Excellent	Very good	Excellent	Excellent	Very good	40P	25	15
	Intermediate: vinyl (2–5 coats)		1.5(×3)								
	Finish: vinyl		1.5								
XIVB	*Primer:* WP-1 vinyl wash primer	10	0.3	Excellent	Very good	Excellent	Excellent	Very good	40P	25	15
	Intermediate: vinyl (hydroxylated) plus (2–4 additional coats)		1.5(×3)								
	Finish: vinyl		1.5								
XV	*Primer:* chlorinated rubber	6	1.5	Excellent	Very good	Excellent	Excellent	Very good	35P	20P	12
	Intermediate: chlorinatcd rubber		3.0								
	Finish: chlorinated rubber		1.5								

(continued)

Table 8-3
(Continued)

System Group	System Description	Surface Preparation (SSPC-SP)	Film Thickness (Dry Mils)	Resistance to Water	Resistance to UV	Resistance to Alkalies	Resistance to Acidic Pollutants	Resistance to Abrasion and Impact	Service in Environment 1B—Dry Exterior (Years)	Service in Environment 2A—Fresh-Water Wet (Years)	Service in Environment 2B—Salt-Water Wet (Years)
XVIA	*Primer:* epoxy/polyamide	6	2.5	Excellent	Fair–Good	Excellent	Very good	Excellent	35P	25P	12
	Intermediate: epoxy/polyamide		2.5								
	Finish: epoxy/polyamide		2.0								
XVIIA	*Primer:* aluminized epoxy mastic	7	5.0	Excellent	Very good	Excellent	Very good	Excellent	10P	7	4
XVIIB	*Primer:* aluminized epoxy mastic	7	5.0	Excellent	Very good	Excellent	Very good	Excellent	25P	15	10
	Finish: aluminized epoxy mastic		5.0								
XVIIC	*Primer:* aluminized epoxy mastic	7	5.0	Excellent	Good	Excellent	Very good	Excellent	25P	13P	8P
	Finish: epoxy/polyamide color coat		5.0								

XVIIIA	*Primer:* moisture-curing urethane aluminum	6	2.5	Excellent	Very good	Very good	Very good	Excellent	10P	7	4
	Intermediate: moisture-curing urethane aluminum		2.5								
	Finish: moisture-curing urethane aluminum		2.5								
XVIIIB	*Primer:* moisture-curing urethane aluminum	6	2.5	Excellent	Excellent	Excellent	Excellent	Excellent	15P	10P	8P
	Intermediate: moisture-curing urethane aluminum		2.5								
	Finish: aliphatic urethane color coat		2.0								
XIX	*Primer:* coal tar epoxy	10	8.0	Excellent	Poor	Excellent	Excellent	Excellent	40P	30P	15
	Finish: coal tar epoxy		8.0								

P=projected service life.

COATING-SYSTEM FAILURE

In the extreme case, postpainting monitoring may, in effect, become the documentation and analysis of a paint-system failure. Unfortunately, the definition of paint failure is elusive, variable, and dependent upon the expectations and standards of the particular observer. Theoretically, any paint-film deterioration mode other than the gradual erosion of the paint system over a number of years might be considered a failure. Certainly, any mode of deterioration that involves intercoat delamination, lateral cohesive separation, excessive cracking, blistering, or widespread pinhole-type corrosion and that dictates special procedures (e.g., total removal of the existing paint system) in subsequent repainting specifications might justifiably be considered a failure of the paint system. Certain types of failure—bubbling in topcoats over inorganic zinc, excessive early chalking, and the like—are often not considered failures at all but are accepted as a "characteristic of the coating" that is to be tolerated. Assessments of this type are quite arbitrary. While excessive chalking in an epoxy topcoat might be allowed, in a vinyl or silicone alkyd, such deterioration would not be allowed.

For their definition of failure, Ray, Henton, and Rideout[17] adapted a criterion involving the failure to achieve an expected or designed service life. Designed service itself, however, is difficult to quantify. In practice, it seems to be only those catastrophic deteriorations of the paint system that occur within a few months or, at most, a few years of application that are universally classed as failures.

Most engineers and nontechnical highway staff do not seem to appreciate the longer service-life possibilities of paint systems today and, notwithstanding the claims of paint companies and specification writers, seem to accept some degree of deterioration within the proposed service life of a paint system. The extent and obviousness of the deterioration will also play a part in the perception of failure. Widespread intercoat delaminations, made obvious by large color differentials but not necessarily accompanied by corrosion or structural deterioration, may cause greater dissatisfaction than more insidious failures of greater consequence.

Failure Analysis

Failure analysis may be carried out by in-house inspectors or engineers but more usually is contracted out to private engineering companies and specialists employed specifically for the purpose, particularly if it seems likely that litigation may become involved. As is noted by Ray, Henton, and Rideout,[17] hard data on specific cases are not always readily accessible, and the details of many coating system failures remain poorly documented even after litigation. Pitfalls of formulation, application, and specification design that are made evident in such cases seem to diffuse gradually into the technology by word of mouth rather than becoming available through widespread publication or more formal distribution channels. This practice is embraced by bridge authorities as much as by paint manufacturers and applicators. Unfortunately, this "Failure has no parent" philosophy does nothing to capitalize on the only positive outcome of failure, the dissemination of technological understanding and system refinement that relates to that failure.

The actual methodology of the analysis of the paint-system failure is vague and poorly documented. While there are many texts describing isolated failure modes,[18–22] terminology,[19,23,24] and prevention, there are few that provide step-by-step instructions as to investigatory procedures. Certain books and papers may highlight specifics and delineate support methodology[25] and analytical techniques.[26,27,28] Works are also available on the forensic examination of paints[26,29] and on case histories of failures,[30,31,32] which may betray some technique. An excellent overview of many of the analytical techniques, including some of the latest instrumentation available for use in failure analysis may be found in the Gardener/Sward *Paint Testing Manual.*[33] An ordered treatise on the what, the where, and the how to of failure analysis, however, does not seem to be available in the literature at this time. KTA Associates in Pittsburgh is currently providing failure analysis instruction courses as well as forum opportunities to discuss particular failures. No specific instructional manual for such courses has been developed to date (K. Trimber, KTA Associates, Pittsburgh, PA, personal communication, 1986).

Ray, Henton, and Rideout[17] conclude that the best failure analysis practice is to list all possible causes of the observed failure and then reduce this list of possibles by a data-collection process. They conclude that, in practice, cause and effect may not be unequivocally established, and that a "most probable mechanism of failure" is the more likely outcome of this failure analysis process. They cite several sources for the generation of possible causes and discuss several bridge-paint-system failures. Other sources of assistance in failure analysis include technical service departments of coatings manufacturers, manufacturers of cleaning and spraying equipment, and materials and maintenance engineers. The input of other experts and engineers in the field (with names obtained from books, papers, and bibliographies on specific or related areas of technology) might also be useful.

The problem with the Ray, Henton, and Rideout approach is that it presupposes that the true failure mechanism will be included in the input from at least one of the sources of information consulted, or that the specific failure mechanism has, in fact, been isolated on other structures. All too often, paint-system failures are found to be not so much related to a single causative factor as to several different factors that, compounded together, produce a failure not seen where any one factor is missing. It is not unusual to find a certain degree of uniqueness in severe coating failures.

In their enumeration of analytical methods that might be instrumental in detection of cause, Ray, Henton, and Rideout cite several viable techniques that are divided into two phases: inspection (discovery) and analysis. In the first phase, inspection reports and logs, weather conditions at the time of application, and so on are reviewed, and, depending on the results of this process, a chemical and/or physical analysis of the various films and interfaces is made. This analysis may include measurements and observations of film thickness, subfilm steel condition, interface conditions, and defect sites not only by means of standard paint-inspection tools but with a wide variety of instrumental analytical techniques.

While a comprehensive treatment of the failure-analysis technique is well beyond the scope of this volume, in view of the relative deficiencies of published material on this aspect of the discipline, an alternative approach to analysis (based on techniques developed by this author and others in the field) is offered here. This procedure is typical of the progression of analysis of cause and mechanism of coating-system failure on bridge structures. It can be divided into six steps.

1. *Failure definition.* The failure must first be defined with accuracy. This may be done by itemizing the following:

a. Manifestations of deterioration
b. Extent and location on structure
c. Patterns of failure that may be related to bridge design, exposure, traffic patterns, and the like

d. Methodology of initiation and propagation (if data is available)

2. *Data collection.* Data relative to the failure must then be collected from the following typical sources:

a. Visual inspection of the failure in the field
b. Laboratory analysis of failure (microscopic examination, film analysis, analysis of interference materials, corrosion products, etc.)
c. Materials batch-testing data
d. Specification
e. Job inspection and postpainting inspection records
f. Climatological records of location during and after application
g. Prior history of specification on other structures
h. Comparison of data pertinent to this structure with that of other successful structures
i. Pertinent change orders, approvals, extra work orders, and general correspondence related to the job

3. *Isolation of anomalies.* All anomalies, differences, and inconsistencies that set the job in question apart from the basic system specification, from good painting practice as it applies to this system, or to the past use of the system on successful structures are then isolated and reviewed against known susceptibilities of this or similar systems in light of the observed manifestations of failure and data from field and laboratory studies. Areas signaled may be related to the following:

a. Specification and change orders
b. Applicator expertise and quality of job execution
c. Bridge design and location
d. Applied system film characteristics (determined substrate quality, film thicknesses, etc.)
e. Climate and season
f. Paint materials
g. Inspection
h. Traffic density and flow
i. Local environment

4. *Identification of specific areas of concern.* From the data collected in Steps 2 and/or 3 above one or more areas of concern will condense as, if not a possible failure mode, then as being most suspect. The analyst then must seek to establish which area, if any, fits a sensible failure mode that could be supported by that combination of data that applies to the particular painting. To do this he will draw from the following sources:

a. Basic scientific principles
b. Experience of researcher
c. Published data of failure manifestations of similar description
d. Literature surveys
e. Consultation with other authorities and experts
f. Consultation with paint suppliers and raw-material suppliers
g. Experimentation

5. *Postulation of possible cause.* The exercise above should provide one or more most-likely-failure scenarios. Each will then be reviewed and attempts to deal with apparently non-supportive or contradictory data will be made by further review and analysis of the particular areas of difficulty. Rectification of such difficulties will lead ideally to a single most-probable cause.

6. *Verification.* As a final step the analyst may seek either to duplicate the failure in the laboratory or in the field. In some cases, it may be possible to predict the incidence of failures on structures painted (with similarly deficient materials or methodologies) after that on which the analysis was performed. Success in either respect is tantamount to proof of the proposed failure mechanism. Inability to duplicate the failure by laboratory model is not necessarily indicative of an inaccurate analysis, and care should be taken to review the appropriateness of test methodology before returning to reevaluate the failure for cause.

The methodology described by Ray, Henton, and Rideout tends to force data into one of several possible failure mechanisms. In this respect, it is somewhat narrow-sighted. The alternative approach offered above would seem to be the reverse of this—that is, to collect all data concerning the particular painting and identify conditions not in keeping with established good practice in specification design, application, formulation, and the like. From these findings and from observations and analysis of interfaces, corrosion products, and bridge-design configurations several most-probable scenarios of failure might emerge, and these might subsequently be reviewed in terms of experience, literature surveys, and the input of others.

Causes of Coating System Failure

Poor surface preparation and inadequate film thickness have been widely held as being the predominant causes of the premature failure of coating systems. An NACE report estimates that some 70% of coating system failures may result solely from poor surface preparation.[34] While there is probably some truth in this, such generalities are dangerous because they foster a preconceived bias against the contractor and may discourage a more complete analysis of the failure. They may even become a convenient hook on which to hang responsibility in cases in which a more definitive interpretation remains elusive. Other mechanisms are by no means rare, and these may be related to inappropriate specifications, faulty materials, climatic conditions during application, other application problems, poor recoating schedules, excessive film thickness, and so on. As noted above, it is not rare that coating failures are found to be unfortunate outcomes of the interaction of several apparently unrelated phenomena.

Failures may be recognized as inherent possibilities in some paint systems because they may be manifestations of certain sensitivities of the chemical composition, film-formation mechanism, or other inherent characteristic of individual vehicle types that lend some definite vulnerability to certain failure modes. Conversely, such may be the properties of the same vehicle system that other modes of failure may be more unlikely in this system than in others. While it is equally misleading to lend too much weight to such considerations, a review of such sensitivities may provide some additional input into the analysis of film failures.

All coating types have failure modes that are inherent to their generic nature and that must be borne in mind when evaluating any specific failure. The list that follows reviews the inherent sensitivities of the most common vehicle types.

Oil, oil/alkyd, and alkyd systems
- Drying problems (sagging, wrinkling, flashing, film contamination) (effect of temperature, lead content, oil length, and barrier pigmentation)
- Wrinkling and poor thorough drying of films of excessive thickness
- Susceptibility to saponification, effect of cathodic alkali at breaks
- Age-induced embrittlement (effect of mechanical- and temperature-induced stress)
- Solvent sensitivity of newly applied films
- Lead paints subjected to sulphide-induced color change
- BLSC color floating in finish

Latex systems
- Early water sensitivity
- Reduced coalescence (crazing and flashing) at low temperature (effect of porous primers and high winds)
- Flash rusting (effect of damp and rusted surfaces)
- Dirt pick up at high temperatures (thermoplastic film softening)
- Embrittlement at low temperatures
- Wet paint subject to freezing
- Poor adhesion to oils and foreign contamination; interference materials, such as chalk, may be particularly threatening to good adhesive security

Inorganic zinc primers
- Surface preparation critical (crazing and flaking on poor surfaces)
- Film thickness critical (mud cracking and flaking at high film builds, especially on poorer surfaces)
- Tendency of high PVC/CPVC ratio films to crack
- Rapid drying—overspray (particularly on high structures or in wind) (effect on film formation and physicals)
- Application difficulties—overreaching
- Bubbling and pinholing of porous films when recoated
- Easy contamination of dark porous films with materials difficult to detect—oil, water, and the like.
- Primer films sensitive to attack at extreme pH ($<6, >10$)
- Climate-controlled cure kinetics (moisture-cured alkyl silicates may be inapplicable for use in desert climates)
- Inadequate cure (especially single-pack inorganics) leading to weakened films if prematurely recoated, with later lateral cohesive failure (zinc splitting)
- Early water sensitivity with alkaline silicate systems
- Unsuitability for recoating with oils and alkyds—delamination of finish coats
- Sensitivity of two-pack systems to component mixing ratios
- Sensitivity of all systems to incomplete homogenization on application (incompletely mixed product affects cathodic protection and physicals)
- Single-pack systems may gas in can

Organic zinc rich primers
- Surface preparation quality affects performance more than adhesion
- Cathodic protection and physical properties are highly dependent on complete homogenization (risk of corrosion and weak porous films); noncontinuous agitation during application will produce heterogeneous films

- High PVC/CPVC ratio films are porous, more prone to mudcracking, and cause bubbling and pinholing of topcoats
- Cure problems related to insufficient access of temperature, air (epoxy esters and uralkyds), water (moisture-curing urethanes), or delayed solvent evaporation (phenoxies, chlorinated rubbers, etc.) may cause lateral cohesive failures (zinc splitting) if recoated too rapidly
- Two- and three-pack systems sensitive to mixing deficiencies
- Fast-drying systems subject to dry-spray problems (effect on film formation and physicals)
- Vertical heterogeneity (intrafilm settling of pigment) in thick films of slow-drying systems cause shiners and color variations, poor performance, and zinc-splitting problems if recoated
- Untopcoated zinc films are susceptible to attack at pH extremes ($<6, >10$)
- Unsuitable for recoating with oils and alkyds—topcoat delamination
- Single-pack systems may gas in can

WASH PRIMERS (WP-1) (DOD-C-15328 D-TYPE)
- Very critical mixing ratios; deviations produce gelation and short pot life
- Formulation very critical; composition of chromate will affect cure, adhesion, and performance
- May not adhere to old paint films
- Critical film thickness (0.3 to 0.5 mils), especially on smooth metal; thick films may disbond
- Performance and adhesion affected by age of mix (no paint should be applied after 8 hours, although gelation does not take place)
- Critical recoat times; films become very hard and are inhospitable for adhesion after several days
- Some finish coats (nonhydroxylated vinyls, acrylics, etc.) will not adhere to WP-1 films and delaminate

VINYLS
- Critical vehicle composition, specific to substrates; selection of vinyl type important; adhesion problems result from incorrect selection
- Adhesion dependent on full solvent evaporation
- Surface preparation critical for high performance
- Powerful solvent systems may attack solvent-sensitive primers and old films
- High sensitivity to pigmentation types and levels (inappropriate pigmentations will result in poor UV resistance, fading and premature chalking)
- Sensitivity to moist surfaces; control of dewpoint and humidity essential during application
- Fast-drying systems sensitive to dry spray, poor film formations, adhesion problems, and aesthetically poor, grainy films with variable gloss (difficulties of application in high temperatures, high winds, or elevated structures)
- High-build films sensitive to bubbling and pinholing over zincs; vacuole formation within films applied by airless spray
- Slightly subject to dehydrochlorination (may require stabilization particularly in low-molecular-weight resin systems)
- Carboxylated vinyls may gel with reactive pigmentation (zinc- and lead-based pigments, etc.)

Chlorinated rubber

- Highly sensitive to dehydrochlorination; stabilization vital—most critical in oxygenated solvent-based systems or zinc primers; possible effects of instability are corrosion of substrates (including zinc and steel), chloride entrapment, color deterioration, gelation, can corrosion, blistering in humid environment; single-pack zinc systems may gas, overheat, and explode without adequate stabilizer; two-pack systems are less vulnerable
- More solvent sensitive than vinyls
- Carrying solvents may attack solvent-sensitive primers and old films
- Sensitive to wet surfaces and high humidity during application
- May embrittle in service or at low temperature
- Fast-drying solvent system may cause dry spraying particularly on high, windy bridges

Epoxies

- Large coating family with wide-ranging property differentials; specifications require critical definitions of epoxy resin, curing agent, and molecular weight
- Thermosetting films of high crosslink density may be brittle and very strong; delamination from poor surfaces possible in service, especially on flexible substrates
- Solvent insensitive; recoating of old epoxy films may require scarification to avoid delamination of new films
- Epoxies chalk and discolor on exposure to UV (leads to aesthetic insufficiencies)
- Old surfaces with heavy chalk will compound intercoat adhesion problems
- Many common systems highly sensitive to application temperature (10°C [50°F] minimum); low-temperature application may cause vehicle stratification, sticky, discolored films, and/or adhesion problems if recoated
- Amino curing agents sensitive to reaction with heavy local carbon dioxide air contamination
- Two-pack systems sensitive to incomplete or inaccurate mixing
- Many systems require preapplication induction period (premature application may result in amine exudation, reduced gloss, and formation of amine carbonate, and consequent adhesion problems between films)
- Finite pot life limits successful application
- Strong solvents may attack solvent-sensitive primers and old films

Urethanes

- Large family of coatings with many different types of components and different curing mechanisms; may be reactive one-component or chemically curing two-component systems; specifications critical
- Single-pack uralkyds are oxidizing systems and have sensitivities similar to alkyds
- Uralkyds are harder than alkyds and may show intercoat adhesion problems
- Single-pack moisture-curing urethanes require moisture from air for cure, sensitive to climate
- Moisture-curing urethanes are sensitive to film thickness and pigmentation types that may restrict moisture access to lower film levels and may produce excessive bubbling and/or pinholing in film
- Moisture-curing urethanes may not cure properly in desert climates and will pick up a great deal of surface contamination

- Moisture-curing and two-pack urethanes sensitive to application on wet substrates or to use in highly humid conditions; reactivity with excess moisture may produce bubbling and pinholing of films
- Moisture-curing urethanes and two-pack urethanes crosslink to form hard films; old films may be difficult to recoat without intercoat adhesion problems; intercoat scarification may be necessary
- Two-pack urethanes are sensitive to inaccuracies in mixing ratios (incorrect mixing may produce cure and performance deficiencies, solvent-sensitive films, frothing, pinholing, reduced aesthetics, and/or increased dirt pick-up)
- Two-pack systems have a finite pot life, and viscosity increase may affect application
- Powerful solvent carriers may attack solvent-sensitive primers and old films
- Fast-drying systems may be subject to dry spray with negative effects on aesthetics and performance

Other causes of coating system failures may relate to local bridge design or maintenance peculiarities, defects in an existing paint film composite to be retained in a spot/recoat specification, or localized contamination.

The failure analysis of coating systems is, in fact, a complex discipline in its own right, and it is one perhaps best relegated to specialists who may not only prepare more accurate interpretations than an in-house engineering staff but who are better qualified and experienced to work with legal counsel in claims prosecution and provide expert testimony in litigation.

Litigation of Coating System Failures

Curiously, state departments of transportation appear more reluctant to pursue restitution in matters involving bridge-coating-system failure than are private owners of other structures. Reasons for this are not readily apparent. Jobs are bonded, and most specifications include phrases relating to a contractor's liability for latent defects. As noted above, however, coating system failures are not always the responsibility of the contractor, and the state's own exposure through poor specification design, lack of inspection, and its acceptance and payment of the job may be considerable. As a result, the cost and difficulties of litigation are often dissuasive. Too often it may be easier to ignore a problem and let it become diluted with time than to incur possible charges in court, from the press, or more certainly from the hierarchy. Also, a certain degree of apathy is often seen in the public sector, perhaps because of the impersonalization of wrongs against public bodies and the nonspecificity of the affront. Only on large structures, where failures are massive and catastrophic, do bridge authorities seem to seek resolution in the courts. Decisions to litigate here are probably forced as much by the sheer costs of the failure and its rectification as anything else, although factors such as public pressure are also possible.

Chapter 9

Costs

Perhaps the most elusive element in the entire discipline of bridge painting is the normalization of costs. The instabilities produced by such factors as inflation and geography notwithstanding, many other obvious and less obvious factors overwhelmingly affect the costs involved and belittle all attempts to reduce cost calculations to some predictability and order. While several sources purport to give indications of the basic costs of paint materials, application, and surface preparation techniques,[1,2] these indications are rarely treated as more than guidelines by most contractors (especially for field maintenance work). Material costs, while dependent on formulation, quantity, and manufacturer, are probably more dependable than the cost of either application or surface preparation, which vary not only with location, design, height, access, intricacies of steel, and the type and condition of existing paint films but also vary widely from contractor to contractor. Furthermore, nontechnical considerations will often have a heavy impact on the bid. Overall bid prices will reflect the makeup of the bidders' list and the manner in which each contractor views his competition. They may be modified at any time by the bidder's present work load and how badly he needs the job, by the availability and proximity of equipment and facilities, and by the economic outlook and the amount of work being let in the same area of the country. There may be a great deal of second-guessing before each contractor finalizes his bid.

Yet the formalization of the process is necessary both for the contractor or fabricator who is trying to get the award and stay solvent and for the bridge authorities with a more fundamental task of long-term planning for bridge maintenance. With process costs escalating dramatically, and with so many different paint-system options now available, decisions concerning long-term maintenance programs now involve large expenditures that mandate a sound economic basis for system selection.

Difficult as the task may be, attempts are being made to bring some order to costing from both viewpoints (contractor and bridge authority). Progress has been made. Many fabricators and contractors are now beginning to utilize complex and not-so-complex computer programs to assist their estimation processes. The bridge-corrosion cost model, designed to assist the bridge engineer in making cost-effective decisions concerning the "when, how, and with what" of bridge painting, is presently being refined into a practical tool.

This chapter reviews costing first from the contractor's perspective and then from that of the bridge engineer.

CONTRACTOR'S ESTIMATION OF JOB COST

The contractor or, in the case of new construction, the fabricator, must base his bid on the cost of labor, materials, and equipment, plus other direct job costs, such as the cost of rigging, bonds and job-insurance premiums, and noncapitalized tools and accessories (e.g., brushes, spray tips, tarps, tape, hoses, etc.). These are added together and overhead and profit are applied as a percentage. In some cases, the cost of small tools may be added as part of the equipment costs; other times that cost may be considered as part of the overhead. Some contractors may include bonding and insurance premiums with payroll taxes and payroll insurance as part of the labor cost. An excellent review of all these costs may be found in the Appendices of FHWA-RD-79-121.[3]

Labor Costs

Labor costs reflect the size of the crew. The exact makeup of the crew varies from job to job and from contractor to contractor, but normally consists of a foreman, blasters and painters (often the same individuals alternating jobs), and helpers who mix paint, tend the blast pots, and so forth. On larger jobs, there are multiple foremen under a single job superintendent and additional operatives, such as traffic-control men, drivers, mechanics, and sweepers. Rigging may be done by a separate crew or, more usually, by the same crew that will later paint the bridge. Specific job descriptions carry set union rates (which are made mandatory as part of the contract documents by most bridge authorities, regardless of the union status of the contractor). The contractor, however, may effectively save money by juggling such job descriptions. To offset this, pay scales are often monitored by the state or bridge authority during the job by direct interviews with the men. On union jobs, premiums of up to one dollar per hour may be assessed for high work, work involving high-performance systems, or even for blasting and spraying operations as opposed to hand-tool and brush-and-roller jobs.

The cost of labor is the most significant component of the contractor's cost. Labor costs consist of the base rate (set by the employer or, more usually, by the local union pay scale) plus the employer's share of Social Security taxes, state and federal unemployment taxes, Workers Compensation, and, possibly, bonding costs. These may add from 15% to 40% to the basic rate, depending on the particular job and the contractor's accident and unemployment records. Union dues and such labor fringe benefits as medical and pension plans, paid vacation time, if required by the contract, can add up to another 15% to 30% of the base rate. Overtime (time-and-a-half and double-time for Sundays and holidays) may be required after eight hours (or even on some union contracts after seven hours) per day. Other incentives, such as bonus plans related to productivity and paid to men or, more usually, to the foremen and/or superintendents, may also be billed into the overall cost per hour of labor. For key men on remote jobs, the contractor may also pay travel or subsistence allowances (as high as $5 to $10 per day for travel time and up to $30 per day for subsistence).

The methodology of dealing with the cost of supervision varies from job to job. On larger jobs with crews of ten and more men, this cost is completely allocated into the labor costs. On small jobs with crews of up to five or six, where the foreman works essentially as one of the crew, there will be no supervisory allocation. On very large jobs with a crew size of twenty and more men, several supervisors, each overseeing about ten men, are allocated into the total labor cost.

Current basic rates for journeymen painters and blasters range from $15 to $23 per hour. The pay scales of helpers, pot tenders, and abrasive handlers are less ($12 to $21 per hour). Foremen are paid up to $2 or more than the highest-paid journeyman. The pay for shop labor in new construction is not as high as field labor and may range from $9 per hour to $14 per hour. Additional data on labor costs may be found elsewhere.[3–6]

Labor costs (and, because of freight, material costs) may show considerable geographical divergency. Brevoort and colleagues[7] show Central and West Coast rates to be as much as 10% or 15% higher than East Coast rates for surface preparation and application for field jobs. In some states, specified department of transportation minimum labor rates are so disproportionately low compared to real current labor rates that men cannot be hired for such minimal pay.

Rigging and Freight Costs

Rigging costs added into labor on large structures over water where nets are required can be considerable (as much as 20% of labor costs). Still higher costs are possible where such specialty devices as platforms and other structures are constructed to facilitate containment methodologies. Pic and cable rigging on small structures is less expensive, as is the use of swing staging, boatswain's chairs, and so on. On simple overpasses, accessible from a road crossing by vehicles or pipe staging, rigging costs are much lower (1%–5% of the labor cost). In shop work, of course, rigging charges do not apply, although handling and freight costs must be added. Freight amounts to approximately $1.50 per loaded mile.

Material Costs

Material costs include the cost of paints and thinners, abrasives, and fuel and utilities, although fuel and utilities may be included with equipment costs. Spent-grit disposal costs may also be added to material costs. While delivered-paint costs vary widely from job to job (depending on the type and quality of coatings used, the contractor's credit standing, and the distance of the job from the paint source), typical figures for various coatings may be found in Tables 9-1 through 9-3. Additional material cost data may be found elsewhere.[3–6] These costs are discussed in more detail later in this chapter.

It is theoretically possible to calculate paint requirements by gallonage (G) from required dry film thickness (t), the decimalized volume solids of a coating (V), and the area to be painted (A), with the equation

$$G = \frac{A \cdot t}{1600\ V} \tag{9-1}$$

But numerous factors may act to modify this theoretical gallonage, principally related to waste and/or variable film thickness in one form or another. No applicator can ever be expected to maintain a film-thickness range narrower than about 1 to 2 mils, except in very low-solids materials, such as certain vinyls. In high-solids systems, film thickness control may be quite erratic. The greater the experience of the applicator, the closer he will average the mean calculated thickness requirements. Waste is another problem, however, and such effects as overspray, particularly with faster drying lacquers or use on open or latticed structures or on high or windy bridges, may substantially increase the gallonage requirements. Brush application is less wasteful than roller application,

Table 9-1
Cost Comparison of Inhibitive Coating Systems for Steel Bridges

System Group	System Description	Typical Volume Solids (%)	Typical Cost per Gallon ($)	Typical Cost per Mil per Square Foot ($)	Typical Cost per Square Foot at Recommended Thickness ($)	Range of Costs per Square Foot of Recommended Surface Preparation ($)	Range of Costs per Square Foot of Application ($)	Range of Costs per Square Foot Total System in Place ($)	Range of System Cost per Square Foot per Year of Expected Service in Environmental Category ($)		
									1B	2A	2B
I	*Primer:* red lead/linseed oil	95	18	0.013	0.026	0.30–0.55	0.15–0.20				
	Intermediate: red lead–iron oxide/oil-alkyd	80	15	0.015	0.022		0.10–0.15				
	Finish: aluminum/alkyd or phenolic	48	15	0.024	0.024		0.10–0.15	0.78–1.12	0.026–0.037	0.052–0.075	0.078–0.112
	Finish: (alternative) BLSC[a]/oil-alkyd	75	15	0.016	0.016		0.10–0.15	0.77–1.11	0.026–0.037	0.052–0.075	0.077–0.111
II	*Primer:* zinc chromate/alkyd	48	16	0.030	0.060	0.30–0.65	0.10–0.15				
	Intermediate: zinc chromate/alkyd	48	16	0.030	0.045		0.10–0.15				
	Finish: aluminum/alkyd or phenolic	48	15	0.024	0.024		0.10–0.15	0.79–1.23	0.032–0.049	0.061–0.095	0.099–0.154
	Finish: (alternative) BLSC/oil-alkyd	75	15	0.016	0.016		0.10–0.15	0.78–1.22	0.031–0.049	0.060–0.094	0.098–0.152
III	*Primer:* BLSC/linseed oil-alkyd	77	16	0.015	0.030	0.30–0.65	0.15–0.20	0.79–1.23	0.026–0.041	0.053–0.082	0.079–0.123
	Intermediate: BLSC/linseed oil-alkyd	75	14	0.015	0.022		0.10–0.15				

(continued)

Table 9-1
(Continued)

System Group	System Description	Typical Volume Solids (%)	Typical Cost per Gallon ($)	Typical Cost per Mil per Square Foot ($)	Typical Cost per Square Foot at Recommended Thickness ($)	Range of Costs per Square Foot of Recommended Surface Preparation ($)	Range of Costs per Square Foot of Application ($)	Range of Costs per Square Foot Total System in Place ($)	Range of System Cost per Square Foot per Year of Expected Service in Environmental Category ($)		
									1B	2A	2B
	Finish: BLSC/linseed oil-alkyd	75	15	0.016	0.024		0.10–0.15				
IV	*Primer:* nontoxic inhibitive/oil-alkyd	63	12	0.017	0.042	0.30–1.00	0.10–0.13				
	Intermediate: nontoxic inhibitive/oil-alkyd	51	11	0.019	0.047		0.10–0.13				
	Finish: nontoxic inhibitive/oil-alkyd	45	18[b]	0.036	0.072		0.10–0.13	0.82–1.55	0.041–0.077	0.082–0.155	0.137–0.258
	Finish: (alternative) aluminum/alkyd or phenolic	48	15	0.024	0.048		0.10–0.13	0.80–1.53	0.036–0.070	0.067–0.127	0.100–0.191
V	*Primer:* inhibitive latex	45	12	0.024	0.048	0.30–1.00	0.13–0.18				
	Intermediate: inhibitive latex	45	12	0.024	0.048		0.13–0.18				
	Finish: noninhibitive latex (2 coats)	39	11	0.025	0.087 (2 coats)		0.26–0.36	1.08–1.90	0.054–0.095	0.108–0.190	0.180–0.317
	Finish: (alternative) noninhibitive latex aluminum/latex	39 37	11 12	0.025 0.029	0.050 0.043		0.26–0.36	1.09–1.91	0.054–0.095	0.109–0.191	0.182–0.318

[a]Basic lead silico chromate.
[b]Silicone alkyd.

Table 9-2
Cost Comparison of Zinc-Based Coating Systems for Steel Bridges

System Group	System Description	Typical Volume Solids (%)	Typical Cost per Gallon ($)	Typical Cost per Mil per Square Foot ($)	Typical Cost per Square Foot at Recommended Thickness ($)	Range of Costs per Square Foot of Recommended Surface Preparation ($)	Range of Costs per Square Foot of Application ($)	Range of Costs per Square Foot Total System in Place ($)	Range of System Cost per Square Foot per Year of Expected Service in Environmental Category ($)		
									1B	2A	2B
VIA	*Primer:* alkyl silicate zinc	65	22	0.030	0.090	0.35–1.35	0.22–0.31	1.09–2.17	0.027–0.054	0.044–0.087	0.073–0.145
	Finish: high-build vinyl	33	13	0.032	0.162		0.19–0.26				
VIB	*Primer:* alkyl silicate zinc	65	22	0.030	0.090	0.35–1.35	0.22–0.31	1.37–2.48	0.034–0.062	0.054–0.099	0.091–0.165
	Intermediate: WP-1 vinyl wash primer	9	11	0.110	0.033		0.16–0.21				
	Finish: high-build vinyl	28	14	0.045	0.225		0.19–0.26				
VIC	*Primer:* alkyl silicate zinc	65	22	0.030	0.090	0.35–1.35	0.22–0.31	1.34–2.46	0.033–0.061	0.054–0.098	0.079–0.145
	Intermediate: epoxy/ polyamide (high build)	60	20	0.030	0.120		0.18–0.25				
	Finish: aliphatic urethane	52	33	0.057	0.114		0.17–0.23				
VID	*Primer:* alkyl silicate zinc	65	22	0.030	0.090	0.35–1.35	0.22–0.31	1.07–2.15	0.027–0.054	0.043–0.086	0.067–0.134
	Intermediate: epoxy/ polyamide (high build)	60	20	0.030	0.150		0.18–0.25				
VIE	*Primer:* alkyl silicate zinc	65	22	0.030	0.090	0.35–1.35	0.22–0.31	1.14–2.22	0.028–0.056	0.046–0.089	0.076–0.148
	Intermediate: high-build aliphatic urethane	60	35	0.052	0.208		0.19–0.26				
VIIA	*Primer:* alkaline silicate zinc	60	30	0.045	0.135	0.35–1.35	0.22–0.31	1.24–2.34	0.031–0.058	0.050–0.094	0.083–0.156
	Intermediate: WP-1 vinyl wash primer	9	11	0.110	0.033		0.16–0.21				
	Finish: vinyl	17	12	0.063	0.095		0.16–0.21				

(continued)

Table 9-2
(Continued)

System Group	System Description	Typical Volume Solids (%)	Typical Cost per Gallon ($)	Typical Cost per Mil per Square Foot ($)	Typical Cost per Square Foot at Recommended Thickness ($)	Range of Costs per Square Foot of Recommended Surface Preparation ($)	Range of Costs per Square Foot of Application ($)	Range of Costs per Square Foot Total System in Place ($)	Range of System Cost per Square Foot per Year of Expected Service in Environmental Category ($)		
									1B	2A	2B
VIIB	*Primer:* alkaline silicate zinc	60	30	0.045	0.135	0.35–1.35	0.22–0.31	1.24–2.34	0.031–0.058	0.050–0.094	0.083–0.156
	Intermediate: vinyl	20	12	0.053	0.053		0.16–0.21				
	Finish: vinyl	25	14	0.050	0.075		0.16–0.21				
VIIIA	*Primer:* phenoxy zinc	38	21	0.050	0.150	0.30–1.00	0.20–0.26	1.35–2.12	0.039–0.061	0.104–0.163	0.135–0.212
	Intermediate: WP-1 vinyl wash primer	9	11	0.110	0.033		0.16–0.21				
	Finish: vinyl	28	14	0.045	0.225		0.19–0.26				
VIIIB	*Primer:* phenoxy zinc	38	21	0.050	0.150	0.30–1.00	0.20–0.26	1.29–2.09	0.037–0.060	0.099–0.161	0.129–0.209
	Intermediate: epoxy/ polyamide (high build)	60	20	0.030	0.120		0.18–0.25				
	Finish: aliphatic urethane	52	33	0.057	0.085		0.17–0.23				
IX	*Primer:* chlorinated rubber zinc	68	22	0.029	0.087	0.30–1.00	0.20–0.26	1.24–2.03	0.035–0.058	0.083–0.135	0.124–0.203
	Intermediate: chlorinated rubber (high build)	35	14	0.036	0.126		0.19–0.26				
	Finish: chlorinated rubber	36	16	0.043	0.086		0.16–0.21				

X	*Primer:* vinyl zinc rich	20	16	0.071	0.142	0.35–1.35	0.21–0.27	1.46–2.55	0.042–0.073	0.073–0.127	0.097–0.170
	Intermediate: vinyl	28	15	0.048	0.144		0.18–0.25				
	Finish: vinyl	28	15	0.048	0.144		0.18–0.25				
XIA	*Primer:* epoxy/polyamide zinc rich	47	23	0.044	0.132	0.35–1.35	0.20–0.26	1.21–2.42	0.035–0.069	0.060–0.121	0.101–0.202
	Intermediate: epoxy/ polyamide	60	20	0.030	0.120		0.18–0.25				
	Finish: epoxy/polyamide	60	20	0.030	0.060		0.18–0.25				
XIB	*Primer:* epoxy/polyamide zinc rich	47	23	0.044	0.132	0.35–1.35	0.20–0.26	1.46–2.54	0.42–0.073	0.073–0.127	0.122–0.212
	Intermediate: epoxy polyamide zinc rich	47	23	0.044	0.066		0.20–0.26				
	Finish: vinyl aluminum	14	11	0.070	0.210		0.19–0.25				
XIC	*Primer:* epoxy/polyamide zinc rich	47	23	0.044	0.154	0.35–1.35	0.20–0.26	1.22–2.31	0.035–0.066	0.061–0.116	0.102–0.192
	Intermediate: epoxy/ polyamide red lead (marine environments only)	49	17	0.031	0.046		0.18–0.25				
	Finish: vinyl toluene/acrylic	50	11	0.020	0.030		0.17–0.22				
XIIA	*Primer:* uralkyd zinc rich (2 coats)	55	17	0.028	0.098	0.35–1.35	0.40–0.52	1.13–2.22	0.038–0.074	0.075–0.089	0.113–0.226
	Finish: vinyl toluene/acrylic	50	11	0.020	0.030		0.17–0.22				
XIIB	*Primer:* moisture-curing urethane zinc	63	35	0.050	0.150	0.30–1.00	0.20–0.26	1.33–2.12	0.038–0.060	0.053–0.085	0.111–0.177
	Intermediate: epoxy/ polyamide	60	20	0.030	0.120		0.18–0.25				
	Finish: aliphatic urethane	52	33	0.057	0.114		0.17–0.23				

(continued)

Table 9-2
(Continued)

System Group	System Description	Typical Volume Solids (%)	Typical Cost per Gallon ($)	Typical Cost per Mil per Square Foot ($)	Typical Cost per Square Foot at Recommended Thickness ($)	Range of Costs per Square Foot of Recommended Surface Preparation ($)	Range of Costs per Square Foot of Application ($)	Range of Costs per Square Foot Total System in Place ($)	Range of System Cost per Square Foot per Year of Expected Service in Environmental Category ($)		
									1B	2A	2B
XIII	*Primer:* galvanizing	100	—	—	—	—	—	1.05–1.70	0.030–0.049	0.042–0.068	0.105–0.170
XIIIA	*Primer:* galvanizing	100	—	—	—	—	—	1.84–3.25	0.046–0.081	0.061–0.108	0.102–0.181
	Intermediate: WP-1 vinyl wash primer	9	11	0.110	0.033		0.16–0.21				
	Finish: vinyl	28	14	0.045	0.225		0.19–0.26				
XIIIB	*Primer:* zinc metallizing	100	—	0.069	0.550	0.50–1.90	0.66–0.93	1.98–3.70	0.050–0.093	0.079–0.148	0.132–0.247
	Finish: vinyl (carboxylated)	20	12	0.053	0.106		0.16–0.21				
XIIIC	*Primer:* zinc metallizing	100	—	0.069	0.550	0.50–1.90	0.66–0.93	2.19–3.96	0.055–0.099	0.088–0.158	0.146–0.264
	Intermediate: WP-1 vinyl wash primer	9	11	0.110	0.033		0.16–0.21				
	Finish: vinyl (hydroxylated)	17	12	0.045	0.126		0.16–0.21				

Table 9-3
Cost Comparison of Barrier Coating Systems for Steel Bridges

System Group	System Description	Typical Volume Solids (%)	Typical Cost per Gallon ($)	Typical Cost per Mil per Square Foot ($)	Typical Cost per Square Foot at Recommended Thickness ($)	Range of Costs per Square Foot of Recommended Surface Preparation ($)	Range of Costs per Square Foot of Application ($)	Range of Costs per Square Foot Total System in Place ($)	Range of System Cost per Square Foot per Year of Expected Service in Environmental Category ($)		
									1B	2A	2B
XIVA	*Primer:* vinyl (carboxylated)						0.80–1.05 (5 coats)	1.67–2.80	0.042–0.070	0.067–0.112	0.111–0.187
	Intermediate: vinyl (2–5 coats)	20	12	0.053	0.40 (5 coats)	0.35–1.35					
	Finish: vinyl										
XIVB	*Primer:* WP-1 vinyl wash primer	9	11	0.110	0.033	0.35–1.35	0.80–1.05 (5 coats)	1.67–2.79	0.042–0.070	0.067–0.112	0.111-0.186
	Intermediate: vinyl (hydroxylated) (plus 2–4 additional coats)	17	12	0.061	0.366 (4 coats)						
	Finish: vinyl										
XV	*Primer:* chlorinated rubber	32	14	0.039	0.058	0.30–1.00	0.48–0.63 (3 coats)	1.09–1.86	0.031–0.053	0.054–0.093	0.091–0.153
	Intermediate: chlorinated rubber	35	14	0.036	0.108						
	Finish: chlorinated rubber	33	16	0.043	0.064						

(continued)

Table 9-3
(Continued)

System Group	System Description	Typical Volume Solids (%)	Typical Cost per Gallon ($)	Typical Cost per Mil per Square Foot ($)	Typical Cost per Square Foot at Recommended Thickness ($)	Range of Costs per Square Foot of Recommended Surface Preparation ($)	Range of Costs per Square Foot of Application ($)	Range of Costs per Square Foot Total System in Place ($)	Range of System Cost per Square Foot per Year of Expected Service in Environmental Category ($)		
									1B	2A	2B
XVIA	*Primer:* epoxy/polyamide	55	18	0.029	0.073	0.30–1.00	0.54–0.75 (3 coats)	1.13–1.96	0.032–0.056	0.045–0.078	0.094–0.163
	Intermediate: epoxy/ polyamide	60	20	0.030	0.075						
	Finish: epoxy/polyamide	60	20	0.030	0.060						
XVIIA	*Primer:* aluminized epoxy mastic	85	24	0.025	0.125	0.30–0.65	0.18–0.25	0.65–1.03	0.065–0.103	0.093–0.147	0.162–0.257
XVIIB	*Primer:* aluminized epoxy mastic	85	24	0.025	0.125	0.30–0.65	0.36–0.50 (2 coats)	0.98–1.40	0.039–0.056	0.065–0.093	0.098–0.140
	Finish: aluminized epoxy mastic	85	24	0.025	0.125						
XVIIC	*Primer:* aluminized epoxy mastic	85	24	0.025	0.125	0.30–0.65	0.36–0.50 (2 coats)	1.02–1.43	0.041–0.057	0.078–0.110	0.127–0.179
	Finish: epoxy/polyamide color coat	80	28	0.031	0.155						

XVIIIA	*Primer:* moisture-curing urethane aluminum	50	25	0.045	0.112	0.30–1.00	0.57–0.78 (3 coats)	1.30–2.12	0.130–0.212	0.186–0.303	0.325–0.53
	Intermediate: moisture-curing urethane aluminum	50	25	0.045	0.112						
	Finish: moisture-curing urethane aluminum	50	25	0.045	0.112						
XVIIIB	*Primer:* moisture-curing urethane aluminum	50	25	0.045	0.112	0.30–1.00	0.19–0.26	1.28–2.09	0.085–0.139	0.128–0.209	0.160–0.261
	Intermediate: moisture-curing urethane aluminum	50	25	0.045	0.112		0.19–0.26				
	Finish: aliphatic urethane color coat	52	33	0.057	0.114		0.17–0.23				
XIX	*Primer:* coal tar epoxy	71	13	0.016	0.131	0.35–1.35	0.40–0.54 (2 coats)	1.09–2.15	0.027–0.054	0.036–0.072	0.073–0.143
	Finish: coal tar epoxy	71	13	0.016	0.131						

which in turn is less wasteful than airless spray. Conventional spray application is more wasteful than airless. Two-pack systems with short pot lives may result in considerable coating loss, particularly on intricate structures with limited production rates. Excessive mil profile and pitted surfaces also depress coverage rates. Spillage and mixing errors can also contribute to increased gallonage requirements, and it is inevitable that the contractor must include substantial loss factors in his material calculations. Usually material losses of 50% for spray, 70% for brush, and somewhere in between for roller applications are used as constants. In Tables 9-1, 9-2, and 9-3, more average loss-adjustment factors of 0.9 for brush, 0.8 for roller, and 0.7 for spray are employed in calculating cost per mil square foot. Inorganic zinc coatings are a special case because of the mechanisms of film formation and film porosity. With those coatings, the calculations above do not readily apply because practical applications without loss provide film thicknesses greater than what might be theoretically expected from a simple volume-solids-based coverage calculation. In Tables 9-1, 9-2, and 9-3 these factors have been built into the calculation by realistically adjusted volume-solids figures.

Sometimes the contractor bases his paint requirement estimations on more simple estimates, such as those noted in the product-data sheet, those reported by Frondistou-Yannas,[3] those derived from the Painting and Decorating Contracting of America's (PDCA) *Estimating Guide*,[1] or the estimates noted by Roebuck and Brevoort.[4] More frequently, the experienced contractor may rely on his own experience using the same or similar materials on similar structures to generate his own material estimates.

The contractor must be wary of discrepancies between the film-building capabilities (without sagging) of a product and the thickness requirement of the specification because these types of divergency can lead to the need for multiple passes, with consequent labor cost inflation.

Solvents for thinning and cleaning are generally estimated to cost about 25% of paint material. Both paint and solvent costs must include an added charge for freight, and, where applicable, sales or use taxes on all materials must be added in.

Similarly, abrasive costs vary with type and grading and with bridge location with respect to the abrasive source. Sand is generally the least costly abrasive, although the use of sand is becoming more widely restricted because of toxicological considerations. Mineral slags, such as H. B. Reed's "Black Beauty," (Reed Mineral's Division of Harsco Corporation, Highland, IN), are relatively inexpensive and less toxic than sand. Recent U.S. Navy restrictions on mineral slags because of toxicological impurities have not yet spread to the bridge-painting industry. Metallic abrasives are much more costly, and are rarely used in field applications unless reclamation and recycling are possible. Finer, particle-sized abrasives are generally more expensive than the larger "utility" grits. Far more important, however, is the effect of abrasive size on production rate and, therefore, on labor costs.

The quantity of abrasive necessary to complete a job is estimated from the assessments of the number of man-days of blasting required. Sand and mineral slag are consumed at rates averaging 0.5 tons per hour, especially on large jobs with bulk equipment. Even with mineral slag, total abrasive cost may be more than total paint cost. Delivery costs for abrasive are relatively high (sometimes 40% or more of the cost of the abrasive itself) and must not be overlooked because they make up a considerable percentage of the delivered-abrasive cost. Sline et al.[7] recommend contracting with a supplier for abrasives at a set price for the duration of the job, if possible. This is not always possible, especially recently, when the popular mineral slag has been in short supply because of competitive roofing-industry markets.

Equipment Costs

Surface preparation and application equipment (including compressors, spray pumps, blast bottles, storage hoppers, trucks, cranes, etc.) are either rented or owned by the contractor. These costs are charged to the job by the contractor or fabricator either in the form of rental fees or compensatory allowances for depreciation (including the cost of equipment delivery). Rental rates vary, but they are published in such publications as *The Rental Rate Blue Book*.[8] Average rental costs for some equipment are noted in Table 9-4. As noted above, the cost of tools, lubricants, fuel, and utilities may also be included as part of the equipment costs, although, in new construction, such costs may be added to the fabricator's overhead. In new construction, such large capitalized pieces as cranes and centrifugal blast equipment are almost always wholly owned by the fabricator. Trucks, however, may be owned, leased, or rented. On larger jobs in the field, the rental of offices, warehouses, and storage yards are also incorporated into the overall rental charges. Such small equipment as hoses, nozzles, helmets, guns, and tarps are included in the overall equipment costs.

Bonding and Insurance Costs

The costs of bid and performance bonds are dependent on the contractor's financial viability and experience record, as are the costs of his property damage and general liability insurance. All are becoming more difficult to obtain and increasingly expensive. Bonding now costs 2% to 4% of the bid price, while insurance costs range between 1% and 5% of the job cost.

Overhead and Profit

Overhead represents that share of the cost of the contractor's or fabricator's operation not directly chargeable to the contract and represents the cost of staff salaries, home-office and base-facility costs (rental or depreciation allowances), utilities, vehicles, equipment, and so on. It is applied as a percentage of the total job cost and is usually determined by dividing annual operating costs by the total chargeable labor, materials, and equipment fees for all operations for the prior year. Overhead percentage factors depend on the size of the contractor's operations and may range from 10% to 25%. Profit in both field and shop work is figured either on total job cost or on total job cost plus overhead. Required profit typically averages approximately 10% of the job cost.

Takeoff

In preparing his bid price, the contractor must "takeoff" the structure to determine the square footage involved. This may be time consuming, especially on older structures where extensive repairs have been made or where the structural design is not totally regular from bent to bent. On more regular structures, takeoff is less of a problem.

In some cases, the contractor may elect to estimate the square footage from the tonage of the structure (sizes of bridge structures are almost universally expressed in tons rather than in square feet). This technique is also used in shop estimates for new construction. A rough rule of thumb used by the industry is that each ton of bridge is equivalent to 250 paintable square feet. Unfortunately, the real square footage per ton will vary considerably from one type of structure to another. Heavy rolled-beam and plate-girder structures may average 80 to 150 square feet per ton, truss members 100 to 300 square feet per ton, and light steel (railings, sways, etc.) 300 to 500 square feet

Table 9-4
Average Monthly Rental Costs for Typical Equipment Used in Bridge Painting Operations (1985)

Office trailer	$ 180
Storage trailer	150
125-CFM compressor	400
365-CFM compressor	800
750-CFM compressor	1,800
7-ton blast bottle	1,000
35-ton abrasive storage hopper	500
Scissor lift	1,000
125-foot boom truck	20,000[a]
Dump truck	1,650
Truck crane	3,500
Swinger bucket loader	2,000
Tugger	360
35 : 1-ratio airless pumps	200
Assorted guns and hoses (500 feet)	200
Rigging (pipe staging, cabling, picks, and platforms)	1,500–2,000[b]
Stake truck	1,650

[a]With operator.

[b]Figures based on painting cost for a 25 bent, 1,250-foot-long, 30-foot-high structure with blast debris contained by tarps.

per ton. Even these estimates are subject to much variation, however, and there is considerable disagreement in the literature concerning assignable square-foot rates to steel type[3,7,9] (see Table 9-5). This author agrees with several authorities who think that the 250-square-foot-per-ton rule is too liberal a figure, particularly as modern cross sections (with stronger steel) are reduced. One contractor (George Campbell, George

Table 9-5
Paintable-Surface-Area-to-Weight Relationship of Bridge Steel

Steel Type	Frondistou-Yannas[a]	Brevoort[b]	Fraser[c]
Extra heavy structural (Plate girders—older structures)	50–100	100	80–110
Heavy structural (Rolled beams—newer structures)	100–150	250	90–150
Medium structural (Truss members)	150–300	200–500	105–190
Light structural and miscellaneous steel (Angled bracings—railings, etc.)	300–500	300–400	

[a]See reference 3, Chapter 9.

[b]See reference 7, Chapter 9.

[c]See reference 9, Chapter 9.

Campbell Painting Corporation, Flushing, NY) reports that he has never found a truss section (even a latticed truss) to have a square-foot-to-weight ratio higher than 150 square feet per ton. Newer sealed boxes, with paintable surfaces only on the exteriors, may have less than the 80-square-feet-per-ton ratios. Because of these variables, it is more usual for the experienced contractor to use the square-foot-to-weight rule as a rough verification of his square footage takeoff. On small, regular overpasses, many experienced contractors quickly assess square footage figures by eye.

Bid Calculation

Armed with known labor rates and material costs once the size of the structure is determined, the contractor can estimate labor and material costs by applying anticipated production rates. Reduced to simple terms, this may be summarized for blasting as follows:

$$C_{bt}=\frac{A\cdot L_r}{B_r} \tag{9-2}$$

where

C_{bt}=total blasting labor cost
A=area of the structure to be blasted
B_r=anticipated hourly production rate of the total blasting work force
L_r=total labor cost rate (in dollars per hour)

Similarly, in painting, costs may be estimated as follows:

$$C_{pt}=\frac{A\cdot L_r}{P_r} \tag{9-3}$$

where

C_{pt}=total painting labor cost
P_r=anticipated hourly production rate of the total painting force

Total abrasive materials cost C_{at} may be calculated

$$C_{at}=\frac{A\cdot a\cdot c_a}{B_r} \tag{9-4}$$

where

a=abrasive consumption per hour in tons
c_a=delivered-abrasive cost per ton

Total paint materials cost C_{pt} for each coat is calculated from the gallonage equation:

$$C_{pt}=\frac{A\cdot t\cdot C_g}{1600\cdot V\cdot K} \tag{9-5}$$

where

C_g=cost per gallon of the paint
K=decimalized modifier reflecting the practical effects of loss, usually 0.70 to 0.90, depending upon application methodology

Obviously, the estimation of production rates is critical to the final cost estimate, and its accuracy determines whether or not the contractor makes a profit. Production rates are heavily influenced by the height and configuration of the structure and, in maintenance work, the type of old coating and condition of the steel. The blasting and priming of open-lattice boxes is much more difficult than solid-sided box members, especially on high trusses. The effects of abrasive loss and overspray raise the materials and insurance costs of such jobs. (For each property-damage claim for overspray or other problems, deductibles are sacrificed and premiums endangered.) The contamination of water crossings on such jobs may also threaten productivity rates through the response of environmental groups and the public to perceived environmental impact.

The condition of the existing steel and paint work can heavily influence production rates, as can climate and season, the type of existing paint, and new profile requirements (especially where large areas of old paint are to be removed). An experienced estimator who is familiar with bridges and old paint systems can approximate blasting, priming, and finishing rates with reasonable accuracy and without too much difficulty. Roebuck and Weismantel[10] report that some contractors, when figuring difficult jobs, will estimate blasting rates by blasting a test patch prior to bidding. Although this practice is undoubtedly an excellent precaution, particularly on older structures or on neglected structures where corrosion appears excessive, it does not appear at all usual in bridge painting.

Productivity rates for the entire job are heavily influenced by the productivity rates for the surface-preparation step. An accurate estimation of this is critical to the accuracy of the bid. Crude figures are provided by Frondistou-Yannas,[3] Roth,[11] and others[12] (see Table 9-6). Depending on the type of structure, steel configuration (flat webs or complex riveted angles), blast quality, containment, and (in spot-cleaning jobs) the amount of cleaning to be done, however, these figures may be wildly inaccurate. Abrasive type and nozzle size are also determining factors in the overall rate of blast production.[13]

Table 9-6
Average Daily Surface Preparation Production Rates per Man on Structural Steel (in square feet)

Surface Preparation Quality	Frondistou-Yannas[a]	Roth[b]	Frenzel[c]
SSPC-SP-7	1600–2000	2300	
SSPC-SP-6	800–1000	1100	
SSPC-SP-10	400–600	650	300–420
SSPC-SP-2	550–850[d]	1000–1200[e]	
SSPC-SP-3	550–850[d]	400–1600[e]	

Note: Rates per man do not include helpers.

[a]On previously coated rusted steel, seven-hour blasting day.

[b]On uncoated steel with mill scale and light rust; adjusted to a seven-hour blasting day.

[c]Steel conditions range from heavily rusted steel with pitting to moderately rusted steel; seven-hour blasting day.

[d]Includes touch-up priming.

[e]Assumes four-hour productive day.

Blasting rates determined by test blasts have generally been conducted on flat stock, over unpainted steel in very good condition, and over relatively limited areas,[14,15] where operator fatigue is not a factor. Better and more comprehensive figures are found in a paper by Hansel that discusses blast-cleaning rates in some detail.[16]

The most important factors affecting productivity rates in blasting are blast quality, mil profile, the condition of the steel, and the nature of the old coating. It may take twice as long to achieve a near-white blast than it would to achieve a commercial blast. Other functions, such as proper operation of equipment and anchor-pattern requirements, also affect productivity. Badly corroded steel with deep pits (such as is found on old structures or in expansion bay areas) will require additional time to clean out the pits, and the presence of mill scale beneath the coating will make for a slower blast than would be the case with coated surfaces that have been blasted previously. Heavy rust scale is severely depressive to high blasting rates and is best removed prior to blasting with hammers and hand tools. Finally, certain coatings respond more easily to blasting than do others. High builds of epoxies and elastomeric materials are particularly difficult to blast, as are tar-based materials. Old oxidizing coatings that readily fracture on blasting are quite easy. Coating thickness, adhesion, and age will also affect the ease of removal. Given all these variables and the criticality of the process, the test blast makes good sense.

The blast-production figures noted in Table 9-6 are very average values. The Frondistou-Yannas figures for rolled girders may be conservative for new steel and too optimistic for such complex, light-steel areas as rails or back-to-back angles. They do, however, show reasonable agreement with Roth's average rates for lightly rusted steel. Note that the blasting day does not entirely consist of production blasting time. Pots must be filled and refilled, rigging moved, and lunch breaks taken. Inspection (and, often, reblasts to required standards) and blow-down are also necessary before the work may be primed in.

Paint-application rates (Table 9-7) are also given by Frondistou-Yannas[3] for both brush and spray. These rates seem high for brush and slightly low for spray application (assuming airless). Figures quoted by Levinson and Spindel[17] seem somewhat more realistic. Again, these figures are averages, and actual application rates will vary widely from paint to paint within a given category, as well as from paint type to paint type. Certain coatings that are more ideally suited to a particular technique will provide a

Table 9-7
Average Daily Paint-Application Production Rates per Man on Structural Steel (in square feet)

	Frondistou-Yannas[a]	Levinson and Spindel[b]	Bayliss[c]
Brush	700	650	X[d]
Roller	1050	1200–2600	
Air spray		2000–6000	
Airless spray	2100	3000–8000	8 times the rate for brush

[a]See reference 3, Chapter 9.
[b]See reference 17, Chapter 9.
[c]See reference 18, Chapter 9.
[d]Rate unspecified.

much better production rate than others. Oil paints are much more suited to brush application, for example, than are high-solids epoxies, and they allow for faster working. Application rates will also vary with film thickness and higher build requirements, particularly higher builds of lower solids lacquers that significantly depress application rates. The necessity for multiple passes to achieve required film build without sagging will similarly increase the cost of application. Touch-up application must also be considered in calculating application costs, especially when areas for touch-up require the movement of scaffolding. Touch-up may be added into the estimated costs as a separate step or built into the overall coating cost as a modification to the application rate.

Generally, in the field, surface preparation is the rate-determining step in the painting process, and priming is often included as part of this. Blasting, blow-down, and priming are done as consecutive operations on the same day. Additional coats are applied later, when time is less critical. Even where some recoat criticality exists (e.g., overcoating coal tar epoxies or wash primers), there is more time than can be allowed before newly blasted steel must be primed. Blasting and priming operations are generally continuous throughout the job, apart from a few days toward completion, when the blast-primed areas are being painted in.

In the shop, where centrifugal blasting techniques speed up the surface preparation process, blasting is still probably the rate-determining step, although there is far less differential between the time it takes to blast and the time it takes to prime in.

Wild Cards and Contingencies

Many of the wild cards that can affect surface preparation rates and seriously confound the contractor's estimate can also affect the rate and the cost of application, although not always to the same extent or in the same manner. These may include weather interruptions, delays caused by paint and equipment problems, delays caused by delivery problems, paint testing, traffic patterns, and so on. Other factors may relate to general accessibility and to the effect on operations of such outside or nontechnical factors as the demands of other trades and contracts on the same bridge, lane closings, enforcibly restricted work days, and community and environmental pressures. Protracted delays in delivery of primers will, of course, prevent blasting operations. Similar delays with finish coats can, at critical times (e.g., when operators must move from one side of the bridge to another, and only one lane can be closed), halt the blasting operation. Traffic flow (both road, rail, and waterway traffic) can have a profound effect on a day's productivity. Lane closings on busy highways are often prohibited during rush hours, and this not only curtails the work day but can, with bad luck, coincide with other delays to result in the loss of a day's blasting. Even on structures where much of the day's work is below the roadway, lane closings will require the removal of compressors, cranes, blast pots, and spray equipment parked on the deck to serve the operations below. On lift and bascule structures over waterways, water traffic may cause repeated work disruptions; on toll structures, maintaining bridge revenue during painting programs will inevitably be a problem.

In new construction, steel deliveries, shop fabrication schedules, finished-goods delivery, and storage problems may all have a dramatic effect on overall production rates.

Containment of blast residues can have a profound effect on productivity and consequent costs. Using vacublast techniques, where some more or less definitive cost factor may be applied (four to five times that of open blasting), the cost of containment is reasonably determinable. In large and small general-area containment enclosures designed for a specific bridge, however, the impact of containment may be widely variable from contract to contract (even on the same structure), and may not only relate to the size and design of the structure but to local politics and the efficiency of local regulatory agencies in enforcing regulations.

On the Tobin Memorial Bridge near Boston, square-foot costs, which included containment of the rails, rose from $5 per square foot in 1978 to approximately $8 per square foot in 1983. The increase was, in part, related to improved containment design, but much of the increase was the contractor's built-in contingency to cover the possible effects of community, media, and political impact on this controversial job.

Within field enclosures, visibility is often poor and, if painting is performed within the enclosures, the difficulties of applying and inspecting zinc rich primers (even when tinted) against near-white blasted steel backgrounds are substantially confounded, even with the use of lights. Large enclosures may also burden the contractor with additional material costs, equipment costs, and even extra steps in the job process.

To avoid downtime and minimize the number of moves when using tarps or solid or semisolid containment devices, the contractor will often try to contain as large an area as possible, working several crews in the same enclosure for several consecutive days. Because the steel is blasted and primed within the same containment structure, work already primed during a previous day's activities is unavoidably badly contaminated with blast residues from the next day's blasting operations. This primed steel must be cleaned thoroughly before subsequent coats may be applied if intercoat adhesion problems are to be avoided. Cleaning procedures involving blowdown or vacuuming may be ineffective on contaminated primer film, and water washing may be necessary. In any case, an added step is included in the job, and the contractor may not always anticipate the need for this step during bidding. Unless inspection is very good, the quality of the job is inevitably jeopardized by inefficiencies in cleaning such areas. The added inspection costs will also add to the price of the job.

While grit disposal is not always monitored by the state or bridge authority, when bona fide testing and documentation for the disposal of spent grit (certainly grit that is classified as a hazardous material) is made part of a pay item, the job cost can increase significantly, depending on the proximity of the nearest (hazardous) waste disposal site to the job. When spent grit is treated as either hazardous or special waste, it must be trucked at extra cost to an approved disposal site, often hundreds of miles from the job. Additional equipment (cranes, scrubbers, hoses, etc.) will also add to the containment costs and increase the overall bid price.

While some contingency may be built into the contractor's price to cover such variables on bid jobs, too much contingency may cost the contractor the award. As a result, the contractor may choose to ignore such possibilities and, if necessary, accept the consequences. The key here seems to be to know the practices of the agency making the award.

Assumed-Cost Calculation

Given his basic labor, materials, and equipment costs, his takeoff, and his anticipated production rates, the contractor will proceed to calculate his assumed cost for the job, adding in his overhead, required profit, and any contingency fees for containment and grit disposal. Simplified examples of this type of exercise are provided in several sources, including Sline et al.[7] and Frondistou-Yannas.[3] Additional examples, including shop work calculations, are provided by Roebuck and Weismantel,[10] and further data of value may be found in the PDCA *Estimating Guide*.[1]

Adjustments before Final Bid

Even after these kinds of calculations have been made, there are often other concerns that may modify the calculated bid price. Almost all bridge-painting contracts are awarded on a low-bid basis. If a contractor is determined to get an award, notwithstanding the above typical calculation, he may have to adjust his calculated bid price accordingly, depending on competition. The prospective contractor often utilizes the prebid meeting to evaluate his competition. Many contractors maintain formal or informal records on competitors' bidding practices, and, based on these, they may adjust their own bid in response to the anticipated bid of a particular competitor. The trick is to stay profitable while making the lowest tender.

The prebid meeting is vital to the successful contractor not only to gauge his competition but to interpret specification package clauses properly. Environmental protection clauses and the degree and quality of inspection can substantially affect the real job costs. In urban areas, containment requirements are increasingly being written into specifications with the deliberate intention of enforcing them only if public pressure should make it necessary. On the other hand, in certain parts of the country, some agencies are enforcing environmental protection and containment, at least to some limited degree. Contractor qualification clauses again, if real, may affect the contractor's perception of his own success in bidding as well as his perception of his rival. He must determine how real such clauses are.

If the contractor must bear the costs of spent-grit disposal, it is to his advantage to know the nature of the existing paint that must be removed, and whether it is likely to test as hazardous. The prebid blast test is of value in this determination, although heavy builds of old paint that are rich in red and white lead will very likely result in a hazardous spent-grit classification. Some general idea may be gained from the thickness of paint present and the total lead content of the film only.

These various specification-writing practices present the contractor with one or more wild cards. He must, therefore, judge carefully the real intentions of the state or bridge authority, not only in terms of his own bid but in terms of his competitors. The net result is likely to be some adjustment in the final bid price, and this, not the accuracy of the takeoff or labor prices, is often the cause of those widely divergent bid ranges that are encountered in open-bidding bridge work.

SHOP WORK ESTIMATING: NEW CONSTRUCTION

Although it is essentially similar to the development of the bid for field work, estimating is less complex for new work, not only because the problems of access, traffic, geographical variations in labor rates, and personnel skills are removed but also because surface preparation processes (centrifugal blasting) and steel surface qualities are less variable and more easily defined from an economic perspective. Production rates are also inevitably faster and more predictable, especially in surface preparation since the widespread adoption of centrifugal-blasting techniques.

Blasting costs with centrifugal techniques may be about one-half or even one-quarter of the equivalent costs for field techniques, although these costs still depend on size, quantity, and steel configuration. In general, the advantages of centrifugal blasting are most clearly demonstrated in the cleaning of complex shapes, railings, ladders, and guard rails, although advantages are still noted in heavy structural steel.[19] Similarly, the cost advantages of centrifugal techniques are emphasized as job size (square feet to be blasted) increases, but these advantages are more obvious with near-white blasts than with commercial blasts. Roebuck and McCage[19] demonstrate that, on relatively small jobs, the cost advantages of centrifugal blasting over air-nozzle techniques may disappear or even reverse.

In shop work, material costs are generally reduced because of recirculated abrasives, reduced paint wastage and overspray, and so on. The fabricator does not have to deal with the environmental concerns of field practice, such as containment of blasting and hazardous grit disposal. Ground work is not only easier and faster than field work, but safety is generally improved and insurance premiums reduced. If the fabricator has indoor facilities, rain days are eliminated and, with heated facilities, the painting season can be greatly expanded in northern latitudes. Not surprisingly, therefore, basic preparation and priming is less expensive in the shop than it is in the field and estimation is easier and more accurate. Additional costs not included in field work are related to the handling and rehandling of new and finished steel and its transportation from shop to remote job sites. The value of storage space prior to shipment (or where multiple coats are shop applied) between coats, must also be considered.

Although one might anticipate that the compromises necessitated in the field to meet job schedules are absent in the shop, this is often far from the truth. The fabricator with limited space may be tempted to move steel out quickly after application and to recoat without allowing enough time for adequate curing of certain primers.

GALVANIZING COSTS

Galvanizing is a shop process that is different from painting.[20] In galvanizing, there are more steps than in the normal preparation and painting of steel, and it is said that more than one-half the costs involved in galvanizing are directly attributable to handling. Zinc costs may amount to less than 25% of the total galvanizing cost (John Malone, Duncan Galvanizing Co., personal communication, 1985).

Galvanizing prices are quoted by the pound of steel not by its square footage. Amounts of zinc are quoted in ounces of zinc per square foot. Both of these practices tend to confound direct comparisons of galvanizing costs with painting costs. ASTM-A-123 requires a minimum of 2 ounces of zinc per square foot (3.4 mils) of surface. Averages for typical bridge structural steel would be 5 to 6 mils, although thicknesses of 4 to 9 mils are by no means unknown. Correlation of steel weight with square footage, as noted in Table 9-5, is not straightforward but will vary with the configuration and size of the steel pieces (i.e., heavy rolled girders, medium structural steel, and light structural materials such as bracings, etc.). Typical prices for galvanizing are shown in Table 9-8.

Large structural pieces are quite difficult for the galvanizer to handle (70–75 feet is a maximum), even with double (end-after-end) dipping. Smaller pieces, on the other hand, particularly those with complex shapes (railings, gratings, etc.) are more effectively handled by hot-dipped galvanizing than by painting. Tube stock, which may be galvanized both inside and out for added protection, is particularly suited to the galvanizing process. As a result, there is a breakpoint at which galvanizing becomes more cost effective than blasting and painting. While this depends on the actual shape and size of the structure and the distance of the galvanizing facility from the job, a breakpoint of 250 to 275 square feet per ton is often quoted, above which the galvanizing is less expensive and below which steel is most cost-effectively blasted and painted.

For these reasons, in new construction and in field work where such small steel sections as railings are to be replaced, the specific utility of galvanizing and duplex

Table 9-8
Average Costs of Galvanizing and Duplex Systems (1985)

	Galvanizing Only	Galvanizing Plus Primer	Galvanizing Plus Coating System
Heavy structural steel (100–200 square feet per ton)	$1.35–1.60 per square foot 0.10–0.12 per pound	$2.15–2.70 per square foot 0.16–0.20 per pound	$2.70–3.20 per square foot 0.20–0.24 per pound
Medium structural steel (200-300 square feet per ton)	0.96–1.20 per square foot 0.12–0.15 per pound	1.45–1.95 per square foot 0.18–0.24 per pound	1.60–2.40 per square foot 0.20–0.30 per pound
Light structural steel (300–400 square feet per ton)	0.86–1.05 per square foot 0.15–0.18 per pound	1.15–1.55 per square foot 0.20–0.27 per pound	1.71–2.30 per square foot 0.30–0.40 per pound

Source: John Malone, Duncan Industries, Inc.

systems is often now being considered. As zinc kettles get larger, we may see more and more of this type of approach being used, particularly in combination with shop- and field-applied metallizing.

METALLIZING COSTS

Metallizing may be either a shop or field process. Because the process involves both surface preparation and coating techniques, the various cost elements beyond the actual cost of materials, fuels, and application equipment (abrasive price, labor rate, cost of access and rigging, containment, etc.) are much the same as those for normal painting, particularly if the metalized surface is sealed or recoated afterward, as is normally the case. Surface preparation is by abrasive blasting to a white-metal (SSPC-SP-5) quality. This is somewhat more intensive than is normal in bridge-painting practice, even with zinc rich systems. White-blasting quality increases costs approximately 30% to 35% over near-white quality.

Zinc metallizing from electric-arc equipment is usually applied at a rate of approximately 0.9 ounce per square foot per mil, although reduced target efficiency (overspray, etc.) reduces this rate by something like 10% (F. M. Velapoldi, TAFA, Inc., personal communication, 1986). Films of 5 to 8 mils are usually applied, and material costs of approximately $1.40 per pound are average for zinc wire. This produces material costs of $0.35 to $0.55 per square foot. Aluminized metallizing that is deposited at the average for zinc wire produces material costs of $0.34 to $0.55 per square foot. Aluminized metallizing that is deposited at approximately 0.25 ounce per square foot per mil is cheaper ($0.20–$0.30 per square foot) at similar film thicknesses, despite the higher cost of wire ($2.35 per pound), because of the lower specific gravity of aluminum.

Metallizing equipment, including the power source (transformer/rectifier) and the control assembly (which monitors both air pressure and the current), together with guns and necessary hoses and cables, sells for approximately $10,000. There appears to be little renting of this type of equipment. Gas-driven metallizing equipment is somewhat lower in cost. The cost of fuel, whether it is electricity or gas, must be added to material costs.

Application rates are governed by the diameter of the feed wire (zinc or aluminum); the larger the diameter, the more material is applied in any given hour. In electric-arc equipment, application rates are also controlled by the size of the electric current (metallizing guns operate with low voltage/high current equipment). At 300 amps, 60 to 65 pounds of material can be deposited per hour; at 100 amps, delivery rates are 15 pounds per hour. At 5 to 8 mils, these rates convert to about 170 to 270 square feet per hour at 300 amps, and 45 to 70 square feet per hour at 100 amps. Production rates in pounds per hour are lower for aluminum, but square-feet-per-hour rates work out to be about the same.

The rates above are in reasonably close agreement with those more theoretical rates quoted by Oechsle and Childs[21] for an oxyacetylene system. Usually, actual application rates with electric-arc guns will be faster than with oxyacetylene and plasma-type equipment.

Additional data on costs and cost calculations of the metallizing process can be found in many of the equipment manufacturer's literature.[22]

These production rates are slower than rates for the application of inorganic zinc, although film thicknesses are higher. Wash primers, phosphate treatments, and vinyl and epoxy sealers that are used in finishing systems are in the same price ranges as are other systems of the same generic type. Film thicknesses of sealers and recoat systems on metallized surfaces may be slightly heavier than over zinc-rich or galvanized films because of the heavier profile of the metallized surface.

DUPLEX-SYSTEMS COSTS

Where duplex systems (galvanized steel painted by the galvanizer in the shop) are concerned, the costs of galvanizing processes are subsequently modified by the addition of the costs of in-shop coating of the steel, even though there are no intermediate preparatory steps, such as solvent washing and cleaning of the galvanized surface, which may be required when new, unweathered galvanized steel sections are subsequently painted in the paint shop or the field. In the duplex process, the various white-rust inhibitive treatments that are used in the finishing of normal galvanized steel are not necessary.

The steel is often coated under better conditions than are normally found in general shop painting, and near-production-line processes have been developed to move the galvanized steel through the painting process within twelve to twenty-four hours of galvanizing. In the commercial duplex process,[23] temperature, humidity, and dew point are controlled so that coatings are applied under optimum conditions, and forced-drying ovens ensure complete and uniform cure. Complete paint systems are applied and include wash-primed/high-build vinyls, as well as both vinyl- and urethane-finished, epoxy-primed systems. In addition, an untopcoated epoxy-primed system is also available from one manufacturer for use under a variety of field-applied topcoats. In general, field thicknesses of coatings in duplex systems are much the same as in equivalent systems used over zinc rich primers.

In addition to the material and application costs involved with duplex systems, therefore, there are also costs for additional handling (including deburring and dross removal), as well as energy costs, which must be included in the final price.

Typical prices for epoxy-primed galvanized duplex systems and epoxy-primed/urethane-finish systems are shown in Table 9-8. Compared to the inorganic zinc-rich/epoxy/urethane systems, these prices are higher even for steel with high square-foot-to-weight ratios, but additional control and resultant performance may improve the long-term cost effectiveness of these systems.

COMPARATIVE COST CONSIDERATIONS IN THE SELECTION OF BRIDGE PAINTING METHODOLOGY

In an exercise so beset with inexactitudes of measurement, inconsistencies in bridge design, material composition, and environmental severity, with so much variability in control, and with insufficiencies in specification design and in the accurate monitoring of service, any hope of real scientific comparison of the cost effectiveness of any two paint systems might seem presumptive if not blatantly absurd. The first half of this chapter has attempted to demonstrate, at least to some degree, the futility of such an undertaking. Even after an honest attempt is made to corral the many variables and to force a viable estimate, the final bid price may be compromised by such foibles as the vague perception of competitive intent and/or the reality of certain specification requirements.

Justification for the Comparisons

If only because of the number of possible permutations involved in coating-system selection and the sporadic appreciation of suitability (as well as value) of coating systems among those whose responsibility it is to direct the painting and maintenance of the nation's bridges, some attempt to harness what order we can is indispensible if viable decisions in selecting optimized painting methodologies are to be made. Few have tried to provide such ordered data, and those who have (Frondistou-Yannas,[3] Roebuck and McCage,[19] and Brevoort and Roebuck[24]) have been widely quoted and re-quoted,[7,10,25,26] and for the most part are again quoted here.

Intelligent and soundly based decisions are needed not only with regard to what paint system should be used but in regard to when to paint and whether painting is, in fact, preferable to such alternative techniques as metallizing or even galvanizing. Complex gratings and, particularly, railings may be coated advantageously and cost-effectively in the shop by such techniques as hot-dipped galvanizing and duplex-system application because of the availability of more cost-effective preparatory and application methods, such as pickling, dipping, and electrostatic application.

As noted by many authorities, including Appleman[26] and Frondistou-Yannas,[3] there is, in the life of all paint systems, a well-defined epoch relating to the extent of deterioration when value in delayed maintenance converts to added costs. It is important, therefore, that maintenance-painting schedules are correlated with such epochs, if maintenance dollars are to be employed effectively.

Available Studies on Costs

Most researchers agree that, given the number of facts and variables that can affect the cost of any job, attempts to make a general quantification of the cost of the different processes and materials that together make up the total job price are condemned to some degree of inaccuracy from the outset. Most researchers, therefore, rely on a statistical approach to balance out discrepancies.

The original Brevoort and Roebuck paper[24] is a simplified guide intended to assist "nonexclusive" coating engineers in making reasonable decisions concerning coating-systems selection based on cost, environment, and service. These authors provide much valuable average data on surface preparation, application, and material costs for four regions of the country, together with service-life estimates in three environments of differing severity, and they calculate cost per year of service. They also discuss such economic concepts as the time value of money and discounted cash flow, and they provide comparative examples of the cost of paint systems in the field and the shop, as well as of two field-applied systems. Much of this material is also used by Roebuck and Weismantel in the *Paint Handbook*.[27] A second paper by Roebuck and Brevoort[4] expands on the first to some extent, enlarging tables to include more paint systems and discussing the necessity for inspection. It updates the cost data from the original 1978 figures to 1981 figures. The 1981 figures are also used by Sline et al. in the *SSPC Manual*.[7] Appleman[26] has subsequently simplified the 1981 figures and further upgraded them by applying an inflation factor to 1984.

Unfortunately, the Brevoort and Roebuck figures are based on industrial plant usage and are not specific to bridges. Field figures represent costs "on the ground," although the authors do offer conversions for work in the air and for structures of more complexity. In the *SSPC Manual*, Sline et al. comment on discrepancies that are apparent between the in-plant service of three-coat alkyd systems and service lives of similar systems on highway bridges in Washington and Massachusetts. Comparisons are also made between an eight- to ten-year plant service life of a three-coat vinyl system and similar systems that have lasted twenty years on water tanks and dam gates. Interviews with many highway department officials across the country would also indicate that Brevoort and Roebuck's service-life estimates are extremely conservative when applied to bridge structures, and that some redefinition of the environmental descriptions is necessary in these different applications.

Surface-Preparation Costs

In this volume, Brevoort and Roebuck's surface-preparation costs are modified somewhat by inflation factors, by considerations from the numerous interviews carried out in the preparation of this volume, and by the specificity of bridge painting. Nevertheless, the figures appear to coincide, for the most part, when applicable divisors are applied.

Table 9-9 shows cost-range estimates for cleaning bridge steel in both shop and field. Shop data presume the use of centrifugal blasting, otherwise shop preparation would approach the figures for simple, easy, accessible steel in the field. The actual cost of centrifugal blasting is also influenced by the sizes of the members and the job itself. Small steel pieces may be manually blasted more cost effectively even in the shop. Field cleaning data are divided into ranges for low steel of simple lines requiring a minimum of rigging (e.g., rolled-beam overpasses) and high steel with considerable intricacy (e.g., latticed-arch trusses). As noted elsewhere, specific factors, such as the addition of containment, may multiply these costs by three, four, or five times, depending upon the extent of the containment. Older structures with heavier scale and highly pitted steel will also cost more to render to a given standard, and the cost per square foot of a spot-blast job will be higher than the same square-footage cost on a full-blast job. Prices in the West are higher than in the Gulf states and the East. Eastern state prices are generally lower than Gulf state prices in the field, but they are higher in the shop. Shop painting in the central states may be somewhat less expensive than in the East, although field costs are more comparable with those of the West.

Application Costs

Application costs (see Table 9-10), like surface preparation costs, vary from one part of the country to another. Again, the West-Coast prices are higher than prices in the East or in the Gulf Coast states, and they are much the same as central states' prices. For field application, prices may be slightly lower in the Gulf states than in the East. Again, the work of Brevoort and Roebuck,[4] with necessary inflation factors applied, is used as a standard. This standard is modified in this volume only as indicated by other reviews and interviews. Figures for brush and roller application of alkyd and oil systems were more difficult to access, and considerable disparity among costs from contractor to contractor was found. As in surface-preparation costs, field work costs more than shop work, and application costs increase with the height and complexity of the structure.

Equally divergent were figures concerning comparable productivity rates from various application methodologies from which application costs might have been derived. Levinson and Spindel[28] estimate roller application to be three times as slow as airless spray and three times as fast as brushing. Frondistou-Yannas[3] considers roller application rates to be one-and-one-half times those of brushing and one-half as fast as spray application. Like all of the above, those prices listed in Table 9-10 are only approxi-

Table 9-9
Range Estimates for Costs of Surface Preparation of Steel Bridges in Shop and Field

		Field Cleaning	
Surface Preparation	Shop Cleaning (New steel)	Low Rolled Beams in Place, Alkyd Coated, 15% Rust	High Complex Truss in Place, Alkyd Coated, 15% Rust
SSPC-SP-1 (Cleaning of old finish)		$0.15–0.30	$0.20–0.35
SSPC-SP-2 (Hand-tool cleaning)	$0.25–0.30	0.45–0.55	0.55–0.65
SSPC-SP-3 (Power-tool cleaning)	0.32–0.36	0.60–0.70	0.75–0.85
SSPC-SP-7 (Brush-blast cleaning)		0.50–0.60	0.60–0.70
SSPC-SP-6 (Commercial-blast cleaning)	0.28–0.32	0.85–1.00	1.00–1.25
SSPC-SP-10 (Near-white-blast cleaning)	0.32–0.37	1.15–1.35	1.40–1.60
SSPC-SP-5 (White-blast cleaning)	0.45–0.55	1.60–1.75	1.90–2.10

Note: Values are expressed as cost per square foot.

mate, and ranges reflect not only variations in geographical location but differences from one contractor to another, one bridge to another, and even one paint to another, within a given coating classification.

Materials Costs

At a time when there are so many different proprietary products being used on bridges under generalized labels, in many cases with little compositional similarity, the institution of order into a review of material costs is necessarily bedeviled. Widely divergent formulations in high-build epoxy mastics, aliphatic urethanes, or high-build vinyls not only foster substantial diversity in mil-per-square-foot costs, but produce considerable price fluctuations as raw-material prices change. In times of work shortage and reduced material demand, competition for a limited dollar may itself bring some stability. Given adequate sales volume, however, larger price differences become manifest.

Throughout the early 1970s and early 1980s, raw-material prices and, perhaps to a lesser extent, paint-material prices fluctuated wildly in response to both general inflation and the instability in price and availability of petroleum-based feed stocks. By 1985, prices were less volatile and the commoditization of such systems as inorganic zinc-rich primers may have added further stability. Fluctuations must be anticipated, however, and these will produce considerable distortion of material costs (such as are used here) relative to one another and, more important, to other costs in the overall discipline. While the dramatic reverberations of large shifts in feed-stock prices eventu-

Table 9-10
Range Estimates for Costs of Coating Application on Steel Bridges in Shop and Field

Application Example	Shop Application	Field Application: Low Simple Rolled-Beam Structure in Place	Field Application: High Complex Truss Structure in Place
Oil/alkyd by brush	$0.13–0.15	$0.17–0.19	$0.21–0.24
Oil/alkyd by roller	0.10–0.12	0.13–0.15	0.16–0.18
Oil/alkyd by spray	0.09–0.10	0.11–0.13	0.13–0.15
Organic zinc rich by spray	0.19–0.21	0.24–0.26	0.29–0.32
Inorganic zinc by spray	0.21–0.23	0.25–0.31	0.31–0.37
High-build (two-pack) epoxy by spray	0.17–0.19	0.20–0.25	0.24–0.30
High-build vinyl by spray	0.18–0.20	0.21–0.26	0.26–0.31
Aliphatic polyurethane finish by spray	0.16–0.18	0.18–0.23	0.22–0.28

Note: Values are expressed as cost per square foot.

ally are dispersed throughout the entire economy, the rate of dispersal is slower in some areas of the economy than in others. This can result in the temporary disproportionate value of coating materials (which are more rapidly affected by feed-stock changes) vis-à-vis labor or equipment costs. It is not inconceivable, in fact, for such disproportionate values to encourage the adoption of a coating system once considered too expensive, because of the relative value of labor or equipment at a given epoch.

In Tables 9-1, 9-2, and 9-3, those paint systems that are most commonly used on bridges (discussed in detail in Chapter 4) are compared using all common criteria from cost-per-gallon to cost of complete in-place coating system per square foot per year of expected service. The range extremes represent high average shop work and field work on typical rolled-girder construction. Shop figures include an 8% allowance for freight and touch-up.

Now generally abandoned, the simplistic cost-per-gallon approach was conceived with paint-material costs alone in mind and took none of the direct costs of application and surface preparation into consideration. Cost-per-gallon estimates made no attempt to incorporate ideas of coating-system value in terms of realized protective service. Moreover, in failing to reduce coating costs to a normalized evaluation criterion based on value per square foot of coverage, the method was totally misleading in evaluating even material costs, except when comparing identical paint systems that, by chance, had equal volume solids. Evaluations based on normalizing material costs adjusting to cost per mil square foot using formulas based on volume solids gave a more realistic comparison, and, when film-thickness multiples are incorporated and the costs of component coats added, a more viable comparison of material costs is provided.

Material costs, on the other hand, represent no more than approximately 10% to 25% of the cost of the in-place coating system. Because the type and degree of surface preparation and application is often variable from system to system, being dictated by the primer if nothing else, it is necessary to consider these costs at the same time as simple material costs in order to achieve viable comparative cost data for in-place systems.

Further refinements consider the expected service that might be provided by the coating system (see Tables 8-1, 8-2, and 8-3) and apply this to obtain a comparative annual cost of bridge protection. Finally, as is discussed later in this chapter, this figure itself may be refined by the use of net-present-value-of-money techniques. If required, the cost of bridge maintenance over the long term, involving several repaintings, may be computed.

Service Life

A meaningful assessment of the service life of a given coating system is another elusive factor in the process of cost comparison. One of the most common questions that the coatings engineer faces from the bridge engineer who must select the coating system is that concerning expected service life. Answers are inevitably vague and couched in caveats. As we have seen throughout this volume, service life is significantly affected by bridge design, geographical location, specific exposure, and by the extent of deicing, spring cleaning, and maintenance practices. Also critical to the life of the system are the accuracy of the original system selection and specification design, the quality of surface preparation, application, and inspection practices, and, in part at least, the condition of old paint systems that are to be repainted. The sheer number of factors and the degree of variability in each is, of course, borne out by the wide differences in system performance that exist across the country and from bridge to bridge even in the same locality. Undoubtedly, as higher performance paint systems of greater complexity have been introduced, service-life variability has (statistically at least) increased compared to those older, more tolerant systems of lower performance that were the norm twenty years ago. Also confounding attempts to assess performance accurately is the lack of real scientific records of field performance and the confusion created by the misrepresentation and the nonspecificity of the data.

Of those data sources on system service that do exist, many are intentionally or unintentionally biased, overly theoretical (based on panel studies, etc.), or are not necessarily applicable to bridge service. Where accurate figures are possible, what constitutes the end of viable service for one observer may be quite different for another. Even such well-intentioned studies as those of Brevoort and Roebuck[4] and Frondistou-Yannas[3] show gross differences from known or reported field service, and, while attempts to rate service in more than one environment help, distortions still exist. Accurate as Frondistou-Yannas's rust curves may be, they are, at least to some degree, specific to those jobs on which they are based. They therefore require additional refinement if they are to provide accurate representation on different structures in different states. It seems probable that, despite variables of specification design, inspection, and job execution, no one set of figures will satisfy a pattern seen in more than one area of a nation with so great a climatological divergency. (In this regard, the necessity for the use of deicing salts is also included in the climatological profile.)

The service figures provided in Tables 8-1, 8-2, and 8-3 and the resultant cost comparisons in Tables 9-1, 9-2, and 9-3 are no more certain or universal than others that are available. They represent judgmental averages of paint-system performance and costs derived from the many other sources reviewed throughout this volume, and from interviews with state transportation department material officials, bridge engineers, paint manufacturers, and other experts throughout the country. They are not only inaccurate they probably can never be accurate, and they are intended for use as a guide only.

Discounted Cash Flow (DCF)

The economic analysis of the comparative values of bridge painting with two or more coating systems must necessarily involve those same accounting procedures that are employed by management in making any large investment decision.[29,30,31] Any discussion of coating-system value must be based on the relative time value of money in maintaining the structure with one system rather than another, taking into account interest rates and, on toll structures, profitability.

DCF analysis is important because the invested dollar (P) accumulates interest (i) and is worth $P(1+i)^n$ if not spent now and instead is invested for n years. Alternatively, P dollars spent n years in the future must be discounted by $1/(1+i)^n$ to determine the money equivalent today. The technique is discussed in more detail elsewhere.[29,30,31] It allows the determination of the value of money at a set epoch and compares the cost of all maintenance methodologies at the same epoch. If, for example, paint system X must be repainted every five years at an increasing total cost to the bridge authority of X, X_5, X_{10}, X_{15}, X_{20}, and X_{25} dollars, while a second system Y must be repainted every ten years at costs of Y, Y_{10}, and Y_{20} dollars, the relative cost of protection over the first thirty years of bridge life may be compared by present-value analysis as follows:

$$PV_x = X + \frac{X_5}{(1+i)^5} + \frac{X_{10}}{(1+i)^{10}} + \frac{X_{15}}{(1+i)^{15}} + \frac{X_{20}}{(1+i)^{20}} + \frac{X_{25}}{(1+i)^{25}} \quad (9\text{-}6)$$

and

$$PV_y = Y + \frac{Y_{10}}{(1+i)^{10}} + \frac{Y_{20}}{(1+i)^{20}} \quad (9\text{-}7)$$

where i is the cost of money. In this analysis, the lesser figure (PV_x or PV_y) is the best alternative.

With these basic data and tools, an appreciation of the probable effects of climate, design, accessibility, and environment for the particular structure in question, and, if available, some comparative history of costs and system service on this structure relative to others, it is possible to render some sort of realistic appreciation of comparative system value, despite the many variables and unknowns.

THE BRIDGE CORROSION COST MODEL (BCCM)

Inevitably, so complex a problem lends itself to some form of computerization. In 1979, a prototype computer model to deal with the problem was developed as part of a government study on corrosion costs.[3,32] The computer model of bridge-corrosion costs was developed by Frondistou-Yannas as part of FHWA Contract DOT-FH-11-9528 and is presented in FHWA-RD-79-121.[3] In its original form, the BCCM is flawed by several assumptions, inaccuracies, and its necessarily simplistic approach to the complex issue; however, it nonetheless remains a brave beginning to the quantification and formalization of data that were, for the most part, regarded as too chaotic, capricious, and quantitatively inextricable.

The model was conceived as a tool to aid the bridge engineer in selecting the most efficient maintenance program for the protection of a given structure in terms of protection method (including, within broad classifications, preferred paint system and film thickness) and optimized painting schedule. As part of its database, the model considers information on performance and costs derived from studies of 2,052 structures in California, Louisiana, Massachusetts, and Washington, as well as from the SSPC, ASTM, NASA, and industrial sources. Unfortunately, most performance data are derived from test panels, although some originate from actual bridges. Only four generalized paint systems are used in the initial model: oleoresinous systems on hand-tool cleaned surfaces, oleoresinous systems on blast-cleaned surfaces, topcoated zinc rich systems (again nonspecific) over near-white surfaces, and vinyl, epoxy, and polyurethane barrier systems on near-white metal. ASTM-A-36 galvanized and weathering steel surfaces are selected, as are new technologies such as metallizing. Environmental factors are generally classified as marine, industrial, rural, and desert. Bridge type is classified in terms of the steel member categories: light, truss, and girder construction. These categories are then assigned square-foot-per-ton values. The classification is primarily based on the determination that corrosion at edges occurs more rapidly than it does on flat plate areas, and rolled girders have a lower edge-to-plate ratio. There is no account taken of the propensity of other steel areas (such as lower flanges) to corrode earlier than webs and the like. Other factors considered are coating thickness and application procedure (shop vs. field and brush vs. spray). Size of the structure (as it affects rigging costs) is also included.

Tests of the BCCM that were performed by the FHWA on structures in Massachusetts, New York, and Virginia show 20% to 25% correlation with the costs of the actual contracts. The model has also been used to demonstrate the value of blast cleaning compared to hand-tool preparation for oil/alkyd system repainting intervals ranging from six to fifteen years. The model identified the optimum repainting interval at ten years for hand-tooled-cleaned steel (5.3¢ per square foot per year) and eleven to twelve years for blast cleaning (4.6¢ per square foot per year). In another study, the FHWA used the model to compare the effects of deferring repainting on cost. Deferring repainting (a spot-prime over hand-tool cleaning plus two recoats) of an oil/alkyd system (reaching the critical 1.0% to 1.5% rust in nine to ten years) for an additional year was found to increase rust to 4% and to increase costs by 18%; a 30% to 50% increase in costs was incurred with a two-year delay. On the other hand, the model showed that on a structure with less than 1% rust, deferred repainting was economically advisable.

In its raw form, the BCCM probably fails as an accurate tool to be universally used in bridge-painting management. Undoubtedly, this results from the fact that the model's design criteria favored ease of use over accuracy, and that the condensation and simplification of so complex a problem must inevitably result in a lack of universal viability until it can be modified by use. The concept is also probably ahead of its time, and the model's viability unavoidably suffers from the lack of more precise performance data (rust curves) on actual structures.

The model has been evaluated critically by Roberts and Davis,[33] who conclude that "the bridge corrosion cost model cannot develop realistic cost data," on even the simplest structure, although it can be modified and, with development, may be transformed into a viable tool for use by bridge authorities responsible for the maintenance of steel bridges. Roberts and Davis's critique is balanced and contains many valuable ideas for refinements based on a sound understanding of the practical criteria that affect the bridge-painting contract.

The BCCM is flexible, however, and is designed so that default values on performance, environmental severity, costs, and so forth may be changed by any given bridge authority. More precise data are likely to be available to that authority on the use of a specific system in a specific area of the country. The user may alter the database with additional information on the discounting rate, film thickness averages more relative to his state's own experience, better rust curves developed from studies in his jurisdiction, and up-to-date figures on the life expectancy of the specified system. Such refinements may lead to quite different adaptations of the model from one area of the country to another.

The original model, in fact, is best considered as no more than a prototype designed for refinement. In some respects, this refinement will have to be basic and new elements will have to be incorporated. As is suggested by Roberts and Davis, items that vary significantly from bridge to bridge must be transferred from the program data file and then input for the individual structure. As presently conceived, the model will allow the specific input of data on average paintable areas of steel per ton, material, equipment, crew size, and productivity. The effects of traffic patterns and control and weather or seasonal work-day modifiers are needed. Expansion of the number of coating systems options also is undoubtedly required for most work. The establishment of defined patterns of containment and grit-disposal costs will certainly require future incorporation, as such containment becomes more organized, in view of the potential magnitude of these effects on final costs.

Chapter 10

Trends, Directions, and Ongoing Research

Unlike the foregoing chapters, which have presented the bridge-painting discipline in its current state, this chapter examines the various aspects of the pertinent technologies in light of present and emerging trends and, with some studied speculation, seeks to anticipate those directions in which the discipline may evolve.

GENERAL TRENDS

Unfortunately, it would appear that, for the near future at least, little improvement can be expected in the overall disparity between what has to be done to address the maintenance painting needs of today's bridges and what can be done with the funds available. Although, over the past decade or so, the problem has been defined with considerable exactitude and while, through the various publications and efforts of such organizations as the SSPC, ASTM, and NACE, as well as through the FHWA, the dissemination of data can be expected to continue and even increase, a full understanding of this complex technology can only be expected to grow slowly within most state departments of transportation and bridge authorities.

At higher levels of state government, where funding originates, appreciation of the needs of the current technology seems static and sadly lacking. Without direct political action, there seems little chance of improvement here in the near future. Yet without some real commitment of facilities, funds, and qualified (and dedicated) personnel in state highway departments, the effort to address the bridge-maintenance problem will inevitably fall short.

Greater federal funding of the required effort will certainly bring much relief. However, in light of the inadequacies in specification design, inspection, prosecution of job control, and postpainting record keeping, it seems inevitable that much of this money will be wasted unless sufficient controls that require appropriate redress of the inadequacies are tied to the funding. Nevertheless, to wait for all states to develop their own efficient response to the needs of the technology will probably be in vain, and some federal funding will have to come quickly if many of our nation's bridges are to be saved.

As a whole, it may be anticipated that system trends will continue to favor the high-performance zinc and barrier systems which are displacing the older oil/alkyd systems.

The transformation will remain a gradual process. In the more rural areas of the country, where environmental concerns are less pressing and where the older lead systems are giving satisfactory performance, change may still not occur for many years. The development of viable, cost-effective methods of deleading existing structures should hasten this trend, although significant breakthroughs in this area are not, from this vantage, anticipated in the near future. Spot repair of lead structures seems destined to be accomplished, in most cases, with lead systems, although the use of the aluminized epoxy mastics is expected to continue to expand as more and more manufacturers add this type of system to their product lines. Less encouragement can be seen for the growth of the aluminized moisture-curing urethane, probably because of the restrictions on high builds from single coats and reported delamination problems.

In general, it is felt that a major trend throughout the industry will be to reduce the number of coats used in bridge painting. Single or, more likely (because of the better probability of holiday reduction, etc.), two-coat high-build systems will gradually succeed the present three- and four-coat methodologies. In the moderate term, this will undoubtedly prolong the present preference for the zinc/vinyl systems over the zinc/epoxy/urethanes. However, zinc/high-build epoxies and zinc/high-build urethanes used judiciously on a single structure with additional coats in highly vulnerable areas may subsequently encroach on the employment of the high-build vinyl system as the major high-performance coating system for new and radically blasted bridges. Uncertainties over the UV stability of vinyl systems, which has caused concern in some circles but which has surely been exaggerated, may dissipate as the qualitative and quantitative parameters of vinyl formulations are realized by specification writers.

Growth in the use of inorganic zinc/epoxy/urethane systems should continue to accelerate as field histories prove the worth of these systems. Recently, the Pennsylvania Department of Transportation has announced construction plans for some twenty-seven new bridges bearing shop-primed 3- to 5-mil DFT inorganic zinc primer coats that are subsequently washed in the field before epoxy midcoating (4 to 6 mils DFT) and finish coating with a 1.5- to 2-dry-mil aliphatic urethane coat. Two other structures are scheduled for similar epoxy/urethane systems built on a zinc-rich moisture-curing urethane primer.

A countertrend away from the newer high-performance systems and other systems that are bound by excellent, if costly, surface preparation requirements may, in some states, slow the aforementioned primary trend. Many factors contribute to this, not the least of which is the increasing divergency among the number of structures requiring attention and the number of dollars available to do the job. In spite of the longer term advantages afforded by the high-performance systems, it becomes increasingly necessary to implement immediate redress of corrosion on many structures and to spread available funds as far as they will go as more bridges reach the point of critical maintenance.

Less widespread but potentially more influential is the increasing public awareness of the environmental impact (whether real or imagined) of sandblasting operations and the increased use of containment methodologies (either local or general) that such awareness has prompted. This, together with the related expense of the disposal of spent grit containing leachable levels of lead and hexavalent chromium salts above the Environmental Protection Agency's (EPA) hazardous threshold, will ironically also add to a reluctance on the part of bridge authorities to undertake radical deleading maintenance procedures.

Recent restrictions by the U.S. Navy[1,2] on the levels of arsenic, lead, cadmium, and beryllium, as well as free silica, gypsum, and carbonates in unused mineral slag abrasives have not yet spread to bridge work, but they could well do so. If nothing else, the Navy's action hoists yet another red flag of environmental concern. Still other indirect environmental factors should not be overlooked. These include a growing reluctance on the part of banks to provide mortgage money without land surveys for hazardous contamination. The proximity of such properties to bridge structures that have recently undergone extensive deleading operations (especially those involving open blasting) raise some interesting questions of liability, which in itself may act as a powerful persuasive against radical-blasting maintenance procedures.

While the industry has anticipated these trends very adequately and is currently evaluating numerous surface-preparation methodologies from which it is hoped a less costly and environmentally compatible procedure will emerge, success in these endeavors has, to date, been minimal.

Supporting such countertrends are the more traditional objections to higher performance systems. These include those state engineers who maintain, often with much justification, that conventional lead/oil/alkyd systems already provide adequate long-term performance. There will also remain those who cite, with only slightly less justification, the relatively higher risks of premature system deterioration with newer coating systems and therefore opt for the more user-friendly traditional systems that have more certain, if more limited, success.

TRENDS IN SURFACE PREPARATION

Research in surface-preparation techniques is discussed in Chapter 3. For the most part, this research is centered on the development of alternative procedures to dry sandblasting in order to mitigate those hazards (from toxic abrasives and atomized paint debris) inherent in the blast process and, if possible, to reduce the cost of the preparatory process through improved productivity. As noted earlier, wet-blasting techniques are the principal practical developments from this approach, although, while more sophisticated research directions are receiving attention, few at this point seem either to satisfy all criteria for successful surface preparation or provide cost-effective resolutions for the difficulties peculiar to large-scale employment on bridges (particularly in the field). In general, the most recalcitrant deficiencies of the newer techniques appear to lie in affording adequate scarification to the steel itself or in achieving effective removal of ionogenic species that congregate tenaciously on corroded bridge-steel surfaces.

Wet Blasting Trends

A pooled SSPC study[3] of the effectiveness of wet-blast cleaning methods concludes that, while wet-blast methods are neither as effective nor as efficient as dry blasting, the wet-abrasive blasting proved quite practical in field applications, providing production rates of 80% to 90% of the dry-blasting rates. Medium- to high-pressure (6,000–20,000 psi.) water blasting with abrasive incorporated is described as being less effective (50% of dry-blasting production) and is quite dangerous because of the very high thrusts and operator fatigue. Yet newer, 35,000-psi. water/abrasive techniques have also been introduced, with back pressures said to be very low. Less effective, although less tiring to operatives, are low-pressure, wet-blasting variations that can remove paint and rust effectively. The report further concludes that all wet-blasting units reduce dust, and, when inhibitors are added, flash rusting can be controlled for several hours.

In other circles, the acceptability of discharging spent water from wet blasting into the environment has been questioned, particularly when inhibitors are used because of the discharge of lead and other toxins into the environment (T. Galvin, Massport, personal communication, 1984).

The SSPC pooled study and summary report[4] contain product directories of water and wet-abrasive-blasting equipment and equipment suppliers.

In another paper,[5] high-pressure water jetting is described as slower than dry blasting in removing mill scale but faster for the removal of energy-absorbent coatings. Surface quality is reported to improve as pressures increase, 20,000 psi. being necessary for white-blast finishes that cannot be achieved by using longer dwell times at 10,000 psi. (which produces a lesser surface quality). Water jetting is also claimed to remove salt debris not stripped by dry blasting. This provides better blistering resistance and less underfilm rust creepage of subsequently applied coatings. When inhibitors were used with water-jetting techniques, no coating failures could be related to chemical residues left on the surface.

The Role of Abrasive Media

Much attention is now being directed toward a more complete understanding of the role of the abrasive media in coating-system performance. While such ionic impurities as chlorides and sulphates in the abrasive have for some time been considered detrimental to performance, in studies by Bleile et al.,[6] the negative effects of abrasives containing copper and/or other less electrochemically active metals (with respect to the substrate) on the flash rusting of newly bared steel and on subsequent paint performance (hot-water blistering) have been shown to be of even greater significance. Theoretically, the results are by no means surprising. Copper slag abrasives have been employed for years with good success, and some further study is advocated by other authorities.[7] The Bleile study also indicates that low-dusting abrasives are generally more productive than high-dusting media and that usage rates (pounds per square foot) for coal slag and mineral grits are generally lower than the rates for smelter slag.

The forementioned concerns of the U.S. Navy over arsenic and other heavy-metal impurities in abrasive-blasting media are receiving increased attention, and the subject has been discussed by industry authorities.[8,9,10] A related medical study linking certain types of slags to changes in lung tissue in animals is described by Stettler.[11]

An increased awareness is also developing of the dangers of ionic impurities on the steel or painted steel with respect to coating-system performance.

The Role of Atmospheric Pollutants

A German study[12] of seventy-eight bridges has revealed the complexity of contamination on bridge structures by inorganic salts, both in terms of the variety of contaminants and their distribution with regard to location, bridge design, and type of crossing. Contamination was found to be heaviest on horizontal members and occurred not only on coating films but between coating layers.

Research into some kind of quantification of the effects of ionogenic impurities (mainly sulphates and chlorides) on the quality of paint-system service is now occupying the attention of several bridge authorities. The development of potential methodologies for the detection and measurement of such contamination in the field is the subject of a study by Avondale Shipyards, New Orleans, Louisiana. Certainly the establishment of a quantitative or even a qualitative (go/no go) surface-cleanliness criterion for acceptable coatings service would be most advantageous, although it seems likely that such a criterion would be dependent upon the type of coating system employed and its film thickness (or its relative permeability to moisture).

A study by Weldon and Cain[13] effectively demonstrates the effects of sulphates and, more particularly, chlorides on the osmotic blistering failure of epoxy/polyamide and vinyl coating systems. Chloride levels of as little as 10 micrograms per square foot and sulphate levels of 20 micrograms per square foot produced blistering within several weeks. Test-strip ion titrators were employed to determine semiquantitative levels of chloride-ion[14] and ferrous-ion content.[15] The difficulties in establishing threshold levels of ionic contaminants is demonstrated by the different threshold levels of ionic material necessary to produce blistering with the two coatings. Notably, when the ion-seeded panels were reblasted before coating, no blistering occurred in subsequent testing.

Bayliss,[16] in summarizing the progress of the International Standards Organization Technical Subcommittee (ISO/TC 35/SC12) in their development of international surface-preparation standards, described similar semiquantitative tests using dipyridyl papers to detect ferrous salts. Baylis also discusses a qualitative technique that uses flash rust spots on an applied white latex paint to reveal the presence of soluble salts on a pitted surface. In a 1985 SSPC paper, several corrosion-indicator papers are also discussed.[17]

The growing awareness of the effects of atmospheric pollutants on the performance of coating systems and the kinetics of corrosion may spawn more practical changes in bridge maintenance. While sand and winter debris are often swept from northern bridges each spring, less attention seems to be paid to ridding the structure of other contaminants. Excessive salting may leave structures in the snow belt so heavily contaminated during the winter that even the color of a bridge is difficult to establish on the inboard rail and roadside box members. Increased adoption of yearly detergent or high-pressure washing schedules, especially along rail areas or beneath the bridge where such accumulations occur, would do much to preserve the life of both the steel and paint system (particularly unprotected, weathering steel in marine environments).

Of particular concern remains the effect of chlorides on the deterioration of weathering steel, where their removal by rain and/or other agents are precluded by bridge design. Studies of related deterioration phenomena and rectification procedures by means of surface preparation and coating-system protection continue to receive an enormous emphasis throughout the industry. Several highway planning and research studies relating to the corrosion of weathering steel are underway in Louisiana,[18] West Virginia,[19] New Jersey,[20] and Michigan. In addition, the SSPC has recently been awarded an FHWA contract to investigate the maintenance coating of weathering steel. Paint systems specifically designed to address the repair of weathered A588 steel in the field are now appearing, and specification design methodologies that address the problems have been suggested.

Additional research focuses on the development and evaluation of alternative deicing media that would eliminate or at least reduce the catastrophic deterioration inherent in the use of chlorides.[20] To this end, studies in the United States are examining the practicality of calcium magnesium acetate.[21,22] In Europe and Canada, urea has also been employed.

Anchor-pattern depth and its effect on coating performance has received increased attention.[23,24,25] The work has been summarized in the *Journal of Protective Coatings and Linings*.[26] The exact significance of anchor-pattern size is still open to considerable discussion among experts, however, and there are conflicting opinions even within the same bridge authorities. In a paper by Keane on the development of surface-preparation specifications by consensus,[27] the effects of mil profile on performance are considered of less consequence than had previously been thought. This conclusion is also reflected in the SSPC report on surface profile, which cautions that the presence of hackles and rogue peaks towering microscopically above the general profile height are much more critical.[28] ASTM has issued a standard for field measurements of anchor patterns.[29]

In two studies by Sandor[30,31] using salt-fog and immersion testing, the value of edge preparation by beveling on coating performance is examined. When the radius of curvature of beveled edges approaches one-half of the plate thickness, coating-system performance at the edge is said to approximate that of the face plate. No advantages were noted when edges were striped before airless application.

Trends in Specification Improvement

Considerable work is underway to improve surface preparation specifications. In an SSPC survey,[17] the SSPC and Swedish standards were criticized for inadequate representation of all groups of cleaning methodologies. Despite the introduction of rotary cleaning and descaling devices,[32] which provide surfaces equivalent to white blast, existing power-cleaning specifications include only the surfaces provided by the traditional tools and cite the removal of loose rust scale and paint only. By 1989, new standards had been issued by SSPC to address this inadequacy. Similarly, standards applicable to water-blasted surfaces are also being developed. New visual standards that would expand the number of rust grades and remove existing abnormalities among the presently adopted Swedish standards, actual practice, and even the SSPC written definitions are also being studied. Finally, a new blast quality, "industrial blast," is being introduced by the SSPC. This blast quality is designed to fit between the present brush-off and commercial-grade blasts.

EMERGING DEVELOPMENTS IN COATINGS TECHNOLOGY

A definitive assessment of trends and directions in coatings technology as it relates to bridge maintenance is unavoidably hampered by the proprietary nature of product development. An appreciation of market needs, overall trends, and the awareness of islands of developing technology, which may satisfy such needs or provide specific improvements in materials available today, does, however, afford some glimpse of what may be down the road.

Coatings Formulation Trends

The fundamental trends in coatings formulation toward 100%-solids systems and toward water are no less pertinent to bridge coatings than to paints for other markets. Trends away from the inhibitive oxidizing system to zinc and/or the barrier system have been noted in other segments of the industry for several decades, and all these trends are encouraged not only by environmental pressures but by economic motives and a need to make more efficient use of chemical feed stocks. Moreover, the constant impact of toxicological concerns over first one solvent and then another exhausts much valuable formulating time; the total elimination of solvents would curtail this necessity. Given the overall trends in bridge painting toward fewer coats and the reduced barrier properties of many water-borne systems presently in use, higher solids systems (toward 100%) would seem to be the most likely objective in the short term. The use of the four-coat latex system in California seems hardly conducive to cost reductions in a labor-intensive practice, and the use of the high-build, one- and two-coat, solvent-based systems of high solids, and consequently lower VOC would, at this time, seem more sensible.

Similarly, overall trends would seem to favor the further development of high-build systems based on thermosetting coatings (such as epoxies and urethanes) instead of the high-molecular-weight thermoplastics. Despite the higher performance profile of the vinyls and chlorinated-rubber systems, the inevitable depression of solids that the high-molecular-weight solution viscosities must bring will change the use strategies of these materials. Lower viscosity polymeric, oligomeric, and even monomeric vinyl, acrylic, and other binders of this type may be expected gradually to displace their high-molecular-weight counterparts, and use strategies will probably involve their combination with thermosetting systems as polyols, coreactants, diluents, or modifiers, instead of using them as lacquers as the sole binder. Already these trends are becoming quite apparent. Acrylics and vinyls are now used with some success as polyols for urethane systems and in hybridized epoxy systems. Vinyl/epoxies also show advantages in protracted pot lives and low-temperature drying properties. Still lower molecular-weight chlorinated-rubber systems are doubtful because of the inherently poor stability against dehydrochlorination in this polymer. Spans coated with vinyl/urethanes and vinyl/epoxies are, however, being tested in both Florida and Maine and are doing well.

Thermosetting polymers that build molecular weight by condensation of low-molecular-weight reactive species after application rather than before seem more easily adaptable to higher solids technologies. High-solids urethane and, particularly, epoxies are receiving (and will continue to receive) much attention from manufacturers. Even with these thermosets, the trends are toward lower molecular-weight liquid reactants at the expense of the more traditional binders. A primary goal of many resin manufacturers even now is to provide systems in which equivalencies favor the highest ratios of the lowest viscosity reactants providing the lowest viscosity resin/hardener combinations that may be applied with least difficulty, without compromise to performance.

Although it seems unlikely that new coating systems based on radically new chemistries can appear in the next five years or so, the application of molecular engineering in the synthesis of coatings resins is already providing many interesting and exciting new applications for coatings formulas that may eventually show up in the bridge-painting market. This type of polymer chemistry, and a less sophisticated but equally promising cold-blending technique at the coating formulating stage, will almost certainly produce even greater diversity in coating-systems types than those now available. Resin manufacturers will be forced by demand and competition to abandon emphasis on some of the more established commodities in favor of the newer, carefully tailored systems. Medium-molecular-weight epoxies with polyamide cures will probably eventually be displaced with liquid epoxies of lower molecular weight, cured with selected proprietary blends using aliphatic amines, cycloaliphatic amines, low-molecular-weight amidoamine blends,[33] and other mixtures. Lower viscosity epoxies based on Bisphenol F[34] or hydrogenated vehicles may even replace Bisphenol A resins. Novel polyols, diluents, and modifiers will probably appear, and all will undergo some kind of natural selection. Inevitably, oligomeric and mono- and multifunctional monomer species will play some part in this revolution.

Some backlash against the use of isocyanates in urethane coating systems may be expected to result from the Bhopal incident. This may both curtail the expansion of urethane finishes and spur the development of alternative light stable topcoats based on new aliphatic epoxies (cycloaliphates, sorbitol-based systems, and triazine-ring–based systems).

While most users are quite familiar with such terms as polyester/urethanes, acrylic/

urethanes, and coal tar/epoxies, such terms as vinyl/urethanes, epoxy/urethanes, and epoxy/acrylics will become more common. Nor is such expanded activity to be limited to actual reactive systems. Experiments with what are apparently diphase systems, such as plasticized thermosets, are also underway. Typified traditionally by coal tar/epoxies, some of the more promising epoxy mastics have utilized this type of technology.

Unavoidably, the drive toward higher solids will involve some compromise in pot life and application, aesthetics and, almost certainly, performance. The response of tighter films of higher crosslink density (inevitable with short-chain, lower molecular-weight reactants) is a reason for concern on bridge structures, where some degree of flexibility is indispensable. Judicious pigmentation and selected application methodologies (two-pack guns, etc.) can compensate for some of these inefficiencies, but it seems likely that some attributes will have to be sacrificed, even if only that degree of universality in use and application that we have come to expect from many systems. It is possible, for example, that new coating systems designed for use in the snow-belt states may not be as acceptable in the different environments of Florida and the Gulf Coast areas, and that increased specificity in formulations may be necessary. The FHWA–SSPC publication *Evaluation of Low Solvent Maintenance Coatings for Highway Structural Steel*[35] includes an excellent bibliography of recent work in this area.

As noted above, the move to water seems less imminent than the move to higher solids. The obvious exceptions do exist, and the alkaline-silicate-based inorganic zinc rich primers, providing perhaps the highest levels of corrosion resistance so far developed, are well established. Water, however, presents significant problems in certain areas of the country where the painting season would inevitably be reduced (compared to that possible with an alkyl-silicate inorganic zinc/vinyl system) because of freezing temperatures and the effect of low temperature on the proper coalescence of latex systems. This trend ignores what will undoubtedly become an increasingly important consideration: the development of painting systems that protract the painting season. There is also a lingering concern over the performance of latex systems on structures where salting is heavy, particularly when one compares the protection afforded by such systems to the protection possible with many of the high-performance solvent systems available today.

Latex dispersions of such thermoplastic polymers as vinyls and acrylics and lately even urethanes provide polymers of much higher molecular weight at much lower viscosities than are possible with conventional lacquers. With polymers plasticized internally by means of copolymerization, films with almost permanent flexibility are possible, with excellent response to those stresses that bedevil the more brittle, less impact-resistant systems (i.e., highly crosslinked thermosets and oxidizing systems prone to age-related autoxidation and chain scission). It is foreseen that as new technologies of emulsion polymerization are developed, latex systems with reduced moisture- and oxygen-transmission properties and with corrosion-resistance properties that rival their high-performance thermoplastic cousins may become available. However, this development may take years and must address the effect of surfactants and incomplete coalescence that, at this time, impair a truly high-performance latex bridge paint.

Significant advances resulting from intense competition for the latex bridge-paint market will be seen in the next decade or so. At this time, styrene acrylics[36] appear to have the inside edge, although urethane dispersions,[37,38] new vinyl technologies, and vinylidene chloride systems (employing a radical new paint-formulation technology[39,40]) deserve some careful attention in the near future. Still newer high-nitrile polymers, when properly cured, have moisture- and oxygen-transmission properties that exceed those of today's solvent-based coatings.

Warness[41] describes Caltrans's formulating experience with latex bridge paints.

Warness notes that not only are the latex polymers themselves critical to performance, but surfactants, defoamers, and wetting-agent selection must be controlled. He reported that by the end of 1984, some 10 million square feet of steel would be protected by the four-coat California latex system. Problem situations (poor blasting, nonuniform and/or inadequate film thickness, and interference materials) seem less related to the type of paint than to the normal deviations from good painting practice. Problems described as being related to the effects of application under cold, damp weather conditions are more definitely related to system type.

Chemically curing water-based systems are presently limited to epoxies. The obvious reactivity of water with polyisocyanates precludes the normal two-pack urethane route. Great strides in such technologies are being made, however, and the realization of effective water-based epoxy bridge systems, with properties that match those of their solvent-based cousins, seems closer than is the case with the latex systems, despite the California experience. After a couple of false starts, latest epoxy technologies utilize emulsifying species[42] that react with the curing agent and, upon curing, are chemically bound within the polymeric matrix and are, therefore, less sensitive to moisture than is the case with the latex systems. Kurnik[43] notes that the presence of water as a carrying solvent is in no manner substantially detractive to the performance of epoxy coatings when they are finally cured. Similar conclusions were originally reached by Masciale with earlier variants of the technology.[44] Khambatta and Warner[45] describe water-borne epoxy systems as well as a cementitious latex system that is said to provide good performance over rusted surfaces. The water-based epoxy technology developed by Celanese Corp. (now Hi-Tek Polymers, Inc.) provides coating systems from a diversity of curing agents with excellent performance characteristics. Water-based alkyds, on the other hand, while used widely in original equipment manufacturer (OEM) applications, have not shown as much acceptance on bridges.

Pigmentation Trends

In pigmentation, trends continue to reflect the efforts to eliminate lead and hexavalent chromium and the swing from the inhibitive primer to the zinc-rich primer. One can expect some continued effort on the part of pigment manufacturers to develop viable lead and chromate replacements,[46] however, because of the size of this market, and also so that they can make more effective use of auxiliary pigment volume in order to make up for the reduced effectiveness of the nontoxic inhibitors. More careful selection of both prime[47] and extender pigmentation[48,49] is to be expected, as well as less flexibility in formulating practice.

With the continued success of zinc-rich primers will inevitably come attempts to extend zinc with aluminum, graphite, and other conductive extenders of various types. As with the inhibitive primers, it is to be expected that other attempts will be made to improve zinc rich performance by means of even more subtle formulation refinements that have little to do with zinc content per se. They are typified by the effects of solvent selection noted by Orr[50] on the wetting properties of zinc primers. Other techniques are now being developed to coat zinc onto spherical and platey inerts in order to reduce the cost (and weight) of zinc-rich systems (K. Lindahl, Fillite, Inc., personal communication, 1985).

New atomization methodologies for the production of zinc pigment, resulting in an elongated sphere, are described in a paper by Leclercq.[51] It is claimed that formulations based on these new zinc pigments are cohesively stronger, without compromise in the level of cathodic protection.

With the greater emphasis on the barrier principle, increased use of both nonleafing and leafing aluminum flake and such other lamella pigments as micaceous iron oxide and stainless steel may be anticipated. This trend is already reflected in the aluminized epoxy mastics and the moisture-curing urethanes.

Environmental Concerns

Where environmental problems or political pressures prohibit the use of lead, but where funding for radical rectification procedures allowing the full use of zinc systems is not available, the aluminized epoxy mastics will continue to score. Without lead, which effectively catalyzes the drying-oil polymerization, the high-wetting, oil-based primers historically used in these applications are nonviable. Unless some innovative procedures for complete conversion of unsaturated vegetable oil without lead appears, in-service structures once repainted with lead and oil combinations may see increasing use of epoxy mastics. The growing concerns over the toxicological properties of synthetic chemicals in general, including solvents and coating resins, may conceivably also spawn some renewed research interest in the adaptation of such natural products as vegetable oils by the coatings industry.[52]

Formulations for Poorly Prepared Substrates

The potential market for a coating system that will provide viable protection for poorly prepared substrates will probably continue to encourage considerable research and development on the part of paint manufacturers. Various estimates on the size of the potential market (in terms of steel area) range from three to seven million square feet.[53,54] Well-controlled evaluations, such as the study by Deanin and Frondistou-Yannas,[53] will probably inhibit the readiness of system-design engineers to give any more credence to the various concoctions based on tannic acids and other rust-converting primers than was afforded them in the past. Most of the viable coatings selected for the Phase II laboratory and field evaluation trials of the Deanin/Frondistou-Yannas study were, in fact, rust-impregnating and emulsifying types. Interestingly, the conclusions of this study concerning the most viable candidates for use over nonblast-cleaned highway steel confirmed the general industry impression that only the epoxy mastics and the aluminum-pigmented moisture-curing urethanes (both considered rust-impregnating types) give such controls as red lead/oil/alkyd and bituminous mastics any real competition. The latest data from this study indicate that the bituminous-material control is the most effective composition.[55]

Research with Zinc Rich Systems

Research with zinc rich systems will probably address those areas of deficiencies that now exist. Cure rate (particularly with the single-pack inorganics), film cohesion, and attempts to eliminate extraordinary recoating procedures, mist coats, tiecoats, and so forth are all targets for potential research. Attempts to further conserve zinc usage in both organic and inorganic zinc rich systems not only by reduced loadings and conductive extenders but by means of more widespread use of anodic-dissolution control techniques that prolong the effectiveness of zinc rich protection (replacing such auxiliary inhibitors now in use as lead and zinc chromate) may be expected. It is, for example, noted that in the electrodeposition of zinc from plating baths, small quantities of cobalt and chromium provide films with reduced zinc corrosion.[56]

Higher solids systems, even in zinc primers, will probably follow the development of such techniques in other coatings (organic) and would be an undoubted consequence of any reduced zinc-binder ratios in solvent-based inorganics. Environmental pressures on such solvent-based inorganics as the alkyl silicates may eventually lead to more widespread use of water-based, alkaline-silicate zinc rich primers. The watchwords in this regard will continue to be "surface preparation" and "recoatability."

The drive toward increased solids and lower viscosities will also foster more research into improved pigmentary–vehicle interfaces by the use of titanate, silane, and zirco-aluminate surface treatments. Such treatments not only produce a marked elevation of the CPVC but have other profound effects on performance properties not apparently related to such CPVC shifts.[57] It may be expected that in the next five years the use of pigmentary surface treatments at both the raw-material manufacturing stage and, to a lesser extent, at the paint-manufacturing stage will see considerable development.

Research in Corrosion Control

Research trends in basic corrosion science as it relates to corrosion protection continue to concentrate on developing an improved understanding of such critical areas as transport phenomena (moisture, oxygen, and ionic materials), the corrosion process beneath the film and at and around holidays (cathodic disbondment, etc.), and the effects of polymeric and morphological heterogeneities. Other studies are more specifically directed toward the prediction of corrosion before it occurs or at least its detection before it becomes obvious. Motivations for such research are to obviate corrosion by means of more specifically directed disruption or hindrance of the isolated processes, thereby prolonging the effectiveness of coatings, and also to substantially reduce the long time spans necessary to evaluate corrosion-control coatings tested by current testing procedures (salt fog, weatherometer, etc.).

More conventional testing methodologies will probably remain in use at least until some of the more scientific techniques can prove themselves. In view of the poor correlation of such tests as the salt-fog test and the Kesternich procedure with long-term field performance, it is possible that customized modifications of these tests will increase. Cycling test methodologies using salt fog or Kesternich with QUV and hot and cold cycling are already being evaluated, and it is probable that such techniques as this will continue. Adaptations of the Prohesion Test methodology (described in Chapter 7) may also receive some attention in view of its apparent success in Europe.

While adhesion[58] and bond uniformity[59] are possibly the most important factors in corrosion prevention by coating films, easy, accurate, quantitative measurements of such film characteristics remain elusive. It should be emphasized that, despite the moisture-transmission rates of a coating film, if adhesion is maintained even in the presence of water, corrosion cannot occur. Present adhesion tests, however, are not only inaccurate but they also generally measure adhesion under dry conditions, not the wet conditions that are critical to the functioning corrosion-resistant coating. Efforts to characterize this important aspect of coatings performance better are being researched: Funke and Zatloukal[60] have examined wet adhesion by comparing the water-absorption–time curves for both supported and nonsupported films. The point of interaction of such curves (where the water absorption of the supported films assumes the same value as the unsupported film), called the "crossover point" by Funke, marks the initiation of water accumulation at the interface. By comparing the relative times different coatings take to achieve crossover, some comparative measure of wet adhesion seems possible. It would seem that increased emphasis on the characterization of wet adhesion might be expected in the coming years.

In a later paper, Funke[61] observes the dichotomy between the advantages of polar and hydrophilic groups for adhesion to metal and their disadvantages in greater sensitivity to water. While this tends to confound the attainment of good adhesion, Funke suggests several possible alternatives for improved wet adhesion even in the presence of such groups. These include the use of high-build systems, surface pretreatment, and/or increased immobility of the polymer backbone in contact with the interface. Papers by Kumins[62] and Dickie[63] on adhesion and de-adhesion and their relationship to corrosion are also of interest.

The profound effects of stress and internal strain on the performance of coating films, while obvious to all having any kind of intimacy with the technology of coating films, have received less-than-deserved attention among those who design coating systems for corrosion protection. Internal stress is built up during solvent release and/or crosslinking because the volume change is restrained due to the restriction of molecular segments by crosslinking and after application by absorption and desorption of moisture; adhesion of the coating at the interface inevitably suffers. Even in single-coat systems, internal stress may be considerable. It may, for example, exceed the tensile strength of the film, in which case the film will inevitably crack or fail cohesively.[64] If it should exceed the adhesive strength of the film, spontaneous delamination from the substrate may occur.[65] The latter danger is of particular concern in thick, thermosetting films (such as high-build epoxies) that are cast from fairly slow solvents. In addition, as noted by Saarnak, Nilsson, and Kornum,[64] internal stresses can increase slowly over quite long time periods (six months after application), and are markedly affected by such external influences as changes in temperature, humidity, UV, the aging process of the film, and the type and cross section of the substrate material.

In such multicoat paint systems as are employed in practical bridge painting, the transfer of stress from one coat to another within the same composite is of particular concern in adhesive systems. This process is manifested in various delamination phenomena, particularly when older systems are recoated with high builds of strong, new systems. Again, in the practical application of strong, highly cohesive finish coats over zinc films of necessarily reduced cohesive strength (particularly when zinc films are incompletely cured), there is significant danger of system disruption. Much value is to be derived from a better understanding of the mechanical properties of paint films and complex paint-system composites.

These considerations become particularly important in an industry in which the trend is toward the application of higher film thicknesses in fewer coats. Using Kendall's[66] equation for the peeling of a coating,

$$\frac{F}{b}=\gamma-t_c U_R \tag{10-1}$$

where

F = peel force applied to the coating
b = width of the coating
t_c = thickness of the coating
U_R = recoverable strain energy stored, by virtue of the internal stress, in unit volume
γ = interfacial work of adhesion

to determine the stress σ necessary to detach an area by the pull-off test, use the following formula:

$$\sigma=\left[\frac{2K}{t_c}(\gamma-t_cU_R)\right]^{1/2} \tag{10-2}$$

where K is the bulk modulus of the coating.

All other things remaining equal, it is apparent that, as film thicknesses increase, the work necessary to enact detachment decreases. A more complete appreciation of these phenomena would seem important if today's trends are to be sustained, and if increased difficulties in adhesion and/or cohesive failure are to be avoided.

Much theoretical work to quantify our understanding of these phenomena has been done by Croll,[67,68,69] Shimbo,[70] Toussaint and D'Hont,[71] Perera and Vanden Eynde[72] at the Belgian Paint Research Institute, and others.[73,74] Various methods of determining internal stress in coatings, including the use of overhanging- and cantilevered-beam distortion techniques, are described by Croll[75] and Perera and Vanden Eynde.[72] Further investigation of these phenomena is expected, and, as appreciation of their relevance grows, it is expected that considerably greater attention will be directed toward this extremely important area of the technology.

Dynamic mechanical testing (DMT) of thermosetting polymers and coatings by means of the torsion pendulum and torsion-braid analysis is also receiving increased interest. Automated test systems of this type are said to provide a viable approach to the accurate investigation of cure kinetics, effects of water uptake, plasticization, and so forth.[76]

A considerable body of research across the world has now been devoted to the investigation and development of potential test methodologies for the evaluation of anticorrosive coatings that use electrical and electrochemical techniques.[77–81] Because corrosion is generally electrochemically induced and because the mechanisms of transport through coating films involve either electrical phenomena or at least can be monitored by electrical property changes in the coating films, this is hardly surprising. The potential value of this type of testing lies in the fact that it should provide not only a comparative and, with luck, a quantifiable measure of the quality of protection but also that some measure of understanding of the actual mechanisms of deterioration and, therefore, protection may also be gleaned. While the literature abounds with many studies of this type, results have been no better than mixed,[58,82] and certainly the proliferation of viable techniques is at this time limited. Tooke and Hurst, in an earlier FHWA report[83] that evaluated apparatus and methods for electrochemical techniques, concluded that, while studies of the dielectric properties of paint films provide rapid data on moisture permeability and while electrical polarization techniques have value in characterizing the inhibitive properties of paint films, the techniques seem limited to research and development or, possibly, diagnostic investigation of the paint-film deteriorations.

Attempts to quantify the performance of coatings on metal by a measurement of impedance (by both DC and AC techniques),[84] capacitance,[76] frequency response, electrical potential, and polarization effects are also receiving considerable attention from many researchers. Past results have been summarized by Leidheiser,[85] Jullien,[86] and Wolstenholme.[87] Methods used to date have involved the study of transport through barrier systems, the detection of water beneath the film, corrosion growth at the interface, the detection of heterogeneities in the polymeric film,[88] and the study of the effects of pigments on the dielectric properties of coatings and on blister growth. Joly and Laout[80] note that polarization resistance measurements may be used in the testing of inhibitive pigments and paints. Callow and Scantlebury and others have done considerable evaluation of electrical impedance in coated metal electrodes.[89–94] The potentiostatic and galvanostatic degradation of zinc rich protection over time have also been examined.[95,96] Lindquist, Mezzaros, and Svenson, using this type of technique, have successfully compared about eight commercial alkyl-silicate and epoxy-polyamide zinc rich primers.[97]

Apart from impedance studies, which generally show that (in barrier systems) as the coating resistance drops, the rate of corrosion increases, most of the electrical studies leave something to be desired, possibly because of the practical difficulty of obtaining truly perfect, defect-free films of uniform thickness. The effects of pores and inhomogeneities on electrical properties of films and subsequent corrosion breakdown of the substrate has been evaluated by Mills and Mayne.[98] Callow and Scantlebury examined selectively dyed paint films under UV illumination and noted the possibility that such heterogeneities may be detectable by means of differences in film fluorescence after dyeing.[99] Film integrity and the presence of pores in coatings have also been investigated using autoradiography.[100] In these more practical experiments, Bayliss and Bray conclude that severe microvoids are entrained by certain application techniques (i.e., airless application), and that the drying times of many commercial systems may be too fast to allow air entrainment to dissipate before film setting. Large vacuoles (voids produced by trapped air) after airless application have also been described in chlorinated-rubber coatings elsewhere.[101]

Increased use of test methodologies involving cathodic disbondment techniques is now evident. This test[102] simulates and accelerates the effect of the natural cathodic polarization (and resultant alkaline condition) of the metal beneath paint films adjacent to defects and corrosion sites by artificially polarizing the substrate.

Other new, nonelectrical techniques for the evaluation of corrosion and anticorrosive coatings that are receiving increasing attention include the use of ellipsometry (the measurement of the growth and/or dissolution of surface films beneath clear coatings by changes in the reflectance of elliptically polarized light)[103] and surface analysis using Auger, Sims, and scanning electron microscopy (SEM) energy-dispersive x-ray analysis.[104] Cunningham and Hansen[105] report the value of electron spectroscopy for chemical analysis (ESCA) and Fourier transform infrared spectroscopy as tools in the early detection of weathering changes in exposed films as well as the promise of these techniques in detecting and understanding the weathering process. Both Fourier transform infrared spectroscopy and Fourier transform nuclear magnetic resonance spectroscopy are discussed in an article by Mathias.[106] Differential scanning calorimetry is being employed to measure the state of absorbed water in polymeric films[107] and radio chemical labeling techniques have been used by several researchers to examine the diffusion phenomenon through paint films.[108,109,110]

Nondestructive tests to signal the onset of corrosion under the film before it becomes manifest are receiving attention from several researchers. Martin and McKnight[111,112,113] at the National Bureau of Standards describe a technique using infrared thermography that holds great promises, particularly when combined with reliability analysis. The latter technique is a philosophical departure from the testing approach traditionally employed in corrosion control by coatings and has evolved from its success in other fields. Traditional testing practice is based on comparative testing of one paint system against another and attempted correlation of the accelerated results with field performance. Problems of reproducibility and lack of quantification so limit this type of evaluation that correlation is often regarded as questionable. In reliability analysis, multiple panels of the same coating are evaluated at different controlled stress levels produced by temperature, humidity, UV, and so on. The frequency of failure at a high stress level is then employed to predict the failure rate under field conditions. The

National Bureau of Standards is also involved in the evaluation of underfilm corrosion and blistering by thermowave-imaging[114] (which, while not as adaptive to field investigations as infrared thermography, appears to be a good laboratory tool).

Acoustic emission, used for years to detect mechanical failure in welds, has recently been applied in the nondestructive evaluation of chemical structure of polymeric paint films, and its relationship to paint performance has been examined.[115] Development of monitoring devices of this type for routine field use could eventually prove invaluable in providing early isolation and, therefore, less costly maintenance of coating systems.

TRENDS IN SPECIFICATION DESIGN

Although relatively isolated at this point, more widespread use of the single-source-responsibility contract in bridge painting can be expected on larger structures run by individual bridge authorities, if costs do not become too prohibitive. The telling factor in this development may well be the first serious failure of a structure painted under this type of contract and the performance of the contractor in upholding his guarantee. On the other hand, success of this type of contract may point out to insurance companies the value of formally underwriting the venture, which could further catalyze the adoption of this procedure. Recent reluctance in general insurance underwriting practices are not, however, encouraging in this regard.

Smaller bridges and state and locally managed structures will probably continue to be let by the present bidding process, although more use of prequalification of contractors and applicators may be tried. If prequalification does grow, it seems almost certain that some form of legal challenge will follow, and apprehension of the ramifications of this should at least temporarily dampen this trend. There are few signs of improvement in any union-led effort to more formally qualify applicators at this point (although the larger contractors do not necessarily deplore accreditation). If the trend toward non-union contracting continues, there may well be some motivation to cultivate accreditation by the unions. Accreditation of contractors through group organizations (witness SSPC's accreditation program) will certainly take up the slack in this regard if the unions fail to take the lead.

The critical nature of coating-system selection and an awareness of the nonuniversality of individual systems is receiving increasing attention. A guide[116] to assist coating-system design engineers in selecting the most suitable coating system for any bridge is presently being developed for the FHWA and is testament to the need for a better understanding of the relationship among the design, environment, and condition of in-service steel bridges and the types of paint systems best suited for their maintenance.

The SSPC is also preparing a guide for maintenance painting[117] that will treat not only the process of establishing a maintenance program but also the repainting of steel structures. Although not specifically directed at bridge painting, this work should provide much valuable assistance in all aspects of the discipline. Subjects are to include: definition of objectives, a structure-condition evaluation and systems-option study, evaluation of corrective-action procedures, preparation of contracts and specifications, securing contracts for labor and materials, as well as actual repainting and inspection, including troubleshooting, postapplication inspection, and record keeping.

The conception of a bridge not as a single structure but as a composite of several component structures, each having different types and degrees of protection, is a relatively new idea in specification design, but one that shows every sign of further development in new specifications.

Specifications must inevitably improve in type and in detail, and, because of the difficulties of batch control by performance alone and the obstacles in the correlation of accelerated testing with field performance, some wider utilization of QPL methodologies or development of compositional forms of inorganic zinc, high-build vinyl, epoxy intermediate and finish coats, and aliphatic urethanes must also eventually evolve, if only to support performance criteria. Performance-control testing increasingly involves the testing of the total coating system instead of simple coatings alone. As more and more zinc rich systems evolve, such full-system performance criteria should be encouraged, because of the incidence of system problems involving bubbling, zinc-splitting failures, and so on.

Improvements in form are likely to evolve by way of transfer from one state to another of successful systems developed by raw-material suppliers, paint companies, and/or consultants, but use of the QPL process as now employed in Florida and Michigan will undoubtedly increase as well.

Stabilization of the present flux and confusion over coating systems that now exist within the industry, and some normalization of the multiformity of specifications that are employed from state to state, must inevitably be beneficial if, as we here conclude, today's best material types are indeed adequate to address the nation's bridge-painting needs. It would seem that, as those most efficient systems (for given specific environments) are proven by reproducible long-term field performance, and that, as comparative data is accurately disseminated, a finite number of well-defined systems with fixed compositional/performance specifications that are refined through use and dependable performance, should be possible. The probability of such order is not apparent, however, at this time.

The process could be fostered by several factors, including increased communication and data dissemination, increased technical awareness of the technologies involved, by the unifying presence of the FHWA and, perhaps, the SSPC and NACE, and by more designed use of federal funding. Most of all, such order is encouraged by an individualized desire among specification writers to avoid the consequences of failure, if not to enjoy the satisfactions of success. Normalization is to some extent discouraged by individual prejudice, competition among the larger paint manufacturers and raw-material suppliers, and by an ever-advancing technology that constantly seeks to introduce "a better system" that is inevitably protected from early competition by proprietary shields. Normalization is qualified to some extent by the disparity in coating-system requirements that come with geographical location, but this limits the degree of normalization (i.e., the number of proven paint systems), not so much the process itself.

Should normalization of specification design not emerge, then it would seem that, if paint and painting-system failures are to be avoided, those who manipulate individual coatings to produce viable coating systems for the protection of steel bridges must develop a much more refined understanding not only of coatings use technology but of the basic technology of coatings and corrosion control in general and of specific products. Furthermore, in this case, strict standards of qualified product and qualified system control would seem mandatory.

TRENDS IN JOB EXECUTION

In job execution, the most significant changes should result from increased regulations and a growing public conception (or misconception) of the effects of materials and processes (e.g., open blasting) on the environment and public safety. Issues of concern that developed in the late 1970s are mushrooming in the late 1980s and spawn a plethora of complications for the painting contractor that will inevitably increase the cost of operations and the price he must levy for a given job. Most obviously, changes in surface preparation that reduce productivity (containment, vacuum blasting, etc.), increase operator fatigue and safety, and raise the cost of equipment should be reflected in increased bids.

Most contractors agree that environmental restrictions will continue to grow, especially in large metropolitan areas, and while these concerns were once limited to isolated geographic pockets, they are gradually likely to spread. Regulations, long existent but rarely levied, are now being enforced by federal, state, and local government agencies, as well as by unions. Grit disposal, blood-lead monitoring, and the control of airborne particulates are all factors with which the contractors are slowly being forced to deal. An 1899 U.S. Coast Guard edict (controlling waterway pollution) is currently being implemented in certain parts of the country, for example.

The advent of OSHA in the 1970s brought about large-scale changes in rigging practices, and a brand new rigging industry continues to grow, where once all rigging was designed and implemented by the contractor himself.

In application, as might be expected, developments will continue to reflect changes in materials technology, and the response to higher solids systems will inevitably continue the trends toward higher airless-pump ratios, air-assisted airless techniques, and the use of multicomponent equipment with metered feeding. Hot spray and electrostatic devices are also expected to increase as is the use of metallizing, with its associated equipment. Air-assisted airless applications that respond to greater control requirements necessary for use with inorganic zinc on complex riveted structures seem to have particular potential. Larger, quieter compressors, sweepers, scrubbers, and articulated access devices will probably also add to the cost of operations. The advantages of air-assisted airless technology, cartridge-powered electrostatic air spray, and electrostatic air-assisted airless applications are discussed by Miller.[118]

Many contractors voice concern over the impact of increased litigation on their operations. For the most part, this involves suits brought by the general public in response to property and personal damage from the contractor's operations, although suits relating to job performance are also increasing. The indirect effect of this is a dramatic increase in insurance premiums and a growing reluctance on the part of insurance companies to underwrite operations.

As noted in Chapter 6, the inconsistencies between specifications and practices regarding contract requirements and the uncertainties over environmental controls on many jobs present the contractor with tremendously difficult estimating problems. While it may be expected that, in the long run, as requirements become more formalized and enforced, this problem may evaporate, most contractors feel that things will get worse before they get better.

With the higher cost of bridge painting, most contractors foresee that state and local engineers will become increasingly aware of the technologies of the process and the materials used. Some contractors note that these authorities are employing the services of outside engineers, consultants, paint specialists, and more inspection forces increasingly to ensure that the value of the money spent is better preserved. The effect of such specialists on job prosecution seems variable and dependent on the level of their individual expertise. The best will assist contract operations, while preserving the bridge authority's interest; the poorest bring no more expertise than would the authorities' own resident, and they may confound operations with trivial concerns.

Throughout the industry, there seems a trend toward nonunionization. Some states require union labor, while others do not. Neither do many local authorities. While all states and bridge authorities usually require union pay scales, not all enforce the equivalent benefit program and many nonunion houses skirt the pay issue by compromising these programs and maintaining a competitive edge over fully unionized shops.

EFFORTS TO IMPROVE INSPECTION

Considerable effort is now underway to redress the deficiencies in inspection through the education process. The bridge paint inspection course that was rewritten for the FHWA by Pinney[119] is proving far more valuable than previous attempts at this material, and the practicality of having experienced inspection personnel rather than academics present this material is undoubtedly a step in the right direction. New general paint-inspection manuals are offered by the American Institute of Steel Construction (AISC) and ASTM, as well as the U.S. Army Corps of Engineers, while inspector accreditation programs are being offered by NACE.

The advent of improved teaching tools and visual aids, such as VCR equipment, should also assist in the training of inspectors, bringing the practicality of shop and field preparation and application into the classroom, as well as generally improving the presentation and understandability of the courses. Such VCR training courses are offered by the Institute of Applied Technology and are distributed through the SSPC. Much wider use of these techniques may be expected in the coming years.

VCR equipment is also being used to support still photography in the actual inspection process. The versatility of VCR cameras, particularly the newer, less cumbersome models for low-light situations and close-up work, is only matched by the indisputable value to the job record that such recordings provide.

Some improvements in inspection instrumentation may also be expected with digital electronic film-thickness gauges, psychrometers, and other equipment. Eventually (as techniques are proven and instrumentation is commercialized), it is hoped that such nondestructive inspection tools as infrared thermography may be introduced.

All of these advances should provide the groundwork for upgrading the inspection force. Unfortunately, the development of the professionalism that is required will necessarily depend on attitudes at the local levels. To attract the type of individuals needed to establish a professional force will require a substantial commitment on the part of the states. Without funding to man and properly equip the required force adequately, the monies and efforts now being directed toward the problem (through education) will be of no avail. A paper by Holl[120] further documents the value of good inspection with several examples.

Paradoxically, the contracting of inspection services to the private sector appears to be increasing and, as more use is made of more complex, high-performance systems, this trend may also be expected to continue, especially if the single-source-responsibility contract becomes more common in bridge painting.

IN-SERVICE MONITORING TRENDS

The corollary of what seems certain to be a continued escalation of both the costs and the complexity of the bridge-painting process is, naturally enough, a need for improved cost effectiveness and a more ordered appreciation of the relative service of different coating systems in different environments and on different types of structures. Much of the present guesswork must be removed from the projections of system service. This inevitably depends on the better and more uniform execution and control of the job process. Accurate assessment of the life of a given system on a given bridge, leading eventually to a more generalized profile of the system's service with respect to different structures (or components of structures) in different but well-defined environments, cannot be achieved without precise documentation of the nature and patterns of deterioration, however.

The formal establishment of deterioration curves is an important key to the quantification of both service and cost, as well as an aid in specification refinement. It is also reasonable to assume that once statistically reliable deterioration curves of system service are established for a given structure, better maintenance programs can be planned at the outset. Additionally, based on a comparison between the actual progression of deterioration curve against the anticipated deterioration curve, early preventive maintenance may be employed to counteract specific or local deficiencies before the deviations between actual and projected values become too great and too costly to adjust without a major painting program. Similarly, the effects of unanticipated environmental changes (e.g., the proximity of new manufacturing plants or the relocation of chemical-storage sites) may be more accurately assessed for changes in rates of deterioration before such effects become too devastating to the paint system or, more importantly, to the structure.

Further evolution of the bridge corrosion cost model and the need for reliable data for the refinement of this tool and for its adaptation to specific states or areas of the country will also encourage more formalized postpainting monitoring programs. Again, the establishment of a specific central databank for receipt and dissemination of the collected information of the disciplines would, because of requirements for more accurate data, generate more programs of formal coating-system service documentation.

Unfortunately, it seems likely that, without both guidance and money from the federal level, controlled programs of quantification of postpainting service monitoring will probably remain restricted to, at best, a handful of more forward-looking states, and the data realized will be too incomplete to be of more than limited local value. This is unfortunate because in many instances the methodology of periodic bridge inspection (if not coating-system inspection) is already in place. What remains lacking is not only quality in such evaluations and the accurate recording of deterioration patterns and quantities but the proper organization of such data and its dissemination.

The ideal scenario would be one in which coating-system service data, retrieved from locally maintained, up-to-date files on each and every structure, are centrally stored and statistically evaluated in terms of location, environment, structural design, and so on. The condensations of these data could then be retrieved from the central databank by local authorities and reapplied in the formulation of more cost-effective maintenance policies and/or in the continual refinement of coating-system specifications and maintenance methodologies. Unfortunately, despite the opportunity for this type of order that our computerized age allows, the realization of such fantasies is a long way off.

Several papers considering coating-system service and costs were presented at the conference "Corrosion 86" in Houston, Texas. These include an interesting paper by Kline and Keyman,[121] which contains much valuable data on the service lives of zinc rich and galvanizing systems, although the data are nonspecific to bridges. The paper also compares several U.S. and European scales for grading paint-system deterioration and discusses maintenance scheduling. A further updating of the work of Roebuck and Brevoort was also presented at the same conference.[122] While the performance figures noted in the revised tables of the newer work still appear inappropriate to practical applications for bridge service, the authors now consider the use of computers in the development of comparative maintenance costs. A paper by Smith[123] also demonstrates the use of computers in the manipulation and retrieval of data required for the management of maintenance painting programs. Like the Roebuck paper, this paper seems more directed toward plant maintenance than bridges. Additional service and cost comparisons are also found in papers by Masciale,[124] Hergenrother,[125] and Mudd and Boyer.[126] The Masciale paper is particularly interesting in that it specifically considers the refurbishment of old coating systems with epoxy mastics.

ANALYSIS AND ORGANIZATION OF COSTS

Improved accuracy and more formal quantification of the costs of coating systems, along with the various individual cost elements of the job process, will be important considerations if the potential value of the bridge corrosion cost model as a tool in the decision-making process of bridge maintenance programs is to be realized. While seemingly elusive, and clearly subject to the inconsistencies of the open-bid process, thanks to the pioneering work of Frondistou-Yannas et al. and those who are shaping the crude model into a more practical tool, the consolidation of cost data into some viable order may be no more difficult than the formalization of system-service profiles. Indeed, because service life is itself an important element in the establishment of cost effectiveness, it is likely that the organization of both cost and service disciplines will progress to some extent simultaneously.

What is again needed is data—from contractors, from coatings manufacturers, from equipment suppliers, and from the analysis of bid packages themselves. Because of the subjectivity and variability of the bid-development process, the more data the better. Armed with increasingly more accurate data, statistically condensed from this input, repeated applications of the BCCM to more and more structures will inevitably shape the product (as revised) into a far more practical tool. Deficiencies in the present model will be isolated and, as these are corrected, the model's efficiency will improve. Certain modifications, such as an extensive broadening of the present coating-system classifications and the transference of items presently included in the program data file to input data for particular structures, are applicable to the general model. Other modifications will probably be more appropriately made to a given state's particular adaptation of the model and reflect peculiarities common to climate, bridge design, and the existing paint systems in that state.

The analysis and organization of cost data can also do much to further a more precise understanding of other aspects of the job process among participating engineers and, by means of subsequent publication, other interested parties. Critical but poorly disseminated data on aspects of job execution that themselves have a direct bearing on painting costs—production rates for blasting and the effects of steel condition or abrasive and nozzle size or air pressure, the relative application rates with different types of coatings and techniques, grit usage, material volumes, and equipment capabilities—all may thus be quantified and better appreciated.

Chapter 11

Conclusions and Recommendations

This volume has attempted to review the state of practice in the selection and use of coating systems on bridges and other steel structures from all vantage points. Consideration has been given not only to the design and performance characteristics of the coating systems themselves but also to their specification, application, and control, all of which strongly influence the success and cost effectiveness of any job. The conclusions and recommendations that follow may be condensed from the whole as being most significant. In outline, the more important areas for continued, new, or renewed emphasis are as follows:

1. Improvements in surface preparation methodologies; safer and less costly removal of toxic paint systems and improved removal of ionogenic material
2. Improved user-friendliness in coatings; development of improved coating systems for nonblasted and marginal bridge surfaces
3. Development of viable, nontoxic coatings with low volatile organic content, and water-based counterparts of present paint systems
4. Improved specification design; education of specification writers and materials engineers, increased communication between field and design engineers, increased use of cost-effective specification-design features and "specific utility"
5. Education and qualification of applicators; development of alternative procurement methodologies
6. Professionalization and training of inspection force
7. Development of reliable, service-predictive testing methods
8. Formalization of postpainting service monitoring
9. Improved quantification of service and cost profiles; further development and refinement of the bridge corrosion cost model
10. Implementation of a central bridge painting databank; development of methodology for collection, processing, and distribution of data
11. Improved nonpainting maintenance; regular cleaning and reduced deicing chloride use
12. Increased employment of structural designs that facilitate the application and performance of coatings, and their inspection; decreased use of lattice work,

open-boxed trusses, and back-to-back angles; built-in rigging and inspection aids

OVERALL TRENDS

In all areas of the discipline, bridge-painting practice seems to be in a state of flux as the industry struggles to keep pace with the technological changes that have been taking place in coatings technology during the past thirty years. The predominant motivations for such change have been the drive for longer, more cost-effective service from coating systems and the increased awareness of the market potential by specialty-coating manufacturers who have, with their more sophisticated "higher performance" coatings, to some extent displaced the smaller specification paint houses from the market. Other factors that have become increasingly important include environmental concerns over lead, hexavalent chromium-based pigments, and photochemical reactive solvents, and the economic inducements of the reduction and elimination of expensive solvents that the swing to high-solids systems and water makes possible. By reducing the necessary coats that make up a typical paint system from three to seven coats to two to three coats, the industry has attempted to reduce the hard dollar costs of paint jobs. These costs have increased not only as a result of labor and materials cost increases but also because of the necessary expense of increased surface-preparation procedures, inspection, and design.

Counter to the general trends toward high-performance coating systems applied over blast-cleaned surfaces in the field are more recent developments in coating systems tailored for application over minimally prepared substrates. This trend reflects and is fueled by increasing concerns over the dangers of airborne atomized-lead-paint contamination of the air and/or, more generally, the costs of the disposal of hazardous waste from the blasting process.

SURFACE PREPARATION

In spite of the limited funding available for bridge maintenance, which acts as something of a deterrent to expensive surface preparation procedures, abrasive blasting is rapidly becoming the norm throughout the industry, particularly in new construction, where centrifugal blasting affords cost-effective excellence in substrate preparation. In field work, abrasive blasting to a minimum of a commercial-blast quality is being adopted in many states in the total rehabilitation of those older bridges that have severely deteriorated existing paint systems and in the spot-blasting maintenance of less deteriorated structures.

Blasting has been fostered not only by the generally increased awareness of the value of the uniform, clean, and profiled substrate and its contribution to the long-term durability of all paint systems but also by the inappropriateness of the older preparatory methodologies to newer paint systems that are based on barrier and/or zinc rich protection. Perhaps the only situations in which blast cleaning is not cost effective are on those bridges where lead-based paint systems utilizing the old TT-P-86A Type I red lead and oil primers are still used in Zones 1B and 2A environments. The extremely long-term protection offered by these systems and by other lead-based systems in many rural parts of the country make their upgrading by means of blast cleaning difficult to justify even from an engineering standpoint.

A newer development particularly suited to field work is the utilization of wet-blasting techniques, although many of these techniques are little more than embryonic at this point. Increased employment of this type of surface preparation may be expected with the development of an increased awareness of the impact of chloride and sulphate nests on the overall performance of coating systems. The limitations of the dry-blast process in the total removal of such contamination from corroding steel and, perhaps more importantly, a sharpened awareness of the hazards of silicosis certainly foster the adoption of the wet-blasting approach. Wet-abrasive blasting and water blasting are also further encouraged by concerns over the hazards (real or perceived) in the removal of lead paints from older structures. The total or partial deleading of the hundreds of thousands of leaded bridges across the country will doubtlessly increase as lead cycles are broken in favor of less toxic or higher performance systems, and, especially in urban areas, it seems certain that the protection of the public from the potential hazards of the process will become an increasingly important factor. The perfection of viable wet abrasive blasting techniques for such deleading operations would, from this vantage, seem a more universal response and probably a less costly one than either complex containment devices or local vacuum-blasting techniques that have been used to date. The disposal of hazardous waste from deleading operations can, in many areas of the country, be extremely expensive. Water-blasting (if not wet-blasting) techniques, which decrease the overall tonnage (if not the concentration) of hazardous waste generated by the deleading process, are also favored because of the consequent reductions in the cost of disposal. These difficulties are also reflected in the resurgence of other nonatomizing preparatory methodologies (hand- and power-tool cleaning) and in the development of equipment to upgrade the quality of surfaces so prepared. While much research is currently underway to explore other methods of preparing steel surfaces for painting, few at this stage seem to show enough practicality for widespread use.

COATING SYSTEMS

The most obvious trend in coating-system use on bridge steel over the last two decades has been the gradual abandonment of the inhibitive alkyd- and oil-based systems in favor of coating systems based on zinc and high-performance barrier vehicles. Nonlead/nonchromate inhibitors to date have not provided the same performance profile as did lead systems, although they are being employed with success on bridges in California and Massachusetts. Optimized formulation techniques doubtlessly upgrade these systems, which remain more universal for employment on older bridges than do bona fide barrier and zinc systems, but the formulating windows are comparatively narrow compared to those of their more toxic progenitors.

While in the late 1980s there are still probably more states painting with lead systems than any of the newer systems, the use of the zinc-based system is increasing, particularly in coastal and industrial areas of the country and where the use of deicing salts is heavy. Many states now employ two different coating systems—one, usually lead based, for spot maintenance in type 2A environments and a second, high-performance system, usually zinc based for type 2B and type 3 environments. Some are experimenting with nonlead inhibitive alkyds and latex systems, aluminized urethanes, and high-build aluminized epoxy mastics in lieu of lead-based systems in spot maintenance and on old complex riveted structures with latticed features and back-to-back angles. Application of zinc systems on such older structures is more difficult because of cleaning difficulties and control of film thickness.

Of the many new systems that have been evaluated over the past twenty-five years, however, the alkyl-silicate-based zinc rich primer (with and without a tiecoat) with a vinyl topcoat has emerged as a successor to the BLSC system as the primary coating system for steel bridges today, although newer zinc systems with aliphatic urethane topcoats (with and without high-build epoxy intermediate coats) are now being seriously examined by several states and bridge authorities. The aliphatic polyurethane has somewhat better UV resistance than the vinyl, although the reputation of the vinyl system in this respect has been undeservedly sullied apparently by the untoward formulating practices of some manufacturers.

Less satisfactory have been developments in new paint systems for spot-cleaned substrates—applications for which traditional lead-based paint systems were singularly suited. Unfortunately, as concerns over hazardous waste disposal and air pollution produced during blasting operations grow, it seems likely that, in the face of the inadequacies of other suitable preparatory techniques, a consequent reluctance to blast will, at least temporarily, generate some increase in the use of spot power-cleaning methodologies. Increased market opportunities for nontoxic coating systems for use over marginally cleaned substrates are already spawning many new coatings directed at this specific application. Claims are made for rust converting coatings, penetrating coatings, and for new nontoxic inhibitive coatings noted above. Unfortunately, data so far indicate that the various penetrants and rust-conversion treatments do not work, and nontoxic inhibitive coatings are less effective over heavily rusted surfaces than were the lead systems they replaced. Most bets seem to be riding on the aluminized epoxy mastics (applied at high builds) to pick up a large portion of this market, and it seems no accident that so many of the larger paint manufacturers have over the past few years included this type of product in their lines. Less effective seems to be the aluminized moisture-curing urethane because of the apparent requirement for high-build application for the protection of rusty surfaces and the limitations of the moisture-curing urethane in this regard.

The abandonment of lead in favor of high-performance systems has not been accomplished without its share of problems, and reports of catastrophic paint failures have induced caution in those who would utilize the newer systems. Such failures as have occurred, however, rarely translate to deficiencies in the actual product. More often than not, the problems relate to insufficiencies in specification design, surface preparation, application, and film thickness. These failures serve to illustrate the gap that exists between the development of the coatings technology and the technical preparedness of the industry to adapt itself to the practical utilization of such products.

In fact, it is the general consensus of most authorities that those coatings available to the industry at this point are indeed quite capable of addressing present needs for long-term protection on bridge structures. If the paint industry itself has been negligent in any aspect of the introduction and shepherding of the new systems through their "teething" stages to repeatable, dependable utility, it is perhaps because of its naïveté in not anticipating at the outset the many and varied deviations from required practice that are met within practical application. The paint companies, which should be expected to know their own products and interactions in a coating system best, sometimes fail to anticipate the nuances of formulation and behavior that may lead to failure in the short or long term. Even this is hardly reprehensible, for these complex systems, with at best a few decades of history, replace relatively simple systems that were originally developed 150 years ago and have, since that time, had much opportunity for study and refinement.

More study might have been made of the response of the newer, proposed systems to the more extreme parameters of possible practice so that peculiar vulnerabilities and mechanisms of system deterioration might be understood better. Armed with such a better understanding, more failures might have been predicted and avoided by tighter specifications and improved communication and education among those who would specify, apply, and control such systems.

While many may consider those paint systems now available as adequate for the job, many of the same motivations that fostered the development of the high-performance system (competition, economics, environmental restrictions, etc.) are expected to continue to fuel new developments, especially in those areas where needs exist. As such, the development of higher solids systems (particularly urethanes and epoxies) leading to simpler, less costly, and more environmentally acceptable one- and two-coat systems may be expected, as may the development of systems with improved performance over wet and/or poorer surfaces and with greater tolerance for application variables. Continued research effort in the area of lead- and chromate-pigment replacement may be expected as manufacturers vie with each other and with the aluminized epoxy mastics for the field maintenance of older structures. Lower solids monomeric and oligomeric hybrids, low-temperature-curing epoxies, and higher solids vinyls are possible, and it seems likely that developments in the water-borne technologies (most immediately, water-borne epoxies) will be significant. As chemically curing thermosets approach 100% solids, severe pot-life–dry-time reversals may increase the use of two-component spray equipment, while higher solids thermoplastics will further increase pump sizes and the use of hot-spray delivery. Some relief in the material costs of such systems as aliphatic urethanes may be expected as the small manufacturers compete in what is certain to become an expanding market.

One very potent wildcard in these developing technologies is the rapidly spreading concern over the toxicological properties of chemicals in general. These concerns are fostered by new government regulations, press reports, and law suits from the outside and by consequent liability concerns from within the industry. For good or bad, the chemical industry has not successfully mustered sufficient lobbying effort to counteract these affronts, and increasingly we are seeing entire raw-material product lines abandoned in the face of these concerns. While in some respects the effect of such pressure is to generate new coatings technologies, there is a growing feeling within the industry that options for new formulations are increasingly restricted, and entire families of chemicals that might address one environmental demand for new technology are themselves threatened by other toxicological concerns before that technology can be completed.

SPECIFICATION DESIGN

While deficiencies are considered more often than not to lie in the coatings implementation process, blame does not lie solely with those who actually apply the systems. Much, for example, could be done to improve the practice of specification writing. All too often bridge-paint specifications are "scissors-and-glue" epics that are stolen from one structure for use on another, without thought or consideration for design, exposure, or other characteristics of the individual structure that may affect the selection of a suitable paint system. Many specification writers blatantly admit to an imprecise understanding of both the materials and processes for which they are charged to prepare specifications. With the acknowledged state of disrepair of the nation's bridges, one must wonder how such individuals maintain their jobs without having to demonstrate a continuing familiarity with trends and advances in their own and allied disciplines, such as coatings technology, design and maintenance engineering, surface preparation and application technology, quality control, environmental regulations, and, where necessary, containment technologies and contract law.

On most smaller structures, the same specification is continually duplicated from bridge to bridge leading in some cases to several system failures of the same type. The monitoring of paint-system service on completed bridges remains an inexact process, and communications between the field and the specification writers are often poor. This leads to a specification refinement process that is slower than it should be and unnecessarily wasteful on account of the repetition of mistakes. Better, more regular, and more frequent communication between specification writers (including materials engineers) and bridge engineers, residents, and inspection crews could do much to accelerate and improve the specification refinement process. The specification writer should come to view his contribution as but part of the overall job-process cycle (see Figure 11-1) and not as a finite task that is divorced from the rest of the job and that, once completed, is never again to be reviewed.

Bridge-paint specifications still seem more concerned with individual coatings than with the paint system as a whole. This is dangerous particularly with today's complex coating systems, where primers, intermediate coats, and finish coats are often chemically and physically much more different from each other than was the case twenty-five years ago. Improved specifications will be required in which not only individual coatings but the total system is considered in detail, and specific attention is directed toward such critically proven areas as cure verification, recoat time frames, the effects of climatic conditions on cure, adhesion, and performance, and film-thickness ranges for each coat and the complete coating system.

Response to deficiencies in specification writing would seem to lie in education and training. Possibly, with a somewhat different emphasis, the type of course now being offered by the FHWA for inspectors would serve well. In addition to the standard review of corrosion, basic corrosion control by coatings, surface preparation, application, and so forth, greater emphasis on coating-system design and composition and the relevant

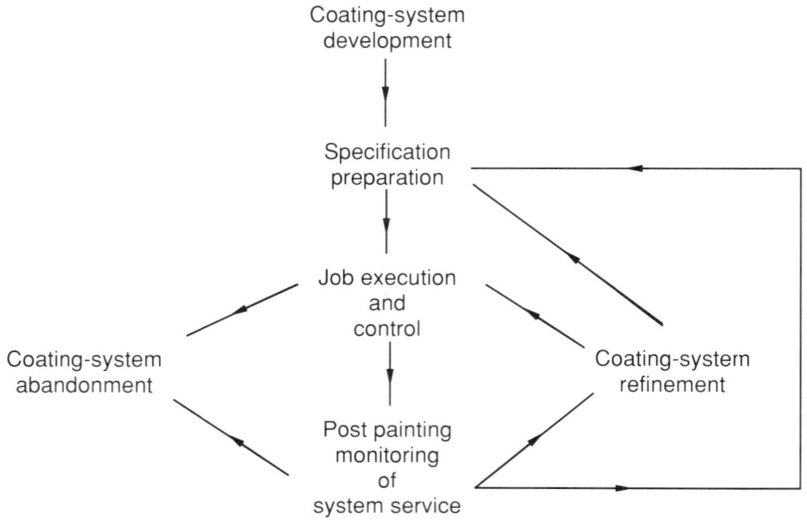

Fig. 11-1. Bridge Painting Job-Process Cycle

elements of paint technology would serve well (along with a refresher course on contract law and the preparation of contract documents) to prepare the specification writer better for his task. The FHWA manual on the selection of coating systems for bridge painting is a start, but the average would-be specification writer requires more expertise in the actual preparation of the technical document itself, and this can only come from a detailed understanding of the strengths and vulnerabilities of the coating system in light of the peculiarities of job execution, bridge design, and environment. Objectives of such courses should be not only to encourage the development and refinement of modern bridge specifications by instruction and example but also to provide a greater understanding of the meaning and ramifications of the many terms, constants, and clauses (now often blindly copied from one document to the next) in terms of the product's composition and its behavior as part of a complete paint system on a given structure.

CONTRACT PROCEDURES

As fair and politically prudent as the U.S. bidding laws are, we are here concerned not so much with the ethics of compensatory advantage or even equality but with engineering excellence (not that the two are incompatible). Without the support of either contractor qualification or a fundamental upgrading of inspection and job-control practices, the "low-bidder-takes-all" philosophy inevitably encourages "low-ball" bidding and the consequent accumulation of profits by the successful bidder through all manner of deceptions and violations of both the specification documents and the mechanics of good painting practice. As utilized in the selection of contractors for bridge painting, the low-bid procedure is self-defeating because, when measured on a cost-per-year-of-service basis, the low-bid job that fails prematurely is more expensive than the same job would have been had the job been performed properly by a well-qualified painting contractor with years of experience in the trade, albeit with a higher tender.

Aside from financial considerations usually related to bondability, there are few qualification criteria that restrict any individual or organization from bidding bridge jobs at this time. On all but the largest jobs (with highest bonding requirements), the large, well-qualified painting contractors of good reputation are often excluded by the open-bidding process and cutthroat practices of many (not all) small nonunion concerns, unless they, too, prostitute their practice and engage in the same sort of conduct as do the small, fly-by-night operators.

Experimentation in such alternative contracting methods as the request-for-proposal-type bid evaluations or single-point-responsibility approaches should be encouraged, particularly where quality-assurance guarantees or postpainting maintenance undertakings are involved.

Contractor Prequalification

Where changes in the contracting methodology from low-bid awards to professional and technical bid-evaluation procedures are too politically explosive, deficiencies inherent in the low-bid procedure might be addressed in other ways. Much, for example, could be done by prequalifying the painting contractors and, no less importantly, the operatives. Both are politically difficult moves, but, in view of the record and rate of bridge deterioration across the nation, exploration of possible ways in which qualification might be accomplished needs little additional justification. While workers in such other trades as electrical work and plumbing have long required certification by means of documented experience as apprentices and journeymen and proctored examinations on theory and practice, the skills of the lowly painter have been judged too basic to warrant such controls. While this may have been true in the 1950s, with today's many and diversified chemical compositions, equipment, and processes, it no longer applies. Adequate training and qualification of painters (with some specific emphasis in bridge work) is essential, if the standard of job execution is to keep pace with the increased complexity of the overall discipline.

Inspection

A second avenue of redress for the practical deficiencies invited by the low-bidder award process is through the implementation of sufficient control. Unfortunately, control, and particularly field inspection, is at this time the one facet of the entire discipline of bridge painting that is universally acknowledged as most blatantly wanting. Most bridge-paint inspectors are ill-trained and lack motivation. There is a lack of professionalism among the ranks of those inspectors employed by most state highway departments and bridge authorities, and only in the private sector are inspectors of real competence predominantly found.

With some notable exceptions, in most highway departments, paint inspection as a whole is regarded as a nuisance chore to be performed alongside other widely diverse duties. In some states, there are no bridge-paint inspectors at all. In others, there is an understanding of the problem, and some token attention may be given by way of short classroom courses and workshop sessions. The FHWA has implemented a formalized bridge-paint inspection course. Unfortunately, in many states, any incentive derived from such courses is immediately frustrated because funding is not available even to equip the would-be inspector with those tools of his trade with which the course has tantalized him.

The problem, in fact, is far too fundamental to be resolved solely by the education of inspectors, although education is certainly needed. Training and education of inspectors can do no good without a conscientious effort on the part of the state departments of transportation to professionalize (and, if necessary, recruit) their bridge-painting inspection force, and that requires both an acknowledgment of the problem and a commitment in terms of time and money to develop such a force.

While, at first sight, shop inspection of new construction might seem somewhat better than the inspection of field work, in practice, the same attitudes prevail and the same difficulties exist. Paint inspection is very often assigned to other inspectors who are mainly responsible for checking welds and other nonpaint-related disciplines.

In the private sector, there are a number of good inspection companies (some specializing in the control of bridge painting), but there are not enough of these to commit to the problem at hand. They are generally employed on larger structures, and, in most cases, they effectively ensure value for the bridge-painting dollar. They may serve effectively to train state inspection personnel by example on actual jobs.

As the paint industry continues to respond to calls for better and better paint systems, we can expect no lessening of the complexity and demands of such systems for precision in use, without compromise. This inevitably translates to a need for sufficient, continuous quality inspection, and, as long as this remains unrealized and/or is addressed with inadequate funding, bridge-paint failures will continue to occur.

MATERIALS CONTROL AND TESTING

Materials control seems not as wanting as is field inspection. State laboratories are generally well equipped, and their staffs generally have adequate expertise with the required analytical and test procedures. Most severe deficiencies relate to the general lack of understanding of paint technology and the formulation process.

With few exceptions, paint chemists are unknown in state laboratories, and this both stifles the generation of in-house specifications and confounds the all-important interpretations of significance in evaluating test data. With the exception of a few specialty concerns, outside testing laboratories are rarely more qualified in paint chemistry than are the state laboratories.

Badly needed are more reliable accelerated-testing methods. The salt-spray test is still widely employed, but most users remain uncomfortable with its value as anything but a screening tool in the early formulation development of new coatings. Weather-ometer studies perhaps satisfactorily reproduce one element (UV) of the normal environment, but they do little to simulate the aggressively corrosive environment that is common on many bridges. Much bridge steel, it may be argued, is never even exposed to UV. No formal test procedures effectively simulate the conditions of most of our Northern structures in the winter months when freeze–thaw cycles and heavy salt contamination can wreak havoc with a deteriorating coating system, and only in a relatively limited number of laboratories are any kind of cycling tests maintained. Little attention has been given to the effects of cycling stresses that are either mechanically (such as by vibrations) or climatologically induced (such as viscoelastic oscillations) on complex steel/multicoat paint composites. The effects of these stresses, particularly in combination with corrodents and the more conventional agents of paint-film deterioration (oxygen, moisture, and UV), need careful evaluation. Few existing test procedures seem well designed both to simulate and intensify (without distortion) all of the destructive agents that have an impact on painted steel bridges, and such is the lack of consistency from environment to environment and location to location across the country that the quest for a universal test methodology may be naïve. Within specific parameters, however, there is surely opportunity for much improvement in this regard.

In view of the inadequacy of correlation with actual field service that is afforded by the most commonly used comparative test procedures (such as salt spray), radically new approaches to short-term testing, typified by reliability analysis and cluster analysis, should be given every opportunity to prove themselves. These new concepts, together with prefailure detection of coating-system distress and/or the initiation of underfilm corrosion (using infrared thermography or electrochemical or acoustic-emission techniques), are welcome and refreshing departures from conventional evaluation techniques.

ASSOCIATED FACTORS

Certain developments, although not directly related to the bridge-painting process, have some bearing on the discipline. These include a greater concern and awareness by bridge-design engineers of ease of painting and future maintenance and the suitability of structural design for the use of the high-performance coating systems. The adoption of simple and continuous beams, with large, flat, uninterrupted areas facilitating the use of zinc systems, and sealed boxes make for easier maintenance than older, riveted trusses, with a profusion of latticed members and intricate designs. Designs to facilitate inspection and rigging during future maintenance (e.g., catwalks, access ports, and permanent movable painting platforms), if possible, will inevitably reduce the cost of painting.

Continued research into such alternative deicing materials as calcium magnesium acetate and urea is also important. While initially more costly, the longer term cost effectiveness of these materials compared to chlorides should be measured in terms of the accelerated deterioration of both structural and reinforcing steel that is caused by the more aggressive electrolytically corrosive deicers.

IN-SERVICE MONITORING

Bridge deterioration is rarely uniform. The cost of maintenance might also be reduced by more careful analysis of the deterioration patterns of a given structure or a group of similar structures as the paint system ages. The correlation of such patterns with exposure, design, traffic flow, and density, as well as with the properties of the coating system, serve well as feedback for new specification design. Through this and such concepts as "specific utility" (the consideration of the bridge not as a simple structure but as a combination of substructures, each with specific requirements of coating type, film thickness, number of coats, etc.) a more uniform deterioration of the structure as a whole might be achieved. This would both lengthen the interpainting service life and render each painting more efficient.

The accurate postpainting documentation of system service and the necessary correlation of areas of distress with bridge design, system type, environment, and job execution factors is still imprecise and often nonexistent, even on test structures. Field trials will always remain a vital step in the proving of any viable coating system. In view of the unpredictability of job execution, further documentation of that process, including surfaces, conditions, and peculiarities existing before and after painting, are essential if any meaningful interpretation of service success or failure is to be gained. All too often, field tests are valueless for want of these records and, where data are available, their interpretation and the distribution of realized conclusions may also be imperfect. The accuracy of analysis of bridge-painting failures is still hampered by insufficient data, poorly qualified investigators, the lack of cooperation from the manufacturers, and inadequate funding. Not the least is the conspiracy of silence that, fueled by embarrassment or concerns over liability, often blanket these failures. On many structures, records of what "went where and when" are not readily available. Consequently, the assessment of service lives and cost per year of service are not known with any accuracy.

SERVICE AND COSTS

The formalization and quantification of cost and service data on coating systems are also a prerequisite to the generation of long-term bridge maintenance programs and to the gradual refinement of such potentially valuable tools as the bridge corrosion cost model. In the long term, we may look forward to the establishment of definitive deterioration profiles by means of a nondestructive methodology, such as periodic thermographic scans of the structure in the field, which, coupled with quality-assurance programs based on reliability-analysis studies, will provide direct-input data. Until then, the simple recording (with or without photographs) of rust percentages on selected structures on a regular basis will do much to provide the more certain required assessment of paint-system service. While accuracy in this is necessary, accuracy is not the immediate problem that getting such a program organized seems to be.

Depending on the quality of job execution, environment, location, and traffic density, the performance of any particular paint system on one structure will necessarily differ from its performance on another. But, given enough data (centrally correlated), mean-value service profiles for each coating system over a broad range of service environments might be established for use in the development of formal maintenance programs.

Similarly, given enough data on enough structures, it may be possible to reduce the chaos in painting costs that results from the open-bidding process and to provide a more definitive appreciation of expected costs. This concept has already been developed to some extent in the BCCM, but more data from structures in all parts of the country are necessary before a more authoritative profile can be established (not only reliable averages of service and costs but also the direction and magnitude of cost/service trends related to geography, environment, specific bridge design features, age, system condition, traffic volume, etc.).

COMMUNICATION

There is, in fact, a continual need for improved communication and data dissemination at all levels of the practice. Misinterpretation of data, misunderstandings and confusion over terminology, inaccuracies in description, premature assumptions, and incomplete or erroneous conclusions are not unfamiliar problems in the reporting of data, and, repeatedly, transmittals (usually, though not always, verbal) of reports and opinions are plagued by distortions of emphasis and meaning.

Even in the published literature, misstatements concerning coatings and coating systems are not unknown, and a plethora of claims and counterclaims (some with appropriate test data) about individual coatings, processes, and equipment abound. Unfortunately, with or without the use of statistics, it is not difficult to design, select, and/or present test data in formats most flattering to a specific product, and it can be difficult for an engineer or specification writer not totally familiar with the technology to discern truth from distortion.

On the other hand, there is a large body of excellent, reliable data in the form of published and unpublished records, studies, reports, journals, papers, articles, books, theses, and opinions from many sources. These relate not only to bridge painting but to associated technologies. They provide a valuable reservoir of data from which all those having any kind of responsibility in the discipline might glean much assistance if they could readily access them. A feasibility study for such a databank has already been completed, although the idea has not been implemented, apparently for want of commitment. Some renewed evaluations of such a databank would seem, from this vantage, a most forward-thinking step. This volume is no more than a partial compendium of such data. In its limited scope and breadth, it can do little justice to the depth and diversity of all the information that is available on the subject to the many who might benefit from its improved accessibility.

Notes

CHAPTER 1

1. Ferguson, P. J. 1976. *Journal of Oil and Color Chemists Association* 59(7):253.
2. De Mare, E. *Bridges of Britain.*
3. Westhofen, W. 1890. *Engineering* February 28, 213.
4. Smith, H. 1899. *Journal of Society of Chemical Industries* 18:1093.
5. Mass, W. B. 1977. *Modern Paints and Coatings* 67(4):31.
6. Bishop, D. M., and F. G. R. Zobel. 1983. *JOCCA* 66(3):67.
7. Bishop, D. M. 1981. *JOCCA* 64(2):57.
8. Navy Formula 84. *Primer, paint, alkyd type*. Federal Specification TT-P-645.
9. Nash, D. N. 1983, September. *Journal of Coatings Technology* 55(704):109.
10. Trade Literature. Ucar solution vinyl resins for coatings. Danbury, CT: Union Carbide Corp.
11. Ray, C. J., L. E. Henton, and F. A. Rideout. 1982. *Improved field reliability of high performance coatings*, Appendix A: Best current technology. FHWA Report RD-82/119.
12. Hare, C. H. 1973. *The Construction Specifier* 26(3):27.
13. Cleveland, J. L. 1978, August. *American Painting Contractor*.
14. Richardson, D. B. 1982. *U.S. News & World Report*, 27 September.
15. Fitzpatrick, M., D. A. Law, and W. C. Dixon. 1981. Deterioration of highway structures and what we can do about it. Paper presented at the First World Bridge Congress, November 3–4, New York, NY.
16. Stevens, L. C. 1983. Problems in disposing of lead based paints. In *Proceedings of Second World Bridge Congress*. Rolla: Univ. of Missouri-Rolla.
17. Tobey, M. 1984. Lead free painting—Mystic River Bridge in Boston. *TRB News* (January–February):2.

CHAPTER 2

1. Hare, C. H. 1978. *Corrosion and the preparation of metallic surfaces for painting*. Federation Society for Coatings Technology Series, Unit 26. Philadelphia, PA: Federation of Societies for Coating Technology.
2. LaQue, F. L. 1974. "Introduction to Corrosion." In *NACE Basic Corrosion Course*, Chapter 2. Houston, TX: National Association of Corrosion Engineers.
3. Green, N. D. 1984. "Corrosion-Related Chemistry and Electrochemistry." In *NACE Basic Corrosion Course*, Chapter 3. Houston, TX: National Association of Corrosion Engineers.
4. LaQue, F. L. 1982. *SSPC Manual*. 2d ed. Vol. 1, Chapter 1.1, 3–9. Pittsburgh, PA: Steel Structures Painting Council.
5. Wilson, C. L., and J. A. Oats. 1968. *Corrosion and the maintenance engineer*. 1–11. New York: Hart.
6. Uhlig, H. H. 1971. In *Corrosion and Corrosion Control*. Chapters 4–5, 37–91. New York: Wiley.
7. LaQue, F. L., and H. R. Copson. 1965. In *Corrosion Resistance of Metals and Alloys*. 2d ed. Chapter 8, 181. New York: Van Nostrand Reinhold.
8. Hackerman, N., and E. S. Snaverly. 1974. "Fundamentals of Inhibitors." In *NACE Basic Corrosion Course*, Chapter 9. Houston, TX: National Association of Corrosion Engineers.
9. Mayne, J. E. O. 1963. "The Mechanism of Protective Action of Paints." In *Corrosion*, Vol. 2, edited by L. L. Shreir, 15.25. New York: Wiley.

10. Knotkova-Cermakova, D., et al. 1982. "Atmospheric Corrosion of Weathering Steels." In *Atmospheric Corrosion of Metals* (ASTM STP 767), edited by S. W. Dean, Jr., and E. C. Rhea, 7–44. Philadelphia, PA: American Society for Testing and Materials.
11. Townshend, H. E., and J. C. Zoccola. 1982. "Eight Year Atmospheric Corrosion Performance of Weathering Steel in Industrial, Rural and Marine Environments." In *Atmospheric Corrosion of Metals* (ASTM STP 767), edited by S. W. Dean, Jr., and E. C. Rhea, 45–59. Philadelphia, PA: American Society for Testing and Materials.
12. Evans, U. R. 1960. *The corrosion and oxidation of metals*. 107. London: Arnold.
13. Hare, C. H. 1982. *SSPC Manual*. Vol. 1, Chapter 1.2, 10–18. Pittsburgh, PA: Steel Structures Painting Council.
14. Paul, S. 1985. *Surface Coatings—Science and Technology*. Chapter 6, 474. New York: Wiley.
15. Leidheiser, H. 1981. *JCT* 53(678):29–39.
16. Hare, C. H. 1979. *Anti-corrosive and barrier and inhibitive primers*. Federation Society for Coatings Technology Series, Unit 27. Philadelphia, PA: Federation of Societies for Coating Technology.
17. Scantlebury, D. J. 1979. "The Mechanism of Corrosion Prevention by Inhibitors in Paint." In *Corrosion Control by Coatings*, edited by H. Leidheiser. Princeton, NJ: Science Press.
18. Appleby, A. J., and J. E. O. Mayne. 1967. *JOCCA* 50:897–910.
19. Mayne, J. E. O. 1946. *J. Soc. Chem. Ind. London* 65:196.
20. Deanin, R., and S. Frondistou-Yannas. 1985, December. *Coatings for non-abrasive cleaned highway steel*. Federal Highway Administration Report RD-86/053.
21. Appleby, A. J., and J. E. O. Mayne. 1976. *JOCCA* 59:69.
22. Dunn, E. J. 1973. *1G-b pigment handbook*. Vol. 1, edited by T. C. Patton, 843. New York: Wiley.
23. Walker, A. C. 1977. Corrosion and coatings. *South Africa* 4(1):27–29.
24. Nylen, P., and E. Sunderland. 1965. *Modern surface coatings*. 693. New York: Wiley.
25. Hare, C. H., and M. G. Fernald. 1984. *Modern Paint and Coatings* 74(4):40.
26. Keane, J. D. 1976. *Modern Paint and Coatings* 66(4):35.
27. Haagenson, K. A., and J. J. Dietzel. 1978. Competitive non lead, non chromate corrosion inhibitive pigments in latex paints: Accelerated tests vs. environmental exposure. Paper presented at European Marketing Conference, March 20–23, Ghent, Belgium.
28. Hare, C. H. 1979. *Modern Paint and Coatings* 69(4):34.
29. Clay, H. F., and J. H. Cox. 1973. *JOCCA* 56:13.
30. Davidson, S. L. 1975. *JOCCA* 58:435–42.
31. Trade Literature. Halox pigments. Pittsburgh, PA: Halox Pigment Division of Hammond Lead Products.
32. Weisberg, H. E. 1968. *Paint and Varnish Production* 58(3):32.
33. Kresse, P. 1979. *Farbe, und Lack*. March, 1981.
34. Hamburg, H. R., and W. M. Morgans. 1979. *Hess's paint film defects*. 345. London: Chapman and Hall.
35. Hare, C. H. 1983. Factors involved in the selection of anti-corrosive coating systems for bridge maintenance. In *Proceedings of Second World Bridge Congress*. Rolla: Univ. of Missouri-Rolla.
36. Johnson, W. C. 1985, December. *JCT* 57(731):67.
37. Hare, C. H. 1976. *White inhibitive pigments in anti-corrosive metal primers*. Stoughton, MA: Clive H. Hare, Inc.
38. Hare, C. H., and M. G. Fernald. 1985. *Industrial and Engineering Chemistry—Product Research and Development* 24:84.
39. Keane, J. D., et al. 1979. *Performance of alternative coatings in the environment*. PACE Phase One Summary Report.
40. Haagen, K. ca. 1970. "Some Aspects of the Use, Application, and Overcoating of Zinc Dust Paints." In *Protecting Steel with Zinc Dust Paints*. Vol. 2. London: Zinc Development Association.
41. Elm, A. C. 1968. *Zinc dust metal protective coatings*. New York: New Jersey Zinc Corp.
42. Snook, R. S. 1984. Corrosion control by zinc thermal spraying. In *Proceedings of Third World Bridge Congress: Coating Systems for Bridges and Steel Structures*. Rolla: Univ. of Missouri-Rolla.
43. Horton, J. B. 1979. "Aluminum-Zinc Alloys as Sacrificial Coatings." In *Corrosion Control by Coatings*, edited by H. Leidheiser, 59. Princeton, NJ: Science Press.
44. See note 11, Chapter 1.
45. Hare, C. H., and S. J. Wright. 1982. *JCT* 54(693):65.
46. Vandemaele, J. ca. 1975. "Zinc Rich Paints from Laboratory Tests to Industrial Application." In *Protecting Steel with Zinc Dust Paints*. Vol. 3. London: Zinc Development Association.
47. Hafford, B. C., W. E. Pepper, and T. B. Lloyd. 1982. *Zinc dust and zinc powder: Their production, properties and applications*. New York: ILZRO, Inc.

48. Tator, K. B. 1975. Zinc rich coatings forge to leadership in common control. In *Proceedings of National Zinc Rich Coatings Conference*. New York: Zinc Institute, Inc.
49. Hare, C. H. 1982. *Modern Paint and Coatings* 72(4):48–52.
50. Evans, U. R. 1944. *Journal of Iron and Steel* 149(1):67–107.
51. Pass, A. 1952. *JOCCA* 35(6):241–61.
52. Elm, A. C. 1956. *Paint and Varnish Production* 46(8):27–30.
53. Lantz, D. S. 1957, September. *WADC Technical Report #52185*. ASTA Document #AD131054.
54. Hare, C. H., M. J. O'Leary, and S. J. Wright. 1983, June. *Modern Paint and Coatings*.
55. Mason, D. G. 1967. *Materials Protection and Performance* 6(5):43–46.
56. Castelli, R., J. Meda, J. Caprain, and M. Damia. 1983. *JCT* 55(70):753–59.
57. Warness, R., and T. L. Shelly. 1978. FHWA-CA-TL-78-21.
58. Tator, K. B. 1976. *Materials Performance* 15(3):9.
59. NACE. Publication 6B173.
60. Young, C. I. "U.K. Experience with Zinc Dust Paints 1." In *Protecting Steel with Zinc Dust Paints*. Vol. 2. London: Zinc Development Association.
61. Lithgow, O. "U.K. Experience with Zinc Dust Paints 2." In *Protecting Steel with Zinc Dust Paints*. Vol. 2. London: Zinc Development Association.
62. U.S. Army Corps of Engineers. Epoxy polyamide zinc rich primer. Specification CE1409E 303a.
63. Berger, D. M. 1975. *Modern Paint and Coatings* 65(6):19.
64. State of California. Zinc rich primer, organic vehicle type. Paint Specifications, 8010-61J-36.
65. Navy Formula 117. 1978. Primer (wash) pretreatment for metals. DOD-P-15328D.
66. Primer coating zinc dust, chlorinated rubber (for steel and galvanized surfaces). Federal Specifications, TT-P-1046.
67. Chlorinated rubber based zinc rich primer. Massport Specification MP-1A.
68. Jitsufuji, M. Zinc Rich Coatings in Japan. In *Protecting Steel with Zinc Dust Paints*. Vol 2. London: Zinc Development Association.
69. Metil, I. 1979. *Modern Paint and Coatings* 69(12):39.
70. Munger, C. G. 1975. *Materials Performance* 14(5):25–29.
71. Nightingall, V. C. J., and A. McDonald. 1949. U.S. Pat. 2,462,763. Dimet Proprietary, Ltd., February 22.
72. Schutt, J. B. 1971. U.S. Pat. 3,620,784. United States as represented by Administrator of NASA, November 16.
73. McMahon, W. M., and C. G. Abba. 1964. U.S. Pat. 3,130,061. American Pipe and Construction Co., April 21.
74. Weldes, H. H., and M. R. Derolf. 1966. U.S. Pat. 3,248,237. Philadelphia Quartz Co., April 26.
75. Munger, C. G. 1982. *SSPC Manual*. Vol. 1, Chapter 4.2, 125–37. Pittsburgh, PA: Steel Structures Painting Council.
76. Berger, D. 1979. *Metal Finishing* 77(4):27–32.
77. Ginsberg, T., and L. G. Kauffman. 1981. *Modern Paint and Coatings* 71(10):138.
78. Munger, C. G. 1984. *Corrosion prevention by protective coatings*. Chapter 6. Houston, TX: National Association of Corrosion Engineers.
79. SSPC Paint 20. 1983. *SSPC Manual*. Vol. 2. Pittsburgh, PA: Steel Structures Painting Council.
80. Hendry, M. C. 1978. *Materials Performance* 17(5):19.
81. Tator, K. B. 1976. *Materials Performance* 15(3):9–16.
82. Munger, C. G. 1977. *Materials Performance* 16(3):33–36.
83. See note 11, Chapter 1.
84. Trade Literature. Zinc-Lock 351 and 352. Louisville, KY: Porter Coatings.
85. Keane, J. D., E. F. Weaver, and J. A. Bruno. 1981. *Topcoats for zinc coatings—Five year report*. Pittsburgh, PA: Steel Structures Painting Council.
86. Bureau of Reclamation. 1983. Specification VR-3.
87. U.S. Army Corps of Engineers. Coal tar epoxy/polyamide. Specification C-200.
88. Kittlberger, W. W., and A. C. Elm. 1952. *Industrial Engineering Chemistry* 44:326.
89. Mayne, J. E. O. 1957. *Official Digest* 24:127, 325.
90. See note 9, Chapter 2.
91. Mayne, J. E. O. 1970. *British Corrosion Journal* 5:106.
92. Mayne, J. E. O., and D. J. Mills. 1975. *JOCCA* 58:155.
93. Mayne, J. E. O., and J. D. Scantlebury. 1970. *British Polymer Journal* 2240.
94. Callow, L. M., and J. D. Scantlebury. 1984. *JOCCA* 67(3):70.
95. Heyes, P. J. 1977. Ph.D. diss., University of Cambridge, Cambridge, England.
96. Corti, H., G. Baro, R. Ferdandez-Prini, and A. J. G. Maroto. 1978. *JOCCA* 61(3):75.

97. Munger, C. G. 1982. *SSPC Manual.* Vol. 1, Chapter 23, 490–519. Pittsburgh, PA: Steel Structures Painting Council.
98. Koehler, E. L. 1981. "Underfilm Corrosion Currents as the Cause of Failure of Protective Organic Coatings." In *Corrosion Control by Organic Coatings*, edited by H. Leidheiser, 87–96. Houston, TX: National Association of Corrosion Engineers.
99. Funke, W. 1983. *JCT* 55(705):31.
100. Hare, C. H., and M. G. Fernald. 1984. *Modern Paint and Coatings* 74(10):138.
101. Trade Literature. Carbomastic 15. St. Louis, MO: Carboline Co.
102. Trade Literature. Desmodur E-21. Pittsburgh, PA: Mobay Chem. Co.
103. Hare, C. H., and M. G. Fernald. 1984. *Amer. Paint Journal* Feb., 37.
104. Kittelberger, W. W., and A. C. Elm. 1947. *Ind. Eng. Chem.* 39(7):876.
105. Maitland, C. C. 1959. Ph.D. diss., University of Cambridge, Cambridge, England.
106. Cherry, B. D., and J. E. O. Mayne. 1961. *Off. Dig.* 33:435,469.
107. Edwards, T. G. 1984. *Materials Performance* 23(2):9–12.
108. Dukes, C. F. 1984. *American Paint and Coatings Journal*, April 30, 42–46.
109. Militello, S. V. 1984. *American Paint and Coatings Journal*, April 30, 51–56.
110. Williams, J. L., and H. C. Grulds. 1984. *American Paint and Coatings Journal*, April 30, 33–38.
111. Mills, D. J., and J. E. O. Mayne. 1981. "The Inhomogenous Nature of Polymer Films and Its Effect on Resistive Inhibitors." In *Corrosion Control by Organic Coatings*, edited by H. Leidheiser, 12–17. Houston, TX: National Association of Corrosion Engineers.
112. Martens, C. R. 1981. *Waterborne coatings.* Chapter 3. New York: Van Nostrand Reinhold.
113. Hare, C. H. 1975. *JCT* 47(605):69.
114. Trade Literature. Haloflex 202. Wilmington, DE: I.C.I.
115. Trade Literature. Rhoplex MV-23. Philadelphia, PA: Rhom and Haas.
116. Caine, M. L. 1983. *Modern Paint and Coatings* 73(9):37.
117. Burgess, A. J., D. Caldwell, and J. C. Padget. 1981. *JOCCA* 64(5):175.
118. Keane, J. D., W. Wettach, and C. Bosch. 1969. *Journal of Paint Technology* 41(533):372.
119. Martel, A. S. 1985. Personal communication.
120. Bruno, J. A., and J. D. Keane. 1981. *Evaluation of low solvent maintenance coatings for highway structural steel.* Pittsburgh, PA: Steel Structures Painting Council.
121. Hare, C. H. 1984. *Examination of blistering failure in submarine escape training tank, Groton, Conn.* Unpublished study. Stoughton, MA: Clive H. Hare, Inc.
122. Gelfer, D. H., and P. Vondorsten. 1981. *Materials Performance* 20(5):23.
123. Starr, T. L., L. E. Henton, and W. S. Lewis. 1984. Environmental factors and the cure of zinc rich coatings: Preliminary results. In *Proceedings of Third World Bridge Congress: Coating Systems for Bridges and Steel Structures.* Rolla: Univ. of Missouri-Rolla.
124. Starr, T. S., J. S. Oechsle, J. A. Lanning, and T. Ginsberg. 1984. *Journal of Protective Coatings and Linings* 1(2):9.
125. See note 11, Chapter 1.

CHAPTER 3

1. Keane, J. D. 1969. *Protective coatings for highway structured steel.* NCHRP Report #74.
2. Leyland, D. S. 1983, July. *A survey of coatings and systems performed by state highway departments.* Technical Paper 83-10. Maine Department of Turnpikes.
3. Rideout, F. A., C. J. Ray, and L. E. Henton. 1982. *Coating systems for old and new steel.* NCHRP Report #4-14.
4. Hudson, J. C. *Corrosion of iron and steel.* London: Chapman Hall.
5. Bigos, J. 1957. *HRB Proceedings* 36:266–81.
6. Brauns, E., H. Determann, and U. Kalla, Stahland, Ewen. 1964. 84:1082–87.
7. Shaw, R. E. 1965. *Corrosion Prevention and Control* 12(3):31–32.
8. Bullet, T. R. 1971. *Corrosion Prevention and Control* 18:8–12.
9. Keane, J. D., and J. A. Bruno. 1966, February. *Surface preparation vs. durability.* Pittsburgh, PA: Steel Structures Painting Council.
10. See note 20, Chapter 2.
11. Barrett, L. D. 1964. *Materials Performance* 4(7):20.
12. Hare, C. H. 1975. *Plant Engineering* 29(21):117.
12a. See note 20, Chapter 2.
13. Bullet, T. R., and A. T. S. Rudram. 1959. *JOCCA* 11:789.

14. Liebman, A. J. 1954. *SSPC Manual.* 1st ed. Chapter 2. Pittsburgh, PA: Steel Structures Painting Council.
15. Cromwall, B., L. Thureson, and V. Victor. 1971. *Centrifugal blasting of steel: Cleaning and cold working.* Swedish Corrosion Industry Bulletin #67.
16. Anonymous. 1978. *Iron Age,* March, 4.
17. Munger, C. G. 1983. *Materials Performance* 22(7):33.
18. See note 15, Chapter 2.
19. Warren, R., et al. 1981. *JCT* 53(683):27.
20. Johnson, W. C. 1984. "Detrimental Materials at the Steel/Paint Interface." In *New Concepts for Corrosion Protection of Steel Structures* (ASTM STP 841), edited by D. M. Berger and R. F. Wint. Philadelphia, PA: American Society for Testing and Materials.
21. Schikorr, G. 1963. *Werkstoffe und Korrosion* 14(2):69–80.
22. Mayne, J. E. O. 1959. *Journal of Applied Chemistry* 9(12):673–80.
23. Igetoft, L. 1983. "Surface cleanliness and durability of anticorrosive paint." In *Proceedings of Second World Bridge Congress.* Rolla: Univ. of Missouri-Rolla.
24. Eriksson, R. 1978. *Werkstoffe und Korrosion* 29:400–403.
25. Steel Structures Painting Council. 1983. Surface preparation standards, SSPC-SP-1–10. In *SSPC Manual.* Vol. 2. Pittsburgh, PA: Author.
26. National Association of Corrosion Engineers. *Surface preparation standards,* NACE 1–4. Houston, TX: Author.
27. Swedish Standards. 1967. SIS 05 59 00.
28. British Standard. 1967. BS 4232.
29. Japan Shipbuilding Research Association. 1975. *Standard of the preparation of steel surface prior to painting.* JSRA.
30. Japanese STD. 1969. *Standard for density of shop painted steel surfaces.* (STPS-1969).
31. Reinigen Door Stralen. 1973. Onderoekingern en Riehtlijnen. *Corrosie Centrum #201.* Netherlands.
32. Din 55928. 1975. Kowosionsschutz von Stahlbauwerken durch Beschichtungen und Uberzuge Teill und Teil 4 Entwurf.
33. Australian Standard 1627. 1974. *Abrasive blast cleaning of steel surfaces.* Part 4.
34. McKelvie, A. N. 1977. *JOCCA* 60:277.
35. McKelvie, A. N. 1981. *JOCCA* 64:509.
36. Trade Literature. Heavy duty roto-peen flap wheel. St. Paul, MN: 3M.
37. Claus, J. 1983. In *Proceedings of Second World Bridge Congress.* Rolla: Univ. of Missouri-Rolla.
38. Calabrese, C., and J. R. Allen. 1978. *Corrosion* 34(10):331–38.
39. National Association of Corrosion Engineers. *Centrifugal wheel blast cleaning of steel plate shapes and fabrication.* Publication 6G-174. Houston, TX: Author.
40. Mallory, A. W. 1977. *Materials Performance* 16(2):11–17.
41. Mallory, A. W. 1982. *SSPC Manual.* 2d ed. Vol. 1, Chapter 2.4. Pittsburgh, PA: Steel Structures Painting Council.
42. Baldwin, B. 1978. *Plant Engineering,* Feb., 116–25.
43. Keane, J. D., J. A. Bruno, and R. E. F. Weaver. *Survey of existing and promising new methods of surface preparation.* Pittsburgh, PA: Steel Structures Painting Council.
44. Borch, E. A. 1982. *SSPC Manual.* 2d ed. Vol. 1, Chapter 2.2. Pittsburgh, PA: Steel Structures Painting Council.
45. Bennett, J. 1982. *SSPC Manual.* 2d ed. Vol. 1, Chapter 2.4. Pittsburgh, PA: Steel Structures Painting Council.
46. Trade Literature. Surface preparation—The foundation for protective coatings. San Francisco: Chemco-Clementina, Ltd.
47. Keane, J. D. 1964. *Zinc shot blasting of structured steel.* Pittsburgh, PA: Steel Structures Painting Council.
48. Sandford, J. E. 1973. *Iron Age,* April 1.
49. Geld, I., L. Deatsch, and F. J. D'Ona. 1968. *Materials Performance,* Aug.
50. Robinson, P. J. 1984. *Materials Performance,* Feb.
51. Bayliss, D. A. 1982. *Assessment of properties of J. blast "Delox" coated abrasive 1T1 anti-corrosive.* 1T1 Anti-Corrosion Ltd. Report #1014. Sept. 10, p. 14.
52. Clark, R. T. 1983. In *Proceedings of Second World Bridge Conference.* Rolla: Univ. of Missouri-Rolla.
53. Peart, J. W., and E. Flounders. 1984, April. *Mineral slag abrasive survey and specification.* U.S. Department of Transportation, Maritime Administration.
54. Bullet, T. R., and D. Dasgupta. 1969. *JOCCA* 52:1099–1129.
55. Weaver, P. E. 1973. *Industrial maintenance painting.* 4th ed. Chapter 3. Houston, TX: National Association of Corrosion Engineers.

56. Keane, J. D., J. A. Bruno, and R. E. F. Weaver. *Surface profile for anti-corrosive paints*. Pittsburgh, PA: Steel Structures Painting Council.
57. O'Leary, J. R. 1980, May. *Durability of coatings applied to steel structures*. West Virginia Department of Transportation Implementation Project #3. Charleston, WV: West Virginia Department of Transportation.
58. Bennett, J. 1982. *SSPC Manual*. 2d ed. Vol. 1, Chapter 2.5. Pittsburgh, PA: Steel Structures Painting Council.
59. NACE. 1972. Surface preparation of steel and other hard materials by water blasting prior to coating and recoating. *Materials Performance* 11(4).
60. Spencer, L. F. 1977. "Abrasive Blasting." In *Metal Finishing Guidebook and Directing*, 61–70.
61. McKelvie, A. N. 1977. Evaluation of various cleaning processes for steel. Paper presented at seminar on Protection of Structured Steel Work, Paint Research Association, Nov. 16–17, Teddington, England.
62. Dasgupta, D., and T. K. Ross. 1971. *British Corrosion Journal* 6(6):237.
63. Appleman, B. R., and J. D. Keane. 1982. *SSPC Manual*. 2d ed. Vol. 1, Chapter 2.8. Pittsburgh, PA: Steel Structures Painting Council.
64. Garroutte, J. D., and M. Hutchinson. 1975. *Hydrocarbon Processing*, May, 95.
65. KUE. 1977. System 9-18. *Polymers, Paint and Color Journal*, March 9, 202.
66. Appleman, B. R., and J. A. Bruno. 1985, July. *Evaluation of the effectiveness of wet blast cleaning methods of surface preparation*. SSPC Report #SSPC8602. Pittsburgh, PA: Steel Structures Painting Council.
67. Trade Literature. Clemtex WAB60031. Houston, TX: Clemtex, Inc.
68. Trade Literature. Clemco wet blast injector system. San Francisco: Clemco Industries.
69. Trade Literature. Rosewell, GA: Hydraire-America Co.
70. Trade Literature. Houston, TX: Aquadyne, Inc.
71. See note 17, Chapter 1.
72. Snyder, M. K., and D. Bendersky. 1983. *Removal of lead based paints*. NCHRP Report #265. Washington, DC: National Research Council.
73. Tobey, M. 1982. In *Proceedings of First World Bridge Congress*. Rolla: Univ. of Missouri-Rolla.
74. Hill, J. W., M. J. Lee, and I. J. Spaulding. 1974. *Optics and Laser Technology* 6(6):267–78.
75. Asmus, J. F. 1981, March. *High energy xenon flash cleaning techniques*. Navy Contract Report #NRC-1037. Maxwell Labs.
76. Anonymous. 1979. Bright ideas—Xenon lamp's intense flash burns off steel truss rust. *Eng. News Record*, May 31, 11.
77. Sandwich, C. J., and T. Briewick. 1976, May. In *Proceedings of the Fourth International Congress of Maritime Corrosion and Fouling*.
78. Shelton, G. T. 1980. Ultrasonic cleaning and vapor. Branson Cleaning Equipment trade literature.
79. Usami, S., and H. Kozu. 1973. *Japan Appl.* 23(749).
80. Fultz, B. S. 1986. *Journal of Protective Coatings and Linings* 3(4):40.
81. Panayappan, R., and D. L. Venezky. U.S. Pat. 4,325,744. United States as represented by the Secretary of the Navy.
82. Conn, A. F., and S. L. Rudy. 1977. *Maritime research center report* 32:127.
83. National Maritime Research Center. Report #NMRC-275-42210-R1.
84. Nicinski, S. A., R. A. May, and M. S. Rice. 1982, January. An investigation into the effects of cavitating water jets on the surface preparation of structural steel prior to painting. Report #FHWA RD 82/002. Alexandria, VA: SEACO, Inc.
85. Tinklenburg, G. 1983. In *Proceedings of the Second World Bridge Congress*. Rolla: Univ. of Missouri-Rolla.
86. McKenzie, M. The corrosion performance of weathering steel in highway bridges. TRRL Report #857.
87. Raska, J. G. 1986. *Journal of Protective Coatings and Linings* 3(3):5.
88. Keane, J. D., et al. 1984. *Remedial painting of weathering steel: State of the art review*. Pittsburgh, PA: Steel Structures Painting Council.
89. Steel Structures Painting Council. 1983. *SSPC Manual*. 2d ed. Vol. 2. Pittsburgh, PA: Author.
90. Sandler, M. H., and S. Spring. 1982. *SSPC Manual*. 2d ed. Vol. 1, Chapter 2.9. Pittsburgh, PA: Steel Structures Painting Council.
91. Spring, S. 1974. *Industrial cleaning*. Melbourne, Australia: Prism Press.

CHAPTER 4

1. Dunn, P. A. 1984. *Polymers, Paint, and Colour Journal* 174(4125):559.
2. See note 65, Chapter 2.

3. See note 11, Chapter 1.
4. Caltrans. Water-borne systems. Specification formulas PWB-135, PWB-137, DA-4498-152-1, and DA-4498-205-1.
5. Morcillo, M. 1984. "Minimum Film Thickness for the Protection of Hot-Rolled Steel: Results after 23 Years of Exposure at Kure Beach, North Carolina." In *New Concepts for Coating Protection of Steel Structures* (ASTM STP 841), edited by D. M. Berger and R. F. Wint. Philadelphia, PA: American Society for Testing and Materials.
6. Steel Structures Painting Council. Paint #2: Red lead, iron oxide, raw linseed oil and alkyd primer. Pittsburgh, PA: Author.
7. State of Missouri. Second coat brown, for steel. Department of Transportation specification 1045.6.
8. State of Alabama. Department of Transportation specification 855.16(e).
9. State of Minnesota. Brown finish. Department of Transportation specification #3522.
10. American Association of State Highway and Transportation Officials. 1982. Specification M69-70.
11. See note 2, Chapter 3.
12. Steel Structures Painting Council. Paint #104: White or tinted alkyd paint. Pittsburgh, PA: Author.
13. State of New Jersey. Basic lead silico chromate brown finish coat. Department of Transportation specification 912.07.
14. TT-P-102D. Oil alkyd (modified) exterior fume-resistant ready-mixed white and tints.
15. State of Oregon Department of Transportation, Highway Division. *Inspection records of White Water Creek Bridge.*
16. See note 1, Chapter 3.
17. State of Alabama. Department of Transportation specification 855.14b.
18. State of Arkansas. Department of Transportation proposed specification 807.60 paint system A(1).
19. Steel Structures Painting Council. 1982. Paint #11: Red iron oxide, zinc chromate, raw linseed oil and alkyd primer. Pittsburgh, PA: Author.
20. Eickhoff, A. J. 1982. *SSPC Manual* Vol. 1, Chapter 4.3, 138. Pittsburgh, PA: Steel Structures Painting Council.
21. State of Missouri. Second coat maroon, basic lead silico chromate. Department of Transportation specification 1045.12.
22. State of New Hampshire Department of Transportation. Finish coat green. NH 708 3.11.
23. Commonwealth of Pennsylvania. Azure blue structural finish coat. Department of Transportation Specification Bulletin #26, 45.
24. Trade Literature. M50 defense in depth specifications and formulas. Hightstown, NJ: National Lead Company.
25. Carter, R. L., et al. 1983. *Modern Paint and Coatings* 73(7):18.
26. State of New Jersey. Graphite paint, finish coat, black. Department of Transportation specification 912.11.
27. Anonymous. 1983, November. *Better Roads.*
28. Commonwealth of Massachusetts. Department of Public Works Specifications M.07.02.41-43.
29. New York and New Jersey Port Authority. Specifications for Goethles Bridge.
30. British Department of Transport. *Use of Department of Transport bridge paint manual.* Technical memorandum (bridges) No. BE 8/77. London: Author.
31. Bruno, J. A., and B. R. Appleman. 1984. *Journal of Protective Coatings and Linings* 1(6):24.
32. Warness, R. *Water based coatings for the protection of steel structures.* FHWA/CA/TL-79/24.
33. Trade Literature. Arolon X-820. Buffalo, NY: Spencer Kellogg, Inc.
34. See note 3, Chapter 3.
35. See note 114, Chapter 2.
36. See note 79, Chapter 2.
37. Berger, D. M. 1984. *Journal of Protective Coatings and Linings* 1(2):20.
38. See note 75, Chapter 2.
39. Trade Literature. Silbond. Westport, CT: Stauffer Chemical Corp.
40. Trade Literature. Chadwicks, NY: Remet Chemical Corp.
41. Medford, W. M., and G. T. Phillips. 1984. Mann's Harbor paint evaluation. In *Proceedings of Third World Bridge Congress: Coating Systems for Bridges and Steel Structures.* Rolla: Univ. of Missouri-Rolla.
42. Porter Coatings. 1981, July. *Protective coatings update.* Vol. 3. Louisville, KY: Author.
43. Trade Literature. Danbury, CT: Union Carbide Corp.
44. State of Virginia. Department of Transportation specification, Section 239.18, Paint 12.
45. Massachusetts Turnpike Authority. Specification for inorganic zinc rich primer/vinyl high build/vinyl acrylic.

46. Steel Structures Painting Council. 1982. Paint system oxide #12.00. *SSPC Manual* Vol. 1, 129. Pittsburgh, PA: Author.
47. Fouad de Fouad. 1978. *Slip behavior of bolted friction type joints with faying type surfaces*. Master's thesis, University of Texas, Austin, TX.
48. See note 88, Chapter 3.
49. See note 81, Chapter 2.
50. See note 85, Chapter 2.
51. See note 95, Chapter 2.
52. See note 84, Chapter 2.
53. Rucker, R. A. U.S. Pat. 3,392,130.
54. Montle, J. F., and D. H. Hasser. 1976. *Materials Performance* 15(8):15.
55. Tinklenberg, G., and J. D. Culp. 1984. *Journal of Protective Coatings and Linings* 1(1):26.
56. State of West Virginia. Department of Transportation specification 711.16.
57. State of Ohio. Department of Transportation vinyl specification 708.18.
58. State of Tennessee. Department of Transportation vinyl specification 910.03.2.
59. State of Missouri. Department of Transportation vinyl specification 1045.16.
60. State of Oklahoma. Vinyl specification 730.04, sec. 3.
61. State of South Carolina. 1979, June. Department of Transportation specification vinyl finish revision.
62. State of Maryland. 1979, July. Department of Transportation vinyl specification.
63. Commonwealth of Massachusetts. Vinyl specification M7.02.03.
64. State of Minnesota. Department of Transportation vinyl finish specification 3529.
65. Martell, R. J. 1984. Vinyl paints and systems. In *Proceedings of Third World Bridge Congress: Coating Systems for Bridges and Steel Structures*. Rolla: Univ. of Missouri-Rolla.
66. Massport. Vinyl Specification MP-14. Boston, MA: Author.
67. Hare, C. H. 1985. Unpublished study. Holbrook, MA: Clive Hare, Inc.
68. State of Michigan. Department of Transportation special provision for complete of structural steel and field repair of damaged coating, Type 3. 5.04(8d).
69. Trade Literature. Acryloid hydroxy terminated acrylics. Philadelphia, PA: Rohm and Haas.
70. Trade Literature. G-cure resins. Minneapolis, MN: Henkel Corp.
71. Trade Literature. Multron, Desmophen, and Desmodur resins. Pittsburgh, PA: Mobay Chemical Co.
72. State of Maine. 1983, June. Special provision, Section 708. Inorganic zinc/epoxy-polyamide/aliphatic urethane specification.
73. State of New Hampshire. 1983. Department of Transportation specification for the painting of the Neil Underwood Bridge, Hampton, NH. Concord, NH: Author.
74. State of Florida. Department of Transportation specification sections 972 and 973.
75. Ramsey, R. R. 1975. Zinc coatings in Florida experience. In *Proceedings of the National Zinc Rich Coatings Conference*. New York: Zinc Institute, Inc.
76. Gelfer, D. H. "The Universal Aspects of Zinc Rich Coatings." In *Protecting Steel with Zinc Rich Paints/2*. London: Zinc Development Association.
77. See note 64, Chapter 2.
78. Massachusetts Turnpike Authority. Specification for phenoxy zinc rich primer/coal tar epoxy/polyamide finish.
79. Massport. Specifications of the painting of the Tobin Memorial Bridge, Chelsea viaduct approach, M.2.038. Boston, MA: Author.
80. American Society for Testing and Materials. B-117, salt spray test.
81. See note 66, Chapter 2.
82. See note 67, Chapter 2.
83. U.S. Army Corps of Engineers. Vinyl zinc rich primer. Specification VZ-108d.
84. State of Oregon Department of Transportation. High-build vinyl finish coat.
85. See note 62, Chapter 2.
86. State of Louisiana Department of Transportation. Specification section 1008.02a.
87. State of Louisiana Department of Transportation. Specification section 1008.02b.
88. State of Louisiana Department of Transportation. Test method for accelerated exposure for zinc rich systems. Designation DOTDTR 503-83.
89. See note 87, Chapter 3.
90. State of Texas Department of Transportation. Zinc rich prime coat system II. Specification #810d.
91. State of Texas Department of Transportation. Vinyl toluene/acrylate finish coats. Specifications #742g (gray) and #744h (green).
92. State of Texas Department of Transportation. Aluminum finish. Specification #702 APE-1.
93. See note 68, Chapter 2.

94. Coburn, S. Hot dip galvanizing for exposed structural and miscellaneous steel. In *NCHRP Synthesis of Highway Practice #112*.
95. Anonymous. 1980. *Galvanized Steel Bridges: A Case History*. Toronto, Canada: Cominco.
96. Hall, J. R. 1964. The case for galvanized bridges. Paper presented at the International Conference on Hot Dip Galvanizing. 473.
97. Van Eijnsberger, J. F. H. 1975, June. Twenty years of duplex systems. *Metalluissenschaf und Technik* 19(6):581.
98. Trade Literature. Colorgalve. Everett, MA: Duncan Galvanizing, Inc.
99. Lwin, M. M. 1984. "Zinc Plus Paint System for Corrosion Protection of a Steel Bridge." In *New Concepts for Coatings Protection of Steel Structures*, edited by D. M. Bergen, and R. F. Wint. p. 53. Philadelphia, PA: American Society for Testing and Materials.
100. American Society for Testing and Materials. Standard practices for the preparation of zinc coated steel surfaces for painting. ASTM D 2092.
101. Oechsle, S. J., and J. N. Childs. 1982. "Thermal Sprayed Coatings." In *SSPC Manual*. Vol. 1, Chapter 20, 456. Pittsburgh, PA: Steel Structures Painting Council.
102. Wood, R. B. 1972. The metallization of the Forth Road Bridge. Paper presented at the symposium on the Protection of Steel Structures by Metal Spraying. London.
103. Jodoin, N., and M. Nadeau. 1980. The biggest on-site metallizing project: The Pierre-Laport Bridge. In *Proceedings of the Ninth International Thermal Spraying Conference* 53. The Hague.
104. See note 42, Chapter 2.
105. See note 86, Chapter 2.
106. U.S. Navy. 1961. Specifications MIL-P-15935B, MIL-P-15936B.
107. Hare, C. H., and R. J. Martel. In publication.
108. Shields, G. D., and W. P. Mayer. 1985. Novel vinyl polyols for high solids coatings. Paper presented at Western Coatings symposium, Feb. 28–March 1, Anaheim, CA.
109. Bishop, R. R. 1976. Zinc coatings and pigments in highway bridge protection. In *Proceedings of Conference on Zinc Coatings in All Their Aspects* 36. Pretoria, South Africa: South African Council for Science and Industrial Research.
110. See note 1, Chapter 1.
111. Pass, A., and J. F. Meason. 1961, April. *Official Digest* Vol. 33.
112. State of Minnesota Department of Transportation. Chlorinated rubber aluminum finish. Specification 3523.
113. Brady, R. F. 1985. "Epoxy Resins for Protective Coatings and Linings." *Journal of Protective Coatings and Linings* 2(11):24.
114. Lee, H., and K. Neville. 1967. *Epoxy Resins*, 2d Ed. New York: McGraw-Hill.
115. Antonino, R. A. 1967, February. *Civil Engineering*.
116. Berger, D. M. 1984. Coating failure of the Newport Rhode Island Bridge. In *Proceedings of the Third World Bridge Congress: Coating Systems for Bridges and Steel Structures* 68. Rolla: Univ. of Missouri-Rolla.
117. Peart, J. W. 1984, May. *Evaluation of rust compatible primers for marine applications*. U.S. Department of Transportation, Maritime Administration.
118. State of Virginia. 1984, April. Aluminum epoxy mastic. Department of Transportation specifications, special provision for section 239.19, #14 Paint.
119. State of Wyoming. Epoxy mastic paint. Department of Transportation specification, section 708.12.
120. State of South Carolina. 1978, July. Aluminum epoxy mastic. Department of Transportation specification.
121. See note 102, Chapter 2.
122. Bracco, J. J. 1984. *Modern Paint and Coatings* 74(8):33.
123. State of Michigan Department of Transportation. 1984, May. Special provision for cleaning and coating of existing painted steel structures constructed of riveted plate girders or trusses, type 2. Specification 504 (13b).
124. Whittier. U.S. Pat. 2,765,288. Issued to Pittsburgh Coke and Chemical.
125. Steel Structures Painting Council. 1983. Paint specification #16: Coal tar epoxy polyamide, black or dark red paint. *SSPC Manual*. Vol. 2, 233.
126. State of New York Department of Transportation. 1984, April. Coal tar epoxy paint system—Cleaning and painting. Specification items 18570.73 and 18570.74.
127. Morawek, R. 1975. *Modern Paint and Coatings* 65(6):43.
128. See note 103, Chapter 2.
129. Hare, C. H., and M. G. Fernald. 1984, February. *American Paint Journal*, p. 39.
130. Trade Literature. TP8A topcoat. Coronado, CA: Coatings Corp. of America.

CHAPTER 5

1. See note 1, Chapter 3.
2. Hare, C. H. 1984. *Journal of Protective Coatings and Linings* 1(5):18.
3. See note 109, Chapter 4.
4. See note 45, Chapter 4.
5. National Association of Corrosion Engineers. 1984. *Structural steel painting—A comparison of various work sites and procedures*. Publication 6H184. Houston, TX: Author.
6. See note 55, Chapter 4.
7. Dillon, C. P. 1975. *Materials Performance* 14(5):29.
8. Bayliss, D. A. 1984. *Journal of Protective Coatings and Linings* 1(3):36.
9. Montle, J. F. 1984. *Journal of Protective Coatings and Linings* 1(5).
10. See note 122, Chapter 2.
11. Schaffer, P., Jr. 1984. Silicate/zinc coatings and the new high-ratio technology. Paper presented at SSPC symposium on Maintenance Painting of Industrial Plants, September 19–20, Pittsburgh, PA.
12. See note 35, Chapter 2.
13. See note 72, Chapter 3.
14. Kline, E. S. 1984. Locking in coatings savings in bridge painting. In *Proceedings of Third World Bridge Congress: Coating Systems for Bridges and Steel Structures*. Rolla: Univ. of Missouri-Rolla.
15. Weismantel, C. E. 1981. *Paint handbook*, Chapters 4–8, and 14. New York: McGraw-Hill.
16. Hall, R. E., et al. *Guide to the preparation of contracts and specifications for the application of protective coatings*. NACE publication 6J162. Houston, TX: National Association of Corrosion Engineers.
17. See note 11, Chapter 1.
18. Baker, J. L. 1972. *Materials Performance* 11(8):23.
19. American Association of State Highway Transportation Officials. 1982. *Standard specification for transportation materials and methods of sampling and testing—Part 1*. Washington, DC: Author.
20. American Association of State Highway Transportation Officials. 1973. *Construction manual for highway bridges and incidental structures*. Washington, DC: Author.
21. Baldino, L. R. 1983. *Plant Engineering* 37:67.
22. Baldino, L. R. 1984. *Plant Engineering* 38:76.
23. Baldino, L. R. 1984. *Plant Engineering* 38:69.
24. See note 1, Chapter 2.
25. Berger, D. M. 1981. *Modern Paint and Coatings* 71(11):61.
26. Munger, C. G. 1984. *Journal of Protective Coatings and Linings* 1(3):61.
27. See note 55, Chapter 3.
28. National Association of Corrosion Engineers. 1984. *Corrosion basics—An introduction*. Houston, TX: Author.
29. Tatton, W. H., and E. W. Dres. 1964. *Industrial paint application*. New York: Hart.
30. Anonymous. 1954. *The ABC's of spray equipment*. Toledo, OH: The DeVilbiss Co.
31. See note 24, Chapter 2.
32. Turner, G. P. A. 1982. *An introduction to paint chemistry*. 2d ed. London: Chapman Hall.
33. Payne, H. F. 1954. *Organic coatings technology* Vols. 1 and 2. New York: Wiley.
34. Martens, C. R. 1974. *Technology of paints, varnishes and lacquers*. Huntington, NY: Krieger.
35. Keane, J. D. 1983. SSPC paint-COM. *SSPC Manual* 3d ed. Vol. 2, Chapter 4. Pittsburgh, PA: Steel Structures Painting Council.
36. Smith, L. 1984. *Journal of Protective Coatings and Linings* 1(5):26.
37. State of West Virginia. Zinc rich system. Department of Transportation specification 711.2.
38. See note 120, Chapter 4.
39. See note 102, Chapter 2.
40. See note 31, Chapter 2.
41. See note 64, Chapter 2.
42. See note 87, Chapter 2.
43. See note 4, Chapter 4.
44. See note 91, Chapter 4.
45. See note 75, Chapter 4.
46. See note 116, Chapter 4.
47. Ramsey, R. R. 1985. *Journal of Protective Coatings and Linings* 2(5):42.
48. Korrosionsschutz von Stahbauten durch Beschichtungen und Uberzuge; Technische Regeln fur Kon-

trollflachen, Gewahrleistung fur den Korrosionsschutz von Staulbauten und Abnahme der Leistung DIN 55 928 T 7.
49. Kolke, F. 1972. *Industrie-Lackier-Betrieb* 40:1.
50. Moore, J. S. 1984. Guarantees for protective coating systems. Paper presented at SSPC symposium on Maintenance Painting of Industrial Plants, September 19–20, Pittsburgh, PA.
51. Hower, H. E. 1985. *Journal of Protective Coatings and Linings* 2(2):22.
52. Pinney, S. G., B. Brock, J. T. Kimball, and W. C. Ashbaugh. 1984. Problem-solving forum. *Journal of Protective Coatings and Linings* 1(6):7.
53. Trimber, K., and Hull. 1984. How to conduct a maintenance painting survey of industrial plants. Paper presented at SSPC symposium on Maintenance Painting of Industrial Plants, September 19–20, Pittsburgh, PA.

CHAPTER 6

1. See note 11, Chapter 1.
2. Sline, M. R., and G. H. Brevoort. 1982. "Comparative Painting Costs." In *SSPC Manual* 2d ed. Vol. 1. Pittsburgh, PA: Steel Structures Painting Council.
3. Manta, L. 1975. How the contractor applicator looks at zinc rich coatings. In *Proceedings of National Zinc Rich Coatings Conference*. New York: Zinc Literature, Inc.
4. Kase, P. E., and C. R. Hess. 1983. Specifications and other problems in shop painting. In *Proceedings of Second World Bridge Congress: Coating Systems for Bridges*. Rolla: Univ. of Missouri-Rolla.
5. Strauss, N. 1982. In *Proceedings of First World Bridge Congress*. Rolla: Univ. of Missouri-Rolla.
6. Brevoort, G. H., and P. N. Thomarios. 1984. Bridge painting costs. Paper #IBC-84-44 presented at the International Bridge Conference, Pittsburgh, PA.
7. Levinson, S. B., and S. Spindel. 1982. "Scaffolding." In *SSPC Manual* 2d ed. Vol. 1, Chapter 5.2. Pittsburgh, PA: Steel Structures Painting Council.
8. Levinson, S. B., and S. Spindel. 1968. *Recent developments in architectural and maintenance painting*. Philadelphia, PA: Fed. Soc. Paint Technology.
9. Tabony, J. S., and C. R. Roussel. 1976, October. *Structural painting*. Louisiana Department of Transportation, Research and Development Section, Training Unit.
10. Anonymous. 1975. *Painting and decorating craftsman's manual and textbook*. 5th ed. Book 4. Falls Church, VA: Painting and Decorating Contractors of America.
11. Brushwell, W. 1982. *Painting and decorating encyclopedia*. Unit 3. Goodheart and Willcox Co., Inc.
12. Fisher, H. 1986. *Journal of Protective Coatings and Linings* 3(3):38.
13. Fox, C. E. 1973. *Materials Performance* 12(10):25.
14. Trade Literature. Covert, MI: Swing-Lo Systems.
15. Anonymous. 1980. *Better Roads* 50(4):22–23.
16. Trade Literature. Snooper. Omaha, NE: Paxton Mitchell Co.
17. Bayliss, D. A. 1984. *Journal of Protective Coatings and Linings* 1(3):36.
18. Wallace, W. J., Jr. 1982. "Shop Painting of Steel in Fabricating Plants." In *SSPC Manual* 2d ed. Vol. 1. Pittsburgh, PA: Steel Structures Painting Council.
19. See note 3, Chapter 3.
20. Leyland, D. S. 1985. *Journal of Protective Coatings and Linings* 2(5):20.
21. See note 39, Chapter 3.
22. See note 46, Chapter 3.
23. See note 59, Chapter 3.
24. Mallory, A. W. 1985. *Journal of Protective Coatings and Linings* 1:14.
25. Munger, C. G. 1984. "Surface Preparation." In *Corrosion Prevention by Protective Coatings*. Houston, TX: National Association of Corrosion Engineers.
26. Anonymous. 1976. *Paint manual*. 3d ed., 97. Washington, DC: U.S. Department of Interior, Bureau of Reclamation.
27. Levinson, S. 1969. Paints and protective coatings. Army Specification TM5-618, Navy Specification Acom-110, and Air Force Specification AFM 85-3. Washington, DC: Department of the Army, the Navy, and the Air Force.
28. Levinson, S., and S. Spindel. 1983. "Paint Application." In *SSPC Manual*. Vol. 2. p. 150. Pittsburgh, PA: Steel Structures Painting Council.
29. Anonymous. 1975. *Painting and decorating craftsman's manual and textbook*. 5th ed. Book 3. Falls Church, VA: Painting and Decorating Contractors of America.
30. Munger, C. G. 1984. "Application of Coatings." In *Corrosion Prevention by Protective Coatings*. Houston, TX: National Association of Corrosion Engineers.

31. Testro H. J. 1965. "The Application of Surface Coating." In *Paint Technology Manual*, Part 4. O.C.C.A. London: Reinhold.
32. Graco, Inc. 1982. *Basic electrostatics*. Minneapolis, MN: Author.
33. See note 101, Chapter 4.
34. Ingram, H. S., and A. P. Shephard. 1964. *Flame spray handbook*. Westbury, NJ: Metco, Inc.
35. Anonymous. 1967. "Recommended Practices for Metallizing." In *Application of Aluminum and Zinc for Protection of Iron and Steel*. American Welding Society.
36. Binks Manufacturing Co. *Compressed airspray gun principles*. Training Division TD-10-1R-2. Franklin Park, IL: Author.
37. Graco, Inc. 1983. *Air spray guns*. Industrial Finishing Training Series. Minneapolis, MN: Author.
38. DeVilbiss Co. 1981. *The ABC's of spray equipment for refinishing*. Toledo, OH: DeVilbiss Co.
39. Binks Manufacturing Co. 1984. *Airspray Manual*. Training Division TD-10-1. Franklin Park, IL: Author.
40. Graco, Inc. Air spray training series. *Air Spray Techniques*. Minneapolis, MN: Author.
41. Binks Manufacturing Co. 1984. Operator techniques. Training Division TD49-IR-2. Franklin Park, IL: Author.
42. Prosser, J. L. 1981. Vacuole formation in high-build coating films. In *Proceedings of the Fifth International Conference in Organic Coating Science and Technologies*, edited by G. D. Parfitt and A. V. Patsis, 231. Westport, CT: Tachnomic.
43. Graco, Inc. 1983. *Graco airless spray systems*. Minneapolis, MN: Author.
44. Binks Manufacturing Co. 1981. *Airless spray manual*. Training Division TD11-1. Franklin Park, IL: Author.
45. Graco, Inc. *Airless spray technique series*. Minneapolis, MN: Author.
46. Kwuok, K. 1984. Understanding air-assisted airless technology and its uses in the finishing market. Paper presented at SSPC symposium, Pittsburgh, PA.
47. Graco, Inc. *Industrial finishing training system*. AA2000 manual, Air-assisted airless spray gun. Minneapolis, MN: Author.
48. Gross, H. 1983. *Materials Performance* 22(10):28.
49. Trade Literature. MZ-7, Inorganic Zinc-Rich. Raritan, NJ: Valspar Corporation.
50. Trade Literature. Carbo Zinc 11. 8th ed. St. Louis, MO: Carboline Co.
51. Clement, K. A. 1985. *Journal of Protective Coatings and Linings* 2(5):30.
52. Starr, T. L. 1986. *Journal of Protective Coatings and Linings* 3(3):22.
53. Keane, J. D. 1982. "SSPC Paint Application Specification #1." In *SSPC Manual*. Vol. 1, 341. Pittsburgh, PA: Steel Structures Painting Council.
54. Weismantel, G. E. 1981. *Paint handbook*, edited by G. B. Charlesworth and G. E. Weismantel, Chapter 18. New York: McGraw-Hill.
55. Tabony, J. S., and C. R. Roussel. 1976, October. *Structural Painting*. In Louisiana Department of Highways, Research and Development Sector, Training Unit.
56. See note 34, Chapter 2.
57. Hess, M. 1965. *Paint film defects*. 2d ed. New York: Van Nostrand Reinhold.
58. Munger, C. G. 1984. "Corrosion Prevention by Protective Coatings." In *Application of Coatings*. Houston, TX: National Association of Corrosion Engineers.
59. Munger, C. G. 1983. "Causes and Prevention of Paint Failure." In *SSPC Manual*. Vol. 2. Pittsburgh, PA: Steel Structures Painting Council.

CHAPTER 7

1. Thompson, S. P. 1982. *Materials Performance* 21:10.
2. Woods, A. L. 1958. *Journal of Protective Coatings and Linings* 1(5):38.
3. Dunkley, F. G. 1976. *JOCCA* 59:39–59.
4. West Virginia Department of Highways. Method of measurement of paint thickness with magnetic gauges. Specification 708.40.00.
5. Florida Department of Transportation. 1979. Structural specification for bridge paint, section 560-7.6
6. Florida Department of Transportation. The painting of structural steel: Inspectors' short course.
7. Pinney, S. G. 1986. *Bridge painting inspection*. Washington, DC: Federal Highway Administration.
8. National Association of Corrosion Engineers. International coating inspector training and certification program. Houston, TX: National Association of Corrosion Engineers.
9. Institute of Applied Technology. 1982. Nuclear quality-assured coating work: Training course for high performance protective coatings inspectors. Washington, DC: Institute of Applied Technology.

10. Bayliss, D. A. 1983. *JOCCA* 66(11):327.
11. Allen Homitz & Associates. *Training course for bridge painting inspectors*. DOT FH-11-9691. Washington, DC: Federal Highway Administration.
12. *Standard guide for paint inspectors*. ASTM D3276-84.
13. American Society for Testing and Materials Subcommittee D.O1.43. 1979. "Inspection." In *Manual of Coating Work for Light Water Nuclear Power Plant Primary Containment and Other Safety Related Facilities*, Chapter 5. Philadelphia, PA: American Society for Testing and Materials.
14. West Virginia Department of Highways. "Durability of Coatings Applied to Steel Structures." In *Inspection of Coatings During and After Application*. (WV-80) 005. Washington, DC: Federal Highway Administration.
15. Munger C. G. 1984. "Corrosion Prevention by Protective Coatings." In *Application of Coatings*, p. 419. Houston, TX: National Association of Corrosion Engineers.
16. See note 34, Chapter 2.
17. See note 57, Chapter 6.
18. See note 58, Chapter 6.
19. See note 59, Chapter 6.
20. Federation of Societies for Coatings Technology. 1978. *Pictorial standards for coating defects*. Philadelphia, PA: Author.
21. National Decorating Products Association. 1980. *Paint problem solver*. St. Louis, MO: Author.
22. National Association of Corrosion Engineers. *Glossary of terms used in maintenance painting*. Task group T-60-12. Houston, TX: Author.
23. LeSota, S., et al. 1978. *Paint and coatings dictionary*. Philadelphia, PA: Federation of Societies for Coating Technology.
24. See note 38, Chapter 6.
25. See note 40, Chapter 6.
26. Binks Manufacturing Co. *Spray painting—Finish problems*. Training Division TD-110-2. Franklin Park, IL: Author.
27. See note 30, Chapter 6.
28. Berger, D. M., and S. E. Mroz. 1976. *Journal of Testing and Evaluation* 4(1):29–39.
29. Berger, D. M. 1977, June. "The Role of Today's Coating Inspector." *American Painting Contractor* 2–4, 6–8, 10–11.
30. Tator, K. B., and K. A. Trimber. 1981. "Coating (Paint) Inspection Instrument Types, Uses and Calibration." *Modern Paint and Coatings* 76(11):54–58.
31. Tator, K. B., and K. A. Trimber. 1982. *SSPC Manual*. 2d ed. Vol. 1, Chapter 6. Pittsburgh, PA: Steel Structures Painting Council.
32. Keane, J. D., J. A. Bruno, and E. F. Weaver. 1976. *Surface profile for anti corrosion paints*. Pittsburgh, PA: Steel Structures Painting Council.
33. Anonymous. 1983. "Measurement of Paint Thickness with Magnetic Gauges." In *SSPC Manual*. Vol. 2. Pittsburgh, PA: Steel Structures Painting Council.
34. U.S. Department of Transportation. 1979. *Bridge Inspection Training Manual 70*. Washington, DC: Federal Highway Administration.
35. Kaminski, W. W., and J. R. Allen. 1984. *Journal of Protective Coatings and Linings* 1(3):30.
36. Roebuck, A. H., and G. H. Brevoort. 1983. *Materials Performance* 23(1):43.
37. See note 35, Chapter 2.
38. See note 47, Chapter 5.
39. Werner, H. M. 1980. *JCT* 52(669):53.
40. Windler, F. J. 1985. *Materials Performance* 24(3):46.
41. Tinklenberg, G. L. 1985. *Journal of Protective Coatings and Linings* 2(5):34.
42. Schlossberg, L. 1964. *Metal Finishing* pp. 57, 93.
43. Appelman, B. R., and P. G. Campbell. 1982. *JCT* 54(686):17–25.
44. Hare, C. H. 1982. *Modern Paint and Coatings* 72(1):50–55.
45. Harrison, J. B. 1979. *JOCCA* 62(1):18.
46. Kesternich, W. 1965. *Werkstoffe und Korrosion* 16(3):193–201.
47. Kutzelnigg, A. 1958. *Werkstoffe und Korrosion* 9(7):429–34.
48. Timmins, F. D. 1979. *JOCCA* 62:131–35.
49. Funke, W. 1984. *JOCCA* 67(3):71.
50. Scott, J. L., and T. E. Andersen. 1976. *JOCCA* 59:404–13.
51. State of Louisiana Department of Transportation. Three coat organic zinc rich system. Materials section qualification procedure #37.

52. See note 11, Chapter 1.
53. Law, D. A., and G. L. Anania. 1977, May. *Acceptance sampling of structural paints*. Report #NYS DOT-ERD-77-RR53. Engineering Research and Development Bureau, New York State Department of Transportation.

CHAPTER 8

1. Reid, H. H. C. 1983. Bridge painting in Britain. In *Proceedings of Second World Bridge Congress*. Rolla: Univ. of Missouri-Rolla.
2. British Department of Transportation. Roads and local transport directorate. Draft departmental standard (BD—/82) for the maintenance of highway structures, procedural, contractual and technical requirements. Bridges and Engineering Standards Division. London, England: Author.
3. British Department of Transportation. Roads and local transport directorate. Draft advice notice (DA—/82) for the maintenance painting of highway structures. Bridges and Engineering Division. London, England: Author.
4. Keane, J. D. 1969. Protective coatings for highway structural steel—Literature search. NCHRP Report #74A.
5. See note 39, Chapter 2.
6. See note 3, Chapter 3.
7. See note 41, Chapter 4.
8. Bishop, R. R. 1979. *British Corrosion Journal* 9(3):149–57.
9. Bruno, J. A., and B. R. Appelman. 1984. *Journal of Protective Coatings and Linings* 1(6):24.
10. Rooney, H. A., A. L. Woods, and T. L. Shelly. *Highway Research Record*, 320, 970:34–40.
11. Leyland, D. 1983. A comparison of new and conventional bridge paint systems used by Maine Department of Transportation. In *Proceedings of Second World Bridge Congress*. Rolla: Univ. of Missouri-Rolla.
12. Appelman, B. R. 1984. *Journal of Protective Coatings and Linings* 1(3):20.
13. See note 2, Chapter 5.
14. British Standards Institute. 1977. "Maintenance." In *Code for Practice for Protective Coating of Iron and Steel Structures Against Corrosion*. London: Author.
15. Roebuck, A. H., and G. H. Brevoort. 1983. *Materials Performance* 22(1):43.
16. Trade Literature. Aliphatic urethane coatings guide. Pittsburgh, PA: Mobay Chemical Corp.
17. See note 11, Chapter 1.
18. See note 34, Chapter 2.
19. See note 57, Chapter 6.
20. See note 58, Chapter 6.
21. See note 59, Chapter 6.
22. McKelvie, A. N., and A. F. Sherwood. 1975. *JOCCA* 58(9):345.
23. See note 22, Chapter 7.
24. See note 23, Chapter 7.
25. Lear, J. B. 1981. *JCT* 53(625):63.
26. Crown, D. A. 1968. *The forensic examination of paints and pigments*. Springfield, IL: Thomas.
27. Vind, H. P., and R. W. Drisko. 1973. *Materials Performance* 12(10).
28. Anderson, D. G., and J. T. Vandeberg. 1981. *Analytical Chemistry* 53:13R–28R.
29. Raaschou Nielsen, H. K. 1984. *JCT* 563(716):21.
30. Tator, K. B. 1978. *American Painting Contractor*, June, 10–14.
31. Montle, J. F. 1977. Analysis of problems in protective coatings application. In *Proceedings of Corrosion '77*. Houston, TX: National Association of Corrosion Engineers.
32. Tator, K. B. 1978. *American Painting Contractor*, Aug., 10–13.
33. Sward, G. G. 1972. *Paint Testing Manual*. Philadelphia, PA: American Society for Testing and Materials.
34. National Association of Corrosion Engineers. *Why coatings fail*. Publication 6D170. Houston, TX: National Association of Corrosion Engineers.

CHAPTER 9

1. Anonymous. *Estimating Guide*. 12th ed. Falls Church, VA: Painting and Decorating Contracting of America.

2. Anonymous. 1977. *The painting and decorating contractor 1977: New construction profile and estimating guide*. New York: McGraw-Hill.
3. Frondistou-Yannas, S. 1980, March. "Coating and Common Costs of Highway Structural Steel." FHWA Report #FHWA-RD-79-121.
4. See note 36, Chapter 7.
5. See note 16, Chapter 8.
6. Haskins, J. E. 1980. *Plant Engineering*, Feb. 7, 73.
7. Sline, M. R., G. H. Brevoort, R. B. Feinberg, and S. S. Oechsle. 1982. "Comparative Painting Costs." In *SSPC Manual*. 2d ed. Vol. 1. Pittsburgh, PA: Steel Structures Painting Council.
8. The Rental Rate Blue Book.
9. Fraser, R. A. 1982. "Painting of Railroad Bridges and Structures." In *SSPC Manual*. 2d ed. Vol. 1. Pittsburgh, PA: Steel Structures Painting Council.
10. Roebuck, A. H., and G. E. Weismantel. 1981. Chapter 10. In *Economics Paint Handbook*, edited by G. E. Weismantel. New York: McGraw-Hill.
11. Roth, R. 1982. "Field Surface Preparation Costs." In *SSPC Manual*. 2d ed. Vol. 2. Pittsburgh, PA: Steel Structures Painting Council.
12. Frenzel, L. M., M. Ginn, and G. V. Spires. 1985. Application of high-pressure water jetting in corrosion control. In *Proceedings of the SSPC Symposium on Surface Preparation: State of the Art*. Pittsburgh, PA: Steel Structures Painting Council.
13. Keefe, K. F. 1970. *Blast off*. San Francisco, CA: Clemco Clemintina Ltd.
14. See note 45, Chapter 3.
15. Appelman, B. R., and J. A. Bruno. 1985. Evaluation of wet blast cleaning units. In *Proceedings of the SSPC Symposium on Surface Preparation: State of the Art*. Pittsburgh, PA: Steel Structures Painting Council.
16. Hansel, D. 1986. Abrasive blast cleaning—A vital foundation for coating cost and success. In *Proceedings of Corrosion '86*. Houston, TX: National Association of Corrosion Engineers.
17. See note 28, Chapter 6.
18. See note 10, Chapter 7.
19. Roebuck, A. H., and D. L. McCage. 1976. *Materials Performance* 5(10):30.
20. See note 94, Chapter 4.
21. See note 101, Chapter 4.
22. Trade Literature. 1984. Thermal spray coating calculation and coating formulas. Technical Data Sheet 1.9.1.1.2.1. Bow, NH: TAFA, Inc.
23. See note 98, Chapter 4.
24. Brevoort, G. H., and A. H. Roebuck. 1980. *Materials Performance* 19(6):24.
25. See note 6, Chapter 6.
26. See note 12, Chapter 8.
27. Roebuck, A. H., and G. E. Weismantel. Chapter 20. In *Paint Handbook*, edited by G. E. Weismantel. New York: McGraw-Hill.
28. Levinson, S. B., and S. Spindel. 1982. "Paint Application." In *SSPC Manual*. Vol. 1. Chapter 5.1. Pittsburgh, PA: Steel Structures Painting Council.
29. Anthony, R. N. 1970. *Management Accounting*. Homewood, IL: Richard D. Irwin, Inc.
30. Davidson, S., and W. L. Weil. 1974. *Handbook of Modern Accounting*. Englewood Cliffs, NJ: Prentice-Hall.
31. Johnson, G. L., and J. A. Gentry. 1974. *Principles of Accounting*. Englewood Cliffs, NJ: Prentice-Hall.
32. See note 15, Chapter 8.
33. Roberts, J. W., and L. H. Davis. 1982. Evaluation of bridge corrosion cost model. Paper 140. In *Proceedings of Corrosion '82*. Houston, TX: National Association of Corrosion Engineers.

CHAPTER 10

1. Anonymous. 1984. Advance Change Notice (ACN) to Naval Ships Technical Manual. Chapter 931, May 29.
2. Anonymous. Abrasive blasting media—Ship hull blast cleaning. MIL-A-22262A. NAVSEA.
3. See note 66, Chapter 3.
4. Appleman, B. R., and J. A. Bruno. 1985. *Journal of Protective Coatings and Linings* 2(8):34.

5. Frenzel, L. M., M. Ginn, and G. V. Spires. 1985. Application of high-pressure water jetting in corrosion control. In *Proceedings of the SSPC Symposium on Surface Preparation: State of the Art*. Pittsburgh, PA: Steel Structures Painting Council.
6. Bleile, H. R., et al. 1985. Specification criteria for abrasive blasting media. In *Proceedings of SSPC Symposium on Surface Preparation: State of the Art*. Pittsburgh, PA: Steel Structures Painting Council.
7. Hansink, J. D. 1985. Problem solving forum. *Journal of Protective Coatings and Linings* 2(10):13.
8. Hansink, J. D., and K. W. Loest. 1985. Abrasive standards and the U.S. Navy—A status report. In *Proceedings of SSPC Symposium on Surface Preparation: State of the Art*. Pittsburgh, PA: Steel Structures Painting Council.
9. Roman, R. J., and D. S. Skeie. 1985. Establishing chemical specifications for blasting grit to be in compliance with OSHA and EPA regulations. In *Proceedings of SSPC Symposium on Surface Preparation: State of the Art*. Pittsburgh, PA: Steel Structures Painting Council.
10. Anonymous. 1985. Testing and specifying slag abrasives for the U.S. Navy shipyards. *Journal of Protective Coatings and Linings* 2(8):22.
11. Stettler, L. F., et al. 1983. "Fibrogenic Effects and Chemical Characterization of Coal and Mineral Slags Used as Sand Substitutes." In *Health Issues Related to Metal and Non-Metallic Mining*, edited by W. L. Wagner et al. London: Butterworth.
12. See note 48, Chapter 6.
13. Weldon, D. G., and T. A. Cain. 1985. Salts: Their detection and influence on coating performance. In *Proceedings of SSPC Symposium on Surface Preparation: State of the Art*. Pittsburgh, PA: Steel Structures Painting Council.
14. Trade Literature. Quantab #1175: Chloride titrator. Elkhart, IN: Ames Division of Miles Laboratories.
15. Trade Literature. EM quant iron test strips #1004-1. Gibstown, NJ: EM Science Division of EM Industries, Inc.
16. Bayliss, D. A. 1985. In *Proceedings of SSPC Symposium on Surface Preparation: State of the Art*. Pittsburgh, PA: Steel Structures Painting Council.
17. Anonymous. 1985. SSPC survey responses on conditions in the industry. *Journal of Protective Coatings and Linings* 2(6):14.
18. Field evaluation of surface modification methods to mitigate rusting and pitting of weathered steel bridge spans. Highway Planning and Research studies HPR84-1ST and HPR84-2ST. Louisiana State University.
19. Arresting the corrosion of (A588) weathering steel. HPR study 66. Charleston, WV: West Virginia Department of Highways.
20. Atmospheric corrosion tests of unpainted low-alloy steels for highway bridges. HPR study HPR7799-5. Trenton, NJ: New Jersey Department of Transportation.
21. Dunn, J. A., and R. U. Schenk. *Alternate deicing chemicals*. FHWA-RD-79-108. Washington, DC: Federal Highway Administration.
22. Chollar, B. H. 1984. *Public Roads* 47(4):113.
23. Fultz, B. S. 1982. *Surface texture (profile) measurement*. U.S. Department of Transportation, Maritime Administration/Avondale Shipyard, Inc. (Available through SSPC, Pittsburgh, PA)
24. National Association of Corrosion Engineers. 1982. Field measurement of surface profile of abrasive blast cleaned steel surfaces. Recommended Practice T-6G-19. Houston, TX: Author.
25. Schwab, L. K., and R. W. Drisko. 1981. *Materials Performance* 20(5):32.
26. Anonymous. 1985. "Surface Texture Profile Management—A Summary." *Journal of Protective Coatings and Linings* 2(9):5.
27. Keane, J. D. 1984. *Journal of Protective Coatings and Linings* 1(1).
28. See note 32, Chapter 7.
29. American Society for Testing and Materials. Standard test method for field measurement of surface profile of blast cleaned steel. ASTM D 4417. Philadelphia, PA: Author.
30. Sandor, L. W. 1983, May. *The Effect of Edge Preparation on Coating Life—Phase One*. U.S. Department of Transportation, Maritime Administration/Avondale Shipyard, Inc. (Available through SSPC, Pittsburgh, PA)
31. Sandor, L. W. 1985, February. *The Effect of Edge Preparation on Coating Life—Phase Two*. U.S. Department of Transportation. Maritime Administration/Avondale Shipyard, Inc. (Available through SSPC, Pittsburgh, PA)
32. See note 36, Chapter 3.
33. Brytus, V. 1984. *Modern Paint and Coatings* 74(10):172.
34. Batzer, H., and E. Leumann. 1982. Recent developments in epoxy chemistry with special reference to

coating problems. In *Proceedings of the Third International Conference on Organic Coatings: Science and Technology*, edited by G. D. Parfitt and A. V. Patsis. Westport, CT: Technomic.
35. See note 120, Chapter 2.
36. Benton, P. 1985. Water-borne coatings for industrial maintenance. In *Proceedings of Corrosion '85*. Houston, TX: National Association of Corrosion Engineers.
37. Dieterich, D. 1982. Polyurethane coatings from aqueous dispersions. In *Proceedings of the Third International Conference in Organic Coatings: Science and Technology*, edited by G. D. Parfitt and A. V. Patsis. Westport, CT: Technomic.
38. Dieterich, D. 1982. Recent developments in aqueous polyurethanes. In *Proceedings of the Sixth International Conference in Organic Coatings*, edited by G. D. Parfitt and A. V. Patsis. Westport, CT: Technomic.
39. Caine, M. L. 1985. Chlorinated vinyl acrylic latexes for steel protection. In *Proceedings of Corrosion '85*. Houston, TX: National Association of Corrosion Engineers.
40. Padget, J. C., and P. J. Moreland. 1983. *JCT* 55(698):39.
41. Warness, R. 1985. Waterborne coatings for corrosion protection of steel highway structures. In *Proceedings of Corrosion '85*. Houston, TX: National Association of Corrosion Engineers.
42. Boyer, C. C. 1985. Waterborne maintenance coatings—A new alternative. In *Proceedings of Corrosion '85*. Houston, TX: National Association of Corrosion Engineers.
43. Kurnik, W. J. 1985. The effect of water as a solvent on the performance of epoxy/amine coatings. In *Proceedings of Corrosion '85*. Houston, TX: National Association of Corrosion Engineers.
44. Masciale, M. *Materials Performance* 11(8):38.
45. Khambatta, J., and W. Warner. 1985. Modern water based coatings stop corrosion and protect the environment. In *Proceedings of Corrosion '85*. Houston, TX: National Association of Corrosion Engineers.
46. Wienand, H., and W. Ostertag. 1984. *Modern Paint and Coatings* 74(11):38.
47. See note 25, Chapter 2.
48. Hare, C. H. 1985. *Modern Paint and Coatings* 75(7):26.
49. See note 38, Chapter 2.
50. Orr, E. W. 1985. *Modern Paint and Coatings* 75(10):173.
51. Leclerq, M. 1986. *JOCCA* 69(6):144.
52. Princen, L. H. 1979. New developments in triglyceride chemistry for plastics and coatings. In *Proceedings of the Third International Conference in Organic Coatings: Science and Technology*, edited by G. D. Parfitt and A. V. Patsis. Westport, CT: Technomic.
53. See note 20, Chapter 2.
54. Patnaik, J. 1982, January. Protective coatings for steel. Report CG-ECV 1-82 (16565). U.S. Coast Guard.
55. Deanin, R., and S. Frondistou-Yannas. 1986. *Journal of Protective Coatings and Linings* 3(8):26.
56. Leidheiser, H. 1978. Corrosion control by coatings. Based on discussions held at Lehigh University, Bethlehem, PA., November 13–15. Report ADA064990. Alexandria, VA: Defense Technical Information Center.
57. Hare, C. H. 1985. *American Paint Journal*, Feb., 40.
58. Funke, W. 1979. *JOCCA* 62:63–67.
59. Bombara, et al. 1983. *Werkstoffe und Korrosion* 33:610–17.
60. Funke, W., and H. Zatloukal. 1978. *Werkstoffe und Korrosion* 84:584.
61. Funke, W. 1985. *JOCCA* 69(9):229.
62. Kumins, C. A. 1982. Sorption and electrical properties of polymers—Adhesion and corrosion effects. In *Proceedings of the Sixth International Conference in Organic Coatings: Science and Technology*, edited by G. D. Parfitt and A. V. Patsis. Westport, CT: Technomic.
63. Dickie, R. A. 1982. Surface studies of de-adhesion of paints by corrosion. In *Proceedings of the Sixth International Conference in Organic Coatings: Science and Technology*, edited by G. D. Parfitt and A. V. Patsis. Westport, CT: Technomic.
64. Saarnak, A., E. Nilsson, and L. O. Kornum. 1976. *JOCCA* 59:427–32.
65. Croll, S. G. 1980. *JCT* 52(665):35.
66. Kendall, K. J. 1973. *Phys. D. Appl. Phys.* 6(15):1782–87.
67. Croll, S. G. 1978. *JCT* 50(638):33.
68. Croll, S. G. 1979. *JCT* 51(659):49.
69. Croll, S. G. 1981. *JCT* 53(672):85.
70. Shimbo, M., et al. 1984. *JCT* 56(713):45.
71. Toussaint, A., and L. D'Hont. 1981. *JOCCA* 64:302–307.

72. Perera, D. Y., and D. Vanden Eynde. 1981. *JCT* 53(677):39.
73. Leidner, J., and R. T. Woodhums. 1974. *Journal of Applied Polymer Science* 18:1639.
74. Corcoran, E. M. 1969. *Journal of Paint Technology* 41(535).
75. Croll, S. G. 1980. *JOCCA* 63:271–75.
76. Roller, M. B., and J. K. Gillham. 1978. *JCT* 50(636):57.
77. Vetere, V., et al. 1978. *JOCCA* 61:419–26.
78. D'Alkaine, C. V., et al. 1980. *JOCCA* 63:23–27.
79. Scantlebury, J. D., and N. Hok. 1979. *JOCCA* 62:89–92.
80. Joly, M., and J. Laout. 1979. Polarization resistance measurements in testing and perfecting anticorrosion organic coatings. In *Proceedings of the Third International Conference in Organic Coatings: Science and Technology*, edited by G. D. Parfitt and A. V. Patsis. Westport, CT: Technomic.
81. Wirth, J. K. 1938. *Chem. Fabrik* 11:455.
82. Koehler, E. L. 1971. Paper presented at the U. R. Evans Memorial Conference. Williamsburg, VA.
83. Tooke, W. R., and D. R. Hurst. 1973, November. Paint characterization by electrical techniques. Technical Report 3, GDOT Research Project 6301. Springfield, VA: National Technical Information Service.
84. Buller, M., et al. 1976. *JOCCA* 59(10):351–55.
85. Leidheiser, H., Jr. 1979. *Progress in Organic Coatings* 7:79–104.
86. Jullien, H. 1973. *Progress in Organic Coatings* 2:99.
87. Wolstenholme. 1973. *Journal of Corrosion Science* 13:521.
88. Mills, D. J., and J. E. O. Mayne. 1981. "The Inhomogeneous Natures of Polymer Films and Its Effect on Resistance Inhibition." In *Corrosion Control by Organic Coatings* 12. Houston, TX: National Association of Corrosion Engineers.
89. Callow, L. M., and J. D. Scantlebury. 1981. *JOCCA* 64(2):83.
90. Callow, L. M., and J. D. Scantlebury. 1981. *JOCCA* 64(3):119.
91. Callow, L. M., and J. D. Scantlebury. 1981. *JOCCA* 64(4):143.
92. Callow, L. M., and J. D. Scantlebury. 1982. *JOCCA* 65(1):11.
93. Callow, L. M., and J. D. Scantlebury. 1985. *JOCCA* 66(3):93.
94. Hepburn, J., et al. 1984. *JOCCA* 67(7):193.
95. Szauer, T., and A. Brandt. 1984. *JOCCA* 67(1):13.
96. Fernadez-Prini, R., and S. Kapusta. 1979. *JOCCA* 62(3):93.
97. Lindquist, S. A., L. Mezzaros, and L. Svenson. 1985. *JOCCA* 68(1):10.
98. See note 111, Chapter 2.
99. Callow, L. M., and J. D. Scantlebury. 1984. *JOCCA* 67(3):73.
100. Bayliss, D. A., and H. Bray. 1981. *Materials Performance* 20(11):29.
101. Paint Research Association. 1979. *PRA reports*. Middlesex, England: Author.
102. Leidheiser, H., and W. Wang. 1981. *JCT* 53(672):77.
103. Ritter, J. J., and J. Kruger. 1980. *Surface Science* 96:364.
104. Larrabee, G. B., and T. J. Shaffner. 1981. *Anal. Chem.* 53:163R–174R.
105. Cunningham, C. P., and C. M. Hansen. 1981, November. *JCT* 53(682):39.
106. Mathias, L. J. 1985. *Modern Paint and Coatings* 75(11):38.
107. Johnson, G. E., et al. 1981. "The State of Water in Polymer Films." In *Corrosion Control by Organic Coatings*, edited by H. Leidheiser, 4. Houston, TX: National Association of Corrosion Engineers.
108. Diachenko, O. R., and B. A. Reka. 1963. *Lakokras Mat.* 50.
109. Wilson, R. W. 1964. *JOCCA* 47:308.
110. Fialkiewicz, A., and M. Szandorowski. 1974. *JOCCA* 57(8):259.
111. Martin, J. W., and M. E. McKnight. 1985. *JCT* 57(724):31.
112. Martin, J. W., and M. E. McKnight. 1985. *JCT* 57(724):39.
113. Martin, J. W., and M. E. McKnight. 1985. *JCT* 57(724):49.
114. Anonymous. 1985. "NBS Research on Protective Coatings for Buildings and Structures." *Journal of Protective Coatings and Linings* 2(3):15.
115. Bahra, M. S., et al. 1984. *JOCCA* 67(5):113.
116. Federal Highway Administration. Coatings guide for in-service steel highway bridges. Study DTFH61-85-R-00082. Washington, DC: Federal Highway Administration.
117. Anonymous. 1985. Association news. *Journal of Protective Coatings and Linings* 2(9):48.
118. Miller, M. 1986. State of the art application technology and the impact on maintenance costs. In *Proceedings of Corrosion '86*. Houston, TX: National Association of Corrosion Engineers.
119. See note 7, Chapter 7.

120. Holl, C. H. 1986. A cost benefit assessment of coatings inspection. Paper 31. In *Proceedings of Corrosion '86*. Houston, TX: National Association of Corrosion Engineers.
121. Kline, H. H., and K. Keyman. 1986. Coating service life and maintenance cost. In *Proceedings of Corrosion '86*. Houston, TX: National Association of Corrosion Engineers.
122. Roebuck, A. H., and G. H. Brevoort. 1986. Coating work costs: Computer application and inspection. In *Proceedings of Corrosion '86*. Houston, TX: National Association of Corrosion Engineers.
123. Smith, T. M. 1986. Data base management and your maintenance painting program. In *Proceedings of Corrosion '86*. Houston, TX: National Association of Corrosion Engineers.
124. Masciale, M. J. 1986. Coating economics in the world of today. In *Proceedings of Corrosion '86*. Houston, TX: National Association of Corrosion Engineers.
125. Hergenrother, R. P. 1986. Urethane coatings: An economical alternative for bridge corrosion control. In *Proceedings of Corrosion '86*. Houston, TX: National Association of Corrosion Engineers.
126. Mudd, R. L., and C. C. Boyer. 1986. High solids coatings—An economical alternative. In *Proceedings of Corrosion '86*. Houston, TX: National Association of Corrosion Engineers.

Index